Progress in Probability

Volume 57

Fractal Geometry and Stochastics III

Christoph Bandt
Umberto Mosco
Martina Zähle
Editors

Springer Basel AG

Editors:

Christoph Bandt
Institut für Mathematik und Informatik
Ernst-Moritz-Arndt-Universität
17487 Greifswald
Germany
e-mail: bandt@uni-greifswald.de

Umberto Mosco
Department of Physics
University of Rome La Sapienza
Via G. Boni 20
00162 Roma
Italy
e-mail: umberto.mosco@uniroma1.it

Martina Zähle
Mathematisches Institut
Friedrich-Schiller-Universität
07740 Jena
Germany
e-mail: zaehle@minet.uni-jena.de

2000 Mathematics Subject Classification: 20H10, 26E25, 28Axx, 37C45, 46E35, 53C65, 60Gxx, 60Jxx

A CIP catalogue record for this book is available from the Library of Congress, Washington D.C., USA

Bibliographic information published by Die Deutsche Bibliothek
Die Deutsche Bibliothek lists this publication in the Deutsche Nationalbibliografie; detailed bibliographic data is available in the Internet at <http://dnb.ddb.de>.

ISBN 978-3-0348-9612-2 ISBN 978-3-0348-7891-3 (eBook)
DOI 10.1007/978-3-0348-7891-3

Originally published by Birkhäuser Verlag Basel in 2004
Softcover reprint of the hardcover 1st edition 2004
Printed on acid-free paper produced from chlorine-free pulp. TCF ∞
ISBN 978-3-0348-9612-2

9 8 7 6 5 4 3 2 1 www.birkhauser-science.com

Contents

Preface vii

Introduction ix

1. Fractal Sets and Measures

Andrzej Lasota, Józef Myjak and Tomasz Szarek
Markov Operators and Semifractals 3

Jacques Lévy Véhel and Claude Tricot
On Various Multifractal Spectra 23

Zhi-Ying Wen
One-Dimensional Moran Sets and the Spectrum of Schrödinger Operators 43

2. Fractals and Dynamical Systems

Albert M. Fisher
Small-scale Structure via Flows 59

Károly Simon
Hausdorff Dimension of Hyperbolic Attractors in $\mathbb{R}^3$ 79

Bernd O. Stratmann
The Exponent of Convergence of Kleinian Groups;
on a Theorem of Bishop and Jones 93

Amiran Ambroladze and Jörg Schmeling
Lyapunov Exponents Are not Rigid
with Respect to Arithmetic Subsequences 109

3. Stochastic Processes and Random fractals

Ai Hua Fan
Some Topics in the Theory of Multiplicative Chaos 119

Peter Mörters
Intersection Exponents and the Multifractal Spectrum
for Measures on Brownian Paths 135

Davar Khoshnevisan and Yimin Xiao
Additive Lévy Processes: Capacity and Hausdorff Dimension 151

4. Fractal Analysis in Euclidean Space

Hans Triebel
The Fractal Laplacian and Multifractal Quantities 173

Jenny Harrison
Geometric Representations of Currents and Distributions 193

Maria Agostina Vivaldi
Variational Principles and Transmission Conditions for Fractal Layers 205

5. Harmonic Analysis on Fractals

Takashi Kumagai
Function Spaces and Stochastic Processes on Fractals 221

Alf Jonsson
A Dirichlet Form on the Sierpinski Gasket,
Related Function Spaces, and Traces 235

Alexander Teplyaev
Spectral Zeta Function of Symmetric Fractals 245

Preface

The conference on Fractal Geometry and Stochastics which took place at Friedrichroda, Germany, from March 17 to 22, 2003, was the third in a series. The previous conferences were held in 1994 and 1998, and their main lectures were published by Birkhäuser, Progress in Probability, Volumes 37 and 46. During recent years the interest in the subject has still been growing, and more profound results were obtained. The size of the conference also increased, with 119 participants from all over the world.

Abstracts of most of the contributions, a list of e-mail addresses and some photos can be found on the web page `www.minet.uni-jena.de/fgs3/`. For this volume, we have asked only the 16 main speakers of the conference to work out their contributions carefully, for a large audience. In order to present new facets of the subject, and new experts working in the field, we decided not to include any authors of the 1994 and 1998 volumes. The papers were chosen to represent the main directions of contemporary research in the area. Most of them are surveys, written in a comprehensible style. A few also contain original results. We hope that non-experts as well as specialists will benefit from the presentations in this book.

We would like to express our gratitude to the Deutsche Forschungsgemeinschaft for their financial support which was essential for organizing the conference.

The Editors

Introduction

Fractal geometry is known for its applications. Irregular phenomena in physics and material science, in biology and medicine, in economy and finance can be described by fractal dimensions and multifractal spectra.

In recent years it turned out that fractals are also at the core of several deep and intricate problems of analysis, mathematical physics and probability theory, as for instance renormalization in dynamical systems and conformal invariance of percolation. Thus some concepts of fractal geometry have been successfully exploited in purely mathematical research. On the other hand, fractal geometry benefitted from the use of analytical and probabilistic methods. Among others, random walks and diffusion processes on fractals are studied by means of Laplace operators and Dirichlet forms. This development is documented in the present book. Fractal analysis and the use of function spaces will be central themes.

Let us briefly review the contents of this volume. Already in the first chapter on general fractal sets and measures, linear spaces play a rôle – as a tool and also as a subject where fractals come in. A. Lasota, J. Myjak and T. Szarek put the well-known construction of self-similar measures into the general setting of Markov operators and derive conclusions for the large class of semifractals. In the paper by Z.-Y. Wen it is shown that the spectra of certain Schrödinger operators are Cantor sets with general self-similarity properties. Methods for determining the dimensions of such sets are worked out. J. Lévy-Véhel and C. Tricot study multifractal spectra arising from Hausdorff dimension and from the principle of large deviations in probability. They define modified spectra which can be estimated more easily in an experimental framework.

Most fractals come from dynamical systems. There are many classes of dynamical systems, each working with specific methods, and Chapter 2 will highlight some important directions. K. Simon clarifies the difficulties of the structure of hyperbolic attractors of differentiable maps and shows how to estimate their Hausdorff dimension. In the case of Kleinian groups acting on hyperbolic space, the Hausdorff dimension of the limit set is better understood. A fundamental theorem says that it coincides with the exponent of convergence of the group. B. Stratmann presents a simplified proof of this basic result. A.M. Fisher describes an approach from ergodic theory to Kleinian limit sets as well as to many other fractals. He studies the magnification flow which arises when we zoom in at a certain point of the fractal. Self-similarity of the fractal implies ergodicity of the flow, and the entropy of the flow equals the Hausdorff dimension of the fractal. A special random walk constructed by A. Ambroladze and J. Schmeling shows that a dynamical system can have positive Lyapunov exponent while the exponent is zero on each arithmetic subsequence.

Random fractals and stochastic processes are studied in Chapter 3. A.H. Fan describes the theory of multiplicative chaos which originated in a model of turbulence, as well as applications to covering and percolation problems. Recent exciting work on the intersection exponents for Brownian paths is reviewed by P. Mörters, who points out the relations to the multifractal structure of certain measures on the paths. D. Koshnevisan and Y. Xiao study the potential theory of additive Levy processes. They find the Hausdorff dimension of the range of the process and of the sets of multiple points.

There are two chapters on fractal analysis. In Chapter 4, the fractals are subsets of Euclidean space, and function spaces on the surrounding $\mathbb{R}^n$ are used. H. Triebel studies the fractal Laplacian of a Radon measure in the plane, and its eigenfunction corresponding to the largest eigenvalue, called Courant function. Connections between the multifractal and Besov characteristics of the given measure and the properties of the Courant function are discussed. In a more abstract approach, J. Harrison considers fractal measures as currents and proves a geometric representation theorem for currents. M.A. Vivaldi investigates transmission problems with fractal layers.

Chapter 5 deals with the analysis on fractal structures without reference to the surrounding space. T. Kumagai studies Dirichlet forms which connect function spaces with diffusion processes on fractals. Heat kernel estimates are given in a very general setting. A. Jonsson considers a special Dirichlet form on the Sierpiński gasket function spaces, and the connections between the corresponding Lipschitz and Besov spaces. The paper by A. Teplyaev also starts with the gasket, observing that the spectrum of the Laplace operator is obtained from a quadratic function. Spectral zeta functions of similar examples are discussed.

Part 1

Fractal Sets and Measures

Progress in Probability, Vol. 57, 3–22

Markov Operators and Semifractals

Andrzej Lasota, Józef Myjak and Tomasz Szarek

Abstract. We show a relationship between the supports of invariant measures with respect to Markov operators and fractals and semifractals defined for Iterated Function Systems (or more generally for suitable multifunctions) without any use of probability theory.

Mathematics Subject Classification (2000). Primary 28A80, 26E25; Secondary 28A78, 60J05.

Keywords. Markov operators, iterated function systems, fractals, semifractals, invariant measure, asymptotic stability.

1. Introduction

Fractals can be obtained as the supports of measures invariant with respect to some asymptotically stable Markov operators acting on the space of Borel measures. Classical results in this area for the case of Markov operators generated by iterated function systems with probabilities were proved by Hutchinson [7] , Barnsley *et al* [2]. The class of sets which can be obtained as supports of invariant measures of asymptotically stable Markov operators contains not only fractals. This fact leads to the notion of semifractals. Roughly speaking, a semifractal is the support of the measure invariant with respect to an asymptotically stable Markov operator.

It is interesting that semifractals can be also constructed by a use of topological limits without any probabilistic tools. What is more, under some natural assumptions both definitions are equivalent.

Since the existence of invariant measure and asymptotic stability of Markov operators are essential for our theory, we start with some results concerning the existence of an invariant measure for Markov operators. Using these general results we will give some criteria for existence of invariant measure and asymptotic stability for Markov operators generated by iterated function systems.

Andrzej Lasota and Tomasz Szarek were partially supported by the state Committe for Scientific Research (Poland) Grant no. Z P03A 031 25.

However, even if a Markov operator P is asymptotically stable, i.e. for every choice of the initial probability measure μ the sequence $(P^n\mu)$ converges weakly to a unique invariant measure μ_*, the corresponding sequence of supports $(\operatorname{supp} P^n\mu)$ needs not converge to the support of μ_*. To study this problem we will introduce the notions of condensing Markov operator and Markov set function. We will show that if P is asymptotically stable, then the sequence of sets $(\operatorname{supp} P^n\mu)$ converges topologically to $\operatorname{supp}\mu_*$ if and only if P is condensing. We will also show that the classical operators appearing in the fractal theory are condensing.

We adjoin to an arbitrary Markov-Feller operator P a Markov set function Γ which maps closed sets into closed sets. This functions has many properties of an iterated function system. In particular, under appropriate assumptions, the iterates $\Gamma^n(A)$ (A a closed set) converge to the support of a measure invariant with respect to P.

The asymptotic stability of Markov operators is a quite strong condition. We will see that Markov operators having unique invariant measure have also some interesting properties. In particular we will prove that every subinvariant with respect to Markov set function closed set has measure μ_* either zero or one.

Finally, it should be noted that all the theorems are stated under quite general assumptions concerning the phase space X, namely X is assumed to be separable (or complete separable) metric space. Thus no compactness conditions are assumed and the theory presented can be applied to systems acting on infinite dimensional phase spaces.

2. Notation and auxiliary results

Let (X,ρ) be a metric space. By $B(x,r)$ (resp. $B^o(x,r)$) we denote the closed (resp. open) ball with center at x and radius r. For a subset A of X, $\operatorname{cl} A$, $\operatorname{diam} A$, and 1_A stands for the closure of A, the diameter of A and the characteristic function of A, respectively. By $\mathbb{R}$ we denote the set of all reals, by $\mathbb{N}$ the set of all positive integers and by $\mathcal{F}$ the family of all closed nonempty subsets of X. Finally, let $\mathbb{R}_+ = [0,+\infty)$ and $\mathcal{F}_0 = \mathcal{F} \cup \{\emptyset\}$.

By $\mathcal{B}$ we denote the σ-algebra of Borel subsets of X and by $\mathcal{M}$ the family of all finite Borel measures on X. By $\mathcal{M}_1$ we denote the space of all $\mu \in \mathcal{M}$ such that $\mu(X) = 1$ and by $\mathcal{M}_s$ the space of all finite signed Borel measures on X. The elements of $\mathcal{M}_1$ are called *probability measures* or *distributions.*

Given $\mu \in \mathcal{M}$ the support of μ is given by the formula

$$\operatorname{supp}\mu = \{x \in X : \mu(B(x,r)) > 0 \text{ for every } r > 0\}.$$

As usual, by $B(X)$ we denote the space of all bounded Borel measurable functions $f: X \to \mathbb{R}$ and by $C(X)$ the subspace of all continuous functions. Both spaces are considered with the supremum norm.

For $f \in B(X)$ and $\mu \in \mathcal{M}_s$ we write

$$\langle f, \mu \rangle = \int_X f(x)\mu(dx).$$

$\mathcal{M}_s$ is endowed with the *Fortet-Mourier norm* given by

$$\|\mu\| = \sup\left\{|\langle f, \mu\rangle| : f \in \mathcal{L}\right\} \quad \text{for} \quad \mu \in \mathcal{M}_s,$$

where $\mathcal{L}$ is the set of all $f \in C(X)$ such that $|f(x)| \le 1$ and $|f(x) - f(y)| \le \rho(x, y)$ for $x, y \in X$.

We say that a sequence (μ_n), $\mu_n \in \mathcal{M}$, converges weakly to a measure $\mu \in \mathcal{M}$ if

$$\lim_{n\to\infty} \langle f, \mu_n \rangle = \langle f, \mu \rangle \quad \text{for every} \quad f \in C(X).$$

It is well known (see [3]) that the convergence in the Fortet-Mourier norm is equivalent to weak convergence.

An operator $P : \mathcal{M} \to \mathcal{M}$ is called a *Markov operator* if

$$P(\lambda_1\mu_1 + \lambda_2\mu_2) = \lambda_1 P\mu_1 + \lambda_2 P\mu_2 \qquad \text{for } \lambda_1, \lambda_2 \in \mathbb{R}_+ \text{ and } \mu_1, \mu_2 \in \mathcal{M}$$

and

$$P\mu(X) = \mu(X) \qquad \text{for } \mu \in \mathcal{M}.$$

A Markov operator P is called *nonexpansive* if

$$\|P\mu_1 - P\mu_2\| \le \|\mu_1 - \mu_2\| \qquad \text{for } \mu_1, \mu_2 \in \mathcal{M}_1$$

and it is called *essentially nonexpansive* if there exists a metric ρ' equivalent to ρ such that P is nonexpansive with respect to Fortet-Mourier norm corresponding to the metric ρ'.

A Markov operator P is called a *Markov-Feller* operator if there is an operator $U : B(X) \to B(X)$ such that:

(i) $\langle Uf, \mu \rangle = \langle f, P\mu \rangle$ for $f \in B(X)$ and $\mu \in \mathcal{M}$;
(ii) $Uf \in C(X)$ for $f \in C(X)$.

The operator U is called *dual* to P.

It can be proved that every nonexpansive Markov operator is a Markov-Feller operator [14].

A measure μ is called *invariant* (or *stationary*) with respect to P if $P\mu = \mu$. A Markov operator P is called *asymptotically stable* if there exists a stationary measure $\mu_* \in \mathcal{M}_1$ such that

$$\lim_{n\to\infty} P^n\mu = \mu_* \qquad \text{for every } \mu \in \mathcal{M}_1.$$

Obviously the measure μ_* satisfying the above condition is unique.

A family of distributions $\widetilde{\mathcal{M}} \subset \mathcal{M}_1$ is called *tight* if for every $\varepsilon > 0$ there exists a compact set $K \subset X$ such that $\mu(K) \ge 1 - \varepsilon$ for every $\mu \in \widetilde{\mathcal{M}}$.

It is well known (see [3]) that every tight sequence of distributions contains a weakly convergent subsequence.

We say that a Markov operator $P : \mathcal{M} \to \mathcal{M}$ is *tight* if for every $\mu \in \mathcal{M}_1$ the sequence of iterates $(P^n\mu)$ is tight.

It is easy to prove the following

Lemma 2.1. *Every tight Markov-Feller operator admits an invariant probability measure.*

Given a set $A \subset X$ and a number $r > 0$ we denote by $\mathcal{N}^o(A,r)$ (resp. $\mathcal{N}(A,r)$) the open (resp. closed) r–neighbourhood of the set A, i.e.

$$\mathcal{N}^o(A,r) = \{x \in X : \rho(x,A) < r\} \quad \text{and} \quad \mathcal{N}(A,r) = \{x \in X : \rho(x,A) \leq r\},$$

where

$$\rho(x,A) = \inf \{\rho(x,y) : y \in A\}.$$

We denote by $\mathcal{C}_\varepsilon$, $\varepsilon > 0$, the family of all closed sets C for which there exists a finite set $\{x_1, x_2, \cdots, x_n\} \subset X$ (ε-net) such that

$$C \subset \bigcup_{i=1}^{n} B(x_i, \varepsilon).$$

Lemma 2.2. *Let (μ_n) be a sequence of distributions such that for every $\varepsilon > 0$ there exists a set $C \in \mathcal{C}_\varepsilon$ with $\mu_n(C) \geq 1 - \varepsilon$ for all $n \in \mathbb{N}$. Then (μ_n) is tight.*

An operator P is called *semi–concentrating* if for every $\varepsilon > 0$ there exist $C \in \mathcal{C}_\varepsilon$ and $\alpha > 0$ such that

$$\liminf_{n\to\infty} P^n \mu(C) > \alpha \quad \text{for} \quad \mu \in \mathcal{M}_1. \tag{2.1}$$

By $\mathcal{C}_\varepsilon^+ = \mathcal{C}_\varepsilon^+(P)$ we denote the family of all $C \in \mathcal{C}_\varepsilon$ such that

$$\inf_{\mu\in\mathcal{M}_1} \liminf_{n\to\infty} P^n \mu(C) > 0.$$

Let $A \in \mathcal{B}$. We say that a measure $\mu \in \mathcal{M}$ is *concentrated* on A if $\mu(X \setminus A) = 0$. By $\mathcal{M}_1^A$ we denote the set of all distributions concentrated on A

Lemma 2.3. *Let P be a nonexpansive Markov operator and let $\varepsilon > 0$. Assume that $A \in \mathcal{B}$ is such that* $\operatorname{diam} A \leq \varepsilon/16$. *Moreover, assume that there exists $\widetilde{\mu} \in \mathcal{M}_1$ such that*

$$\liminf_{n\to\infty} P^n \widetilde{\mu}(A) > 0.$$

Then there exists $C \in \mathcal{C}_\varepsilon$ such that

$$P^n \mu(C) \geq 1 - \varepsilon/2 \qquad \text{for all } n \in \mathbb{N} \text{ and } \mu \in \mathcal{M}_1^A.$$

Lemma 2.4. *Let P be a nonexpansive and semi–concentrating Markov operator. Then for every $\varepsilon > 0$ there exist a finite sequence of Borel sets $A_1, \ldots, A_k$ with* $\operatorname{diam} A_i \leq \varepsilon$, *$i = 1, \ldots, k$ and a measure $\mu_0 \in \mathcal{M}_1$ such that*

$$\bigcup_{i=1}^{k} A_i \in \mathcal{C}_\varepsilon^+ \quad \text{and} \quad \liminf_{n\to\infty} P^n \mu_0(A_i) > 0 \quad \text{for } i = 1, \ldots, k.$$

The proofs of these lemmas can be found in [16] and [18].

A multifunction $F : X \to X$ is a subset of $X \times X$ such that for every $x \in X$ the set $F(x) = \{y : (x, y) \in F\}$ is nonempty. For $A \subset X$ we define

$$F^{-}(A) = \{x \in X : F(x) \cap A \neq \emptyset\}.$$

A multifunction F is called *measurable* if the set $F^{-}(A)$ is Borel for every open subset A of X and it is called *lower semicontinuous* (shortly l.s.c.) if the set $F^{-}(A)$ is open for every open subset A of X.

In the next lemma we recall some known properties of l.s.c. multifunctions.

Lemma 2.5. *The following conditions are equivalent:*

(i) *F is l.s.c.;*

(ii) *$F(\operatorname{cl} A) \subset \operatorname{cl} F(A)$ for every $A \subset X$;*

(iii) *for every sequence (x_n), $x_n \in X$, we have*

$$\lim x_n = x \quad \textit{implies} \quad F(x) \subset \operatorname{Li} F(x_n);$$

(iv) *for every sequence (x_n), $x_n \in X$, we have*

$$\lim x_n = x \quad \textit{implies} \quad F(x) \subset \operatorname{Ls} F(x_n).$$

A set $A \subset X$ is called *subinvariant* (resp. *invariant*) with respect to a multifunction F if $F(A) \subset A$ (resp. $F(A) = A$).

Let (A_n) be a sequence of subsets of a metric space X. The *lower bound* $\operatorname{Li} A_n$ and the *upper bound* $\operatorname{Ls} A_n$ are defined by the following conditions. A point x belongs to $\operatorname{Li} A_n$ if for every $\varepsilon > 0$ there is an integer n_0 such that $A_n \cap B(x, \varepsilon) \neq \emptyset$ for $n \geq n_0$. A point x belongs to $\operatorname{Ls} A_n$ if for every $\varepsilon > 0$ the condition $A_n \cap B(x, \varepsilon) \neq \emptyset$ is satisfied for infinitely many n. If $\operatorname{Li} A_n = \operatorname{Ls} A_n$, we say that the sequence (A_n) is topologically convergent and we denote this common limit by $\operatorname{Lt} A_n$.

Observe that $\operatorname{Li} A_n$ and $\operatorname{Ls} A_n$ are always closed sets. We recall that $\operatorname{Li} A_n = \operatorname{Li}(\operatorname{cl} A_n)$, $\operatorname{Ls} A_n = \operatorname{Ls}(\operatorname{cl} A_n)$ and that $\operatorname{Li} A_n \subset B$ provided that $A_n \subset B$ for sufficiently large n and B is a closed set. Moreover, $\operatorname{Lt} A_n = \operatorname{cl} \bigcup_{n=1}^{\infty} A_n$ if A_n is an increasing sequence.

In the case when X is a compact space, $\operatorname{Lt} A_n = A$ if and only if the sequence (A_n) converges to A in the sense of the Hausdorff distance.

3. Existence of invariant measures for Markov operators

The classical proof of the existence of an invariant measure for Markov operators defined on compact spaces goes as follows: first we construct a positive invariant functional defined on the space of all continuous and bounded functions $f : X \to \mathbb{R}$ and then using the Riesz representation theorem, we obtain the desirable invariant measure (see [4], [6]). The first existence result in the case of locally compact spaces was given in [14] by using the concept of nonexpansiveness and lower bound technique. The existence of an invariant measure for Markov operators on Polish spaces has been established quite recently (see [15-18]). The proofs of the last results are based on the concept of tightness and suitable concentration properties of Markov operators. Here we give some examples of such results and we will use these results for proving the asymptotic stability of iterated function systems.

Theorem 3.1. *Every nonexpansive and semi–concentrating Markov operator P admits an invariant measure.*

Proof. Fix $\varepsilon > 0$ and set $\tilde{\varepsilon} = \varepsilon/16$. By virtue of Lemma 2.4 there exists a sequence of Borel sets $A_1, \ldots, A_k$ with $\operatorname{diam} A_i \le \tilde{\varepsilon}$ for $i = 1, \ldots, k$, and a measure $\mu_0 \in \mathcal{M}_1$ such that $A = \bigcup_{i=1}^k A_i$ belongs to $\mathcal{C}_{\tilde{\varepsilon}}^+$ and

$$\liminf_{n\to\infty} P^n\mu_0(A_i) > 0 \qquad \text{for } i = 1, \ldots, k.$$

By Lemma 2.3, for every $i \in \{1, \ldots, k\}$ there exists a set $C_i \in \mathcal{C}_\varepsilon$ such that

$$P^n\nu(C_i) \ge 1 - \varepsilon/2 \qquad \text{for all } n \in \mathbb{N} \text{ and } \nu \in \mathcal{M}_1^{A_i}.$$

Evidently the set $C = \bigcup_{i=1}^k C_i$ belongs to $\mathcal{C}_\varepsilon$. Moreover, we have

$$P^n\nu(C) \ge 1 - \varepsilon/2 \qquad \text{for all } n \in \mathbb{N} \text{ and } \nu \in \bigcup_{i=1}^k \mathcal{M}_1^{A_i}. \tag{3.1}$$

Since $A \in \mathcal{C}_{\tilde{\varepsilon}}^+$, there exists $\tilde{\alpha} > 0$ such that

$$\liminf_{n\to\infty} P^n\mu(A) > \tilde{\alpha} \quad \text{for all } \mu \in \mathcal{M}_1. \tag{3.2}$$

Define

$$\eta = \sup\Big\{\gamma \ge 0 : \liminf_{n\to\infty} P^n\mu(C) \ge \gamma \text{ for all } \mu \in \mathcal{M}_1\Big\}.$$

We claim that $\eta \ge 1 - \varepsilon/2$. To see this, assume that on the contrary $\eta < 1 - \varepsilon/2$. Set $\alpha = \tilde{\alpha}/k$. Clearly

$$\eta > \frac{\eta}{1-\alpha} - \frac{\alpha}{1-\alpha}\Big(1 - \frac{\varepsilon}{2}\Big).$$

Choose $\gamma > 0$ such that

$$\eta > \gamma > \frac{\eta}{1-\alpha} - \frac{\alpha}{1-\alpha}\Big(1 - \frac{\varepsilon}{2}\Big).$$

By the definition of η we have

$$\liminf_{n\to\infty} P^n\mu(C) \ge \gamma \qquad \text{for } \mu \in \mathcal{M}_1. \tag{3.3}$$

Fix $\mu \in \mathcal{M}_1$. From (3.2) it follows that there are $n_0 \in \mathbb{N}$ and $i \in \{1, \ldots, k\}$ such that

$$P^{n_0}\mu(A_i) > \alpha.$$

Define

$$\nu(B) = \frac{P^{n_0}\mu(B \cap A_i)}{P^{n_0}\mu(A_i)} \qquad \text{for } B \in \mathcal{B}$$

and

$$\tilde{\mu}(B) = \frac{1}{1-\alpha}\big(P^{n_0}\mu(B) - \alpha\nu(B)\big) \qquad \text{for } B \in \mathcal{B}.$$

Observe that $\nu \in \bigcup_{i=1}^{k} \mathcal{M}_1^{A_i}$, $\widetilde{\mu} \in \mathcal{M}_1$ and

$$P^{n_0}\mu = (1-\alpha)\widetilde{\mu} + \alpha\nu.$$

By the last equality and conditions (3.1) and (3.3) we obtain

$$\begin{aligned}\liminf_{n\to\infty} P^{n_0+n}\mu(C) &\geq (1-\alpha)\liminf_{n\to\infty} P^n\widetilde{\mu}(C) + \alpha \liminf_{n\to\infty} P^n\nu(C)\\ &\geq (1-\alpha)\gamma + \alpha(1-\varepsilon/2) > \eta.\end{aligned}$$

Since $\mu \in \mathcal{M}_1$ is arbitrary, we have

$$\liminf_{n\to\infty} P^n\mu(C) \geq (1-\alpha)\gamma + \alpha(1-\varepsilon/2) > \eta \quad \text{for } \mu \in \mathcal{M}_1, \tag{3.4}$$

which contradicts the definition of η. Finally, one can easily check that for every $\mu \in \mathcal{M}_1$ the sequence $(P^n\mu)$ is tight. An application of Lemma 2.1 completes the proof. □

Remark 3.2. From condition (3.4) and the fact that $\eta \geq 1-\varepsilon/2$ it follows that if P is a nonexpansive and semi-concentrating Markov operator, then for every $\varepsilon > 0$ there exists $C \in \mathcal{C}_\varepsilon$ such that

$$\liminf_{n\to\infty} P^n\mu(C) \geq 1-\varepsilon/2.$$

A nonexpansive and semi–concentrating Markov operators can have a lot of invariant measures. However, the family of all invariant measures is tight. To formulate this result we need the following notation.

For $\mu \in \mathcal{M}_1$ we set

$$\Omega(\mu) = \big\{\nu \in \mathcal{M}_1 : \|P^{n_k}\mu - \nu\| \to 0 \text{ for some } (n_k) \subset (n)\big\}.$$

Theorem 3.3. *Let $P : \mathcal{M} \to \mathcal{M}$ be a nonexpansive and semi–concentrating Markov operator. Then*

(i) *$\Omega(\mu) \neq \emptyset$ for every $\mu \in \mathcal{M}_1$,*

(ii) *$\widehat{\Omega} = \bigcup_{\mu\in\mathcal{M}_1} \Omega(\mu)$ is tight.*

Proof. (i) Fix $\varepsilon > 0$ and $\mu \in \mathcal{M}_1$. As in the proof of Theorem 3.1 one can verify that there exists a set $C \in \mathcal{C}_\varepsilon$ such that

$$\liminf_{n\to\infty} P^n\mu(C) > 1-\varepsilon/2.$$

By the Ulam theorem we can find a compact set $K \subset X$ such that

$$P^n\mu(K \cup C) > 1-\varepsilon \qquad \text{for } n \in \mathbb{N}.$$

Since $K \cup C \in \mathcal{C}_\varepsilon$, by Lemma 2.1 the sequence $(P^n\mu)$ is tight. From this and the Prokhorov Theorem it follows that $\Omega(\mu) \neq \emptyset$.

(ii) To prove the tightness of $\widehat{\Omega}$ fix $\varepsilon > 0$. By Remark 3.2 there exists a sequence $(C_j)_{j\geq 1}$ of subsets of X with $C_j \in \mathcal{C}_{\varepsilon/2^j}$ such that

$$\liminf_{n\to\infty} P^n\mu(C_j) \geq 1-\varepsilon/2^j \quad \text{for } \mu \in \mathcal{M}_1 \text{ and } j \in \mathbb{N}.$$

Define

$$K = \bigcap_{j=1}^{\infty} \mathcal{N}(C_j, \varepsilon/2^j)$$

and observe that K is compact. We are going to show that $\widetilde{\mu}(K) > 1 - \varepsilon$ for all $\widetilde{\mu} \in \widehat{\Omega}$. Fix $\widetilde{\mu} \in \widehat{\Omega}$ and let $\mu \in \mathcal{M}_1$ be such that $\widetilde{\mu} \in \Omega(\mu)$. Let $(n_k)_{k\geq 1}$ be a sequence of integers such that $\|P^{n_k}\mu - \widetilde{\mu}\| \to 0$ as $k \to \infty$. Then by the Alexandrov theorem we have

$$\begin{aligned}\widetilde{\mu}\left(\mathcal{N}(C_j, \varepsilon/2^j)\right) \geq \widetilde{\mu}\left(\mathcal{N}^o(C_j, \varepsilon/2^j)\right) &\geq \limsup_{k\to\infty} P^{n_k}\mu\left(\mathcal{N}^o(C_j, \varepsilon/2^j)\right) \\ &\geq \limsup_{k\to\infty} P^{n_k}\mu(C_j) \geq 1 - \varepsilon/2^j \qquad \text{for } j \in \mathbb{N}.\end{aligned}$$

Hence

$$\widetilde{\mu}(X \setminus K) \leq \sum_{j=1}^{\infty} \widetilde{\mu}\left(X \setminus \mathcal{N}(C_j, \varepsilon/2^j)\right) < \sum_{j=1}^{\infty} \varepsilon/2^j = \varepsilon,$$

which completes the proof of (ii). □

4. Iterated Function Systems

An *Iterated Function System* (shortly IFS) is given by a family of continuous transformations

$$w_i : X \to X, \quad i \in I.$$

Further, let

$$p_i : X \to X, \quad i \in I.$$

be a given family of continuous functions such that

$$p_i(x) \geq 0 \quad \text{and} \quad \sum_{i\in I} p_i(x) = 1 \quad \text{for} \quad x \in X,$$

where $I = \{1, \ldots, N\}$. The family$\{(w_i, p_i) : i \in I\}$ is called an *iterated function system with probabilities.*

We consider some special Markov operators describing the evolution of measures due to the action of randomly chosen transformations. A simple but somewhat unprecise description of this process goes as follows. Choose $x_0 \in X$. When an initial point x_0 is chosen, we randomly select from the set I an integer i in such a way that the probability of choosing i is $p_i(x_0)$. When a number i_0 is drawn we define $x_1 = w_{i_0}(x_0)$. Having x_1 we select i_1 according to the distribution $p_1(x_1), \ldots, p_N(x_1)$ and we define $x_2 = w_{i_1}(x_1)$ and so on. Denoting by μ_n, $n \in \mathbb{N}$, the distribution of x_n, i.e. $\mu_n(A) = \text{prob}\,(x_n \in A)$, we define P as the transition operator such that $\mu_{n+1} = P\mu_n$. It is easy to verify that this operator must be of the form

$$P\mu(A) = \sum_{i\in I} \int_{w_i^{-1}(A)} p_i(x)\mu(dx). \tag{4.1}$$

Moreover, P is a Feller operator and its adjoint operator U is given by the formula

$$Uf(x) = \sum_{i\in I} p_i(x) f\big(w_i(x)\big). \tag{4.2}$$

Now assume that

$$\sum_{i\in I} |p_i(x) - p_i(y)| \le \omega\big(\rho(x,y)\big) \qquad \text{for } x, y \in X, \tag{4.3}$$

where $\omega : \mathbb{R}_+ \to \mathbb{R}_+$ is a continuous function. The function ω is called a modulus of continuity. Further, we assume that ω satisfies the *Dini condition*, i.e. $\omega : \mathbb{R}_+ \to \mathbb{R}_+$ is a nondecreasing and concave function such that

$$\int_0^a \frac{\omega(t)}{t} dt < \infty \qquad \text{for some } a > 0.$$

We will assume the *average contractility condition*

$$\sum_{i=1}^{N} p_i(x) \rho\big(w_i(x), w_i(y)\big) \le r\rho(x,y) \qquad \text{for } x, y \in X, \tag{4.4}$$

where $r < 1$.

Lemma 4.1. *Assume that an iterated function system* $\{(S_i, p_i) : i \in I\}$ *satisfy conditions* (4.3) *and* (4.4). *If the modulus of continuity* ω *defined by* (4.3) *satisfies the Dini condition, then the Markov operator* P *given by* (4.1) *is essentially nonexpansive.*

Sketch of the proof. If ω satisfies the Dini condition then

$$\sum_{n=1}^{\infty} \omega(r^n t) < \infty$$

for $r < 1$ and $t \ge 0$. It is easy to check that the function

$$\varphi(t) = t + \sum_{n=0}^{\infty} \omega(r^n t)$$

is a solution of the inequality

$$\omega(t) + \varphi(r(t)) \le \varphi(t).$$

Moreover φ is a nondecreasing and concave function, $\varphi(0) = 0$ and $\varphi(t) > 0$ for $t > 0$. Set $\rho_\varphi(x,y) = \varphi\big(\rho(x,y)\big)$ for $x, y \in X$ and observe that (X, ρ_φ) is a Polish space. To prove the nonexpansiveness in the new space it is enough to show that $U(\mathcal{L}_\varphi) \subset \mathcal{L}_\varphi$, where $\mathcal{L}_\varphi$ denotes the family of all continuous functions f such that $|f(x)| \le 1$ and $|f(x) - f(y)| \le \rho_\varphi(x,y)$ for all $x, y \in X$. □

Theorem 4.2. *Let an iterated function system* $\{(S_i, p_i) : i \in I\}$ *satisfy conditions* (4.3) *and* (4.4). *Assume that there exists* $\delta > 0$ *such that*

$$\sum_{i\in J(x,y)} p_i(x) p_i(y) \ge \delta \qquad \textit{for } x, y \in X, \tag{4.5}$$

where $J(x,y)$ is the set of all $i \in I$ such that

$$\rho\big(w_i(x), w_i(y)\big) \leq r\rho(x,y). \tag{4.6}$$

Then the Markov operator P given by (4.1) is semi–concentrating. As a consequence, P has an invariant distribution.

Proof. Fix $x_0 \in X$ and define $V(x) = \rho(x, x_0)$. We have

$$\begin{aligned} UV(x) &= \sum_{i\in I} p_i(x)\rho\big(w_i(x), x_0\big) \\ &= \sum_{i\in I} p_i(x)\rho\big(w_i(x), w_i(x_0)\big) + \sum_{i\in I} p_i(x)\rho\big(w_i(x_0), x_0\big) \\ &\leq rV(x) + b, \end{aligned}$$

where $b = \max_{i\in I} \rho\big(w_i(x_0), x_0\big)$. From [18, Corollary 2.4.1] it follows that there exists a bounded Borel set $Y \subset X$ such that

$$\liminf_{n\to\infty} P^n\mu(Y) > 1/2 \qquad \text{for } \mu \in \mathcal{M}_1. \tag{4.7}$$

To show the semi–concentrating property for P, fix $\varepsilon > 0$ and choose an integer m such that

$$r^m \operatorname{diam} Y < \varepsilon.$$

Now fix $c \in X$ and define

$$C = \bigcup_{i_1,\dots,i_m\in I} B\big(w_{i_m,\dots,i_1}(c), \varepsilon\big),$$

where

$$w_{i_m,\dots,i_1} = w_{i_m} \circ \ldots \circ w_{i_1}.$$

Let $\delta > 0$ be such that (4.5) is satisfied and let $\delta_* = \delta/N$. Further, for arbitrary $i_1, \dots, i_m \in I$ denote by $Y_{i_1,\dots,i_m}$ the set of all $x \in Y$ such that

$$w_{i_m,\dots,i_1}(x) \in B\big(w_{i_m,\dots,i_1}(c), r^m\rho(x,c)\big) \tag{4.8}$$

and

$$p_{i_1}(x) \cdot \ldots \cdot p_{i_m}\big(w_{i_{m-1},\dots,i_1}(x)\big) \geq \delta_*^m. \tag{4.9}$$

By an induction argument, using (4.5) and (4.6), one can verify that

$$Y = \bigcup_{i_1,\dots,i_m\in I} Y_{i_1,\dots,i_m}.$$

We claim that P satisfies condition (2.1) with $\alpha = \delta_*^m/(2N^m)$. For this, fix $\mu \in \mathcal{M}_1$. According to (4.7), we may choose an integer n_0 such that

$$P^n\mu(Y) > 1/2 \qquad \text{for } n \geq n_0.$$

Thus, for every $n \geq n_0$ there exist $k_1^n, \dots, k_m^n \in I$ such that

$$P^n\mu(Y_{k_1^n,\dots,k_m^n}) \geq 1/(2N^m). \tag{4.10}$$

On the other hand, for arbitrary $n \ge n_0$ we have

$$\begin{aligned} P^{n+m}\mu(C) &= \langle 1_C, P^{n+m}\mu\rangle = \langle U^m 1_C, P^n\mu\rangle \\ &= \sum_{i_1,\dots,i_m\in I} \int_X p_{i_1}(x)\cdot\ldots\cdot p_{i_m}\big(w_{i_{m-1},\dots,i_1}(x)\big) 1_C\big(w_{i_{m-1},\dots,i_1}(x)\big) P^n\mu(dx) \\ &\ge \int_{Y_{k_1^n,\dots,k_m^n}} p_{i_1}(x)\cdot\ldots\cdot p_{i_m}\big(w_{i_{m-1},\dots,i_1}(x)\big) 1_C\big(w_{i_{m-1},\dots,i_1}(x)\big) P^n\mu(dx). \end{aligned}$$

Now, by (4.8), (4.9) and (4.10) we obtain

$$P^{n+m}\mu(C) \ge \delta_*^m/(2N^m) = \alpha \qquad \text{for } n \ge n_0,$$

whence condition (2.1) follows. Since $\varepsilon > 0$ is arbitrary, this completes the proof of the semi–concentrating property. The existence of an invariant distribution with respect to P follows immediately from Lemma 4.1 and Theorem 3.2 □

To prove the next result we need the following

Proposition 4.3. *Let P be a nonexpansive Markov operator. Assume that for every $\varepsilon > 0$ there is a number $\alpha > 0$ with the following property: for every $\mu_1, \mu_2 \in \mathcal{M}_1$ there exist a Borel set A with* $\operatorname{diam} A \le \varepsilon$ *and an integer n_0 such that*

$$P^{n_0}\mu_j(A) \ge \alpha \quad \text{for} \quad j = 1, 2. \tag{4.11}$$

Then

$$\lim_{n\to\infty} \|P^n(\mu_1 - \mu_2)\| = 0 \quad \text{for} \quad \mu_1, \mu_2 \in \mathcal{M}_1. \tag{4.12}$$

Proof. See [18, Theorem 5.4]. □

Theorem 4.4. *Under the hypotheses of Theorem 4.2 the Markov operator P given by (4.1) is asymptotically stable.*

Proof. Fix $\varepsilon > 0$. According to Theorem 3.2 there is a compact set $K \subset X$ such that

$$\bar\mu(K) \ge 4/5 \qquad \text{for } \bar\mu \in \widehat{\Omega}. \tag{4.13}$$

Choose an integer m such that

$$r^m \operatorname{diam} K \le \varepsilon/3 \tag{4.14}$$

and for arbitrary $i_1, \dots, i_m \in I$ and $x \in K$ define

$$\Pi_{i_1,\dots,i_m}(x) = p_{i_1}(x)\cdot\ldots\cdot p_{i_m}\big(w_{i_m,\dots,i_1}(x)\big).$$

Now, for every $x \in K$ and $i_1, \dots, i_m \in I$ we define the open neighbourhood $O_{i_1,\dots,i_m}(x)$ of x in the following way:

If $\Pi_{i_1,\dots,i_m}(x) > 0$, we set

$$\begin{aligned} O_{i_1,\dots,i_m}(x) = \Big\{ y \in X:\ & \rho\big(w_{i_m,\dots,i_1}(x), w_{i_m,\dots,i_1}(y)\big) < \varepsilon/3 \\ & \text{and } \Pi_{i_1,\dots,i_m}(y) > \Pi_{i_1,\dots,i_m}(x)/2 \Big\}. \end{aligned} \tag{4.15}$$

If $\Pi_{i_1,\ldots,i_m}(x) = 0$, we set

$$O_{i_1,\ldots,i_m}(x) = \Big\{ y \in X : \rho\big(w_{i_m,\ldots,i_1}(x), w_{i_m,\ldots,i_1}(y)\big) < \varepsilon/3 \Big\}$$

Now, for $x \in K$ we define

$$O_x = \bigcap_{i_1,\ldots,i_m \in I} O_{i_1,\ldots,i_m}(x).$$

Clearly $\{O_x : x \in K\}$ is an open covering of K. Since K is a compact set, there are $x_1, \ldots, x_q \in K$ such that

$$K \subset \bigcup_{k=1}^{q} O_{x_k}. \tag{4.16}$$

Set $G = \bigcup_{k=1}^{q} O_{x_k}$. Let $\delta > 0$ be such that (4.5) holds and let $\delta_* = \delta/N$. We claim that P satisfies condition (4.11) with $\alpha = \delta_*^m/(4q)$. In fact, let $\mu_1, \mu_2 \in \mathcal{M}_1$. Set $\mu_0 = (\mu_1 + \mu_2)/2$. By Theorem 3.3 the set $\Omega(\mu_0)$ is nonempty. Let $\bar{\mu} \in \Omega(\mu_0)$ and let $(n_k)_{k \geq 1}$ be such that

$$\lim_{k\to\infty} \|P^{n_k}\mu_0 - \bar{\mu}\| = 0.$$

Since G is open, the Alexandrov theorem implies

$$\liminf_{k\to\infty} P^{n_k}\mu_0(G) \geq \bar{\mu}(G). \tag{4.17}$$

From conditions (4.13), (4.16) and (4.17) it follows that for n sufficiently large

$$P^n\mu_0(G) = \big(P^n\mu_1(G) + P^n\mu_2(G)\big)/2 \geq 3/4$$

and consequently

$$P^n\mu_j(G) \geq 1/2 \qquad \text{for } j = 1, 2. \tag{4.18}$$

From the last inequality and (4.16) it follows that there exist $\widetilde{n} \in \mathbb{N}$ and $k_1, k_2 \in \{1, \ldots, q\}$ such that

$$P^{\widetilde{n}}\mu_1(O_{x_{k_1}}) \geq 1/(2q) \quad \text{and} \quad P^{\widetilde{n}}\mu_2(O_{x_{k_2}}) \geq 1/(2q). \tag{4.19}$$

Write for simplicity $\mathcal{O}_1 = O_{x_{k_1}}$ and $\mathcal{O}_2 = O_{x_{k_2}}$. By an induction argument, using (4.4) and (4.5), one can show that there are $i_1, \ldots, i_m \in I$ such that

$$\rho\big(w_{i_m,\ldots,i_1}(x_{k_1}), w_{i_m,\ldots,i_1}(x_{k_2})\big) \leq r^m \rho(x_{k_1}, x_{k_2}) \leq \varepsilon/3 \tag{4.20}$$

and

$$\Pi_{i_1,\ldots,i_m}(x_{k_j}) \geq \delta_*^m \quad \text{for } j = 1, 2. \tag{4.21}$$

Define

$$A = A_1 \cup A_2 \quad \text{where} \quad A_j = w_{i_m,\ldots,i_1}(\mathcal{O}_j), \;\; j = 1, 2.$$

According to (4.15), (4.20) and (4.21) we have $\operatorname{diam} A \leq \varepsilon$ and

$$\Pi_{i_1,\ldots,i_m}(x) \geq \delta_*^m/2 \tag{4.22}$$

for arbitrary $x \in \mathcal{O}_1 \cup \mathcal{O}_2$. Now, using (4.19) and (4.22) we obtain

$$\begin{aligned} P^{m+\tilde{n}}\mu_j(A) &= \langle U^{m+\tilde{n}} 1_A, \mu_j \rangle = \langle U^m 1_A, P^{\tilde{n}}\mu_j \rangle \\ &\geq \sum_{i_1,\dots,i_m=1 \in I} \int_X \Pi_{i_1,\dots,i_m}(x) 1_{A_k}(w_{i_m,\dots,i_1}(x)) P^{\tilde{n}}\mu_j(dx) \\ &\geq \delta_*^m P^{\tilde{n}}\mu_j(\mathcal{O}_j)/2 \geq \delta_*^m/(4q) \end{aligned}$$

for $j = 1, 2$, which proves condition (4.8) (with $n_0 = m + \tilde{n}$). Since, by Theorem 4.2, the operator P admits an invariant distribution, an application of Proposition 4.3 completes the proof. □

5. Markov set function

Using the Alexandrov Theorem it is easy to verify the following

Proposition 5.1. *Assume that a sequence of measures (μ_n), $\mu_n \in \mathcal{M}$, converges weakly to a measure $\mu \in \mathcal{M}$. Then*

$$\operatorname{Li} \operatorname{supp} \mu_n \supset \operatorname{supp} \mu.$$

A sequence of measures (μ_n), $\mu_n \in \mathcal{M}$, is called *condensed* at a point $x \in X$ if for every $\varepsilon > 0$ there is $\eta > 0$ such that

$$\inf \{\mu_n(B(x,\varepsilon)) : n \in \mathbb{N}_\eta\} > 0,$$

where

$$\mathbb{N}_\eta = \{n \in \mathbb{N} : B(x,\eta) \cap \operatorname{supp} \mu_n \neq \emptyset\}.$$

We say that a sequence (μ_n) is *condensed on X* if it is condensed at every point $x \in X$.

Proposition 5.2. *Assume that a sequence of measures (μ_n), $\mu_n \in \mathcal{M}$, converges weakly to a measure $\mu \in \mathcal{M}$. Then the following conditions are equivalent:*

(i) *(μ_n) is condensed on X;*
(ii) $\operatorname{Lt} \operatorname{supp} \mu_n = \operatorname{supp} \mu$.

Sketch of the proof. (i) $\Rightarrow$ (ii). By virtue of Proposition 5.1 it is sufficient to verify that $\operatorname{Ls} \operatorname{supp} \mu_n \subset \operatorname{supp} \mu$. For a contradiction, suppose that $x \in \operatorname{Ls} \operatorname{supp} \mu_n \setminus \operatorname{supp} \mu$ and let $\varepsilon > 0$ be such that $\mu(B(x,\varepsilon)) = 0$. Since the sequence (μ_n) is condensed on X and $x \in \operatorname{Ls} \operatorname{supp} \mu_n$, there exist $\alpha > 0$ and n_0 such that $\mu_n(B(x,\eta)) \geq \alpha$ for $n \geq n_0$. Now the Alexandov Theorem leads to a contradiction.

To prove (ii) $\Rightarrow$ (i), fix $x \in X$ and $\varepsilon > 0$. If $x \in \operatorname{supp} \mu$, then $\mu(B^o(x,\varepsilon)) > 0$ and by the Alexandrov Theorem $\mu_n(B^o(x,\varepsilon)) > \mu(B^o(x,\varepsilon))/2$ for n sufficiently large. This implies that (μ_n) is condensed at x. If $x \notin \operatorname{supp} \mu_n$, then condition (ii) implies that $x \notin \operatorname{Ls} \operatorname{supp} \mu$. Thus there exist $\eta > 0$ and $n_0 \in \mathbb{N}$ such that $\operatorname{supp} \mu_n \cap B(x,\eta) = \emptyset$ for $n \geq n_0$, which again implies that (μ_n) is condensed at x. □

Proposition 5.3. *Let P be a Markov-Feller operator and $\mu_1, \mu_2 \in \mathcal{M}$. Then the following implications hold:*

$$\operatorname{supp}\mu_1 \subset \operatorname{supp}\mu_2 \quad \textit{implies} \quad \operatorname{supp}P\mu_1 \subset \operatorname{supp}P\mu_2$$

and

$$\operatorname{supp}\mu_1 = \operatorname{supp}\mu_2 \quad \textit{implies} \quad \operatorname{supp}P\mu_1 = \operatorname{supp}P\mu_2.$$

Proof. The inclusion $\operatorname{supp}\mu_1 \subset \operatorname{supp}\mu_2$ is equivalent to the following condition

$$\langle f, \mu_1\rangle > 0 \quad \text{implies} \quad \langle f, \mu_2\rangle > 0,$$

for every $f \in C(X)$, $f \geq 0$. From this observation and equality $\langle f, P\mu\rangle = \langle Uf, \mu\rangle$ the first implications follows. The second one is an immediate consequence of the first. □

Let a Markov-Feller operator $P : \mathcal{M} \to \mathcal{M}$ be given. We adjoint to P a function $\Gamma : \mathcal{F}_0 \to \mathcal{F}_0$ by the formula

$$\Gamma(A) = \operatorname{supp}P\mu, \tag{5.1}$$

where $\mu \in \mathcal{M}$ be an arbitrary measure such that $\operatorname{supp}\mu = A$.

According to Proposition 5.3 the function Γ is well defined. It is called a *Markov set function* corresponding to P. If $A = \{x\}$ for simplicity we write $\Gamma(x)$ instead of $\Gamma(\{x\})$. Evidently

$$\Gamma(x) = \operatorname{supp}P\delta_x.$$

Some simple properties of Γ are summarized in the following

Proposition 5.4. *The Markov set function Γ corresponding to the Markov-Feller operator P has the following properties:*

(i) $\Gamma(\emptyset) = \emptyset$;
(ii) $\Gamma(A) \subset \Gamma(B)$ *for* $A, B \in \mathcal{F}$, $A \subset B$;
(iii) $\Gamma(A) = \operatorname{cl}\bigcup_{x\in A}\Gamma(x)$ *for* $A \in \mathcal{F}$;
(iv) *If the sequence* (x_n) *converges to* x *then*

$$\operatorname{Li}\Gamma(x_n) \supset \Gamma(x).$$

Moreover, using the induction argument, it is easy to prove

Proposition 5.5. *Let P be a Markov-Feller operator. Then for every $\mu \in \mathcal{M}$ and $n \in \mathbb{N}$ we have*

$$\operatorname{supp}P^n\mu = \Gamma^n(\operatorname{supp}\mu). \tag{5.2}$$

In particular, if μ is invariant with respect to P then

$$\Gamma(\operatorname{supp}\mu) = \operatorname{supp}\mu. \tag{5.3}$$

Theorem 5.6. *Let P be a Markov-Feller operator and Γ be the corresponding Markov set function. Assume that P is asymptotically stable and μ_* is the invariant measure with respect to P. Then*

$$\operatorname{Li}\Gamma^n(A) \supset \operatorname{supp}\mu_* \quad \textit{for every} \quad A \in \mathcal{F} \tag{5.4}$$

and

$$\operatorname{Lt}\Gamma^n(A) = \operatorname{supp}\mu_* \quad \textit{for every} \quad A \in \mathcal{F},\ A \subset \operatorname{supp}\mu_*. \tag{5.5}$$

Proof. Let $A \in \mathcal{F}$ and $\mu \in \mathcal{M}$ be such that $\operatorname{supp}\mu = A$. Using (5.1) and (5.2) it is easy to verify that

$$\Gamma^n(A) = \operatorname{supp} P^n\mu \quad \text{for} \quad n \in \mathbb{N}. \tag{5.6}$$

Since $(P^n\mu)$ converges weakly to μ_*, Proposition 5.1 and relation (5.6) imply (5.4).

For $A \subset \operatorname{supp}\mu_*$ by Proposition 5.4 (ii) and equality (5.3) we have

$$\Gamma^n(A) \subset \Gamma^n(\operatorname{supp}\mu_*) = \operatorname{supp}\mu_*.$$

Consequently

$$\operatorname{Ls}\Gamma^n(A) \subset \operatorname{supp}\mu_*,$$

which proves (5.5) and completes the proof of Theorem 5.6. □

Theorem 5.7. *Let P be a Markov operator and Γ the corresponding Markov set function. Assume that P has an unique invariant probability measure μ_*. Then*

$$\mu_*(D) = 0 \quad \textit{or} \quad \mu_*(D) = 1$$

for every closed set $D \subset X$ such that $\Gamma(D) \subset D$.

Proof. Let D be a closed set such that $\Gamma(D) \subset D$. From Proposition 5.3 and the definition of Γ it follows that $\operatorname{supp} P\mu \subset D$ for every $\mu \in \mathcal{M}$ with $\operatorname{supp}\mu \subset D$. Let μ_0 be a measure given by formula

$$\mu_0(A) = \mu_*(A \cap D) \quad \text{for} \quad A \in \mathcal{B}.$$

Observe that $\operatorname{supp} P\mu_0 \subset D$. For $A \in \mathcal{B}$ we have

$$\begin{aligned}\mu_0(A) &= \mu_*(A \cap D) = P\mu_*(A \cap D) = \langle 1_{A\cap D}, P\mu_*\rangle = \langle U1_{A\cap D}, \mu_*\rangle \\ &\geq \langle U1_{A\cap D}, \mu_0\rangle = \langle 1_A - 1_{A\setminus D}, P\mu_0\rangle = \langle 1_A, P\mu_0\rangle = P\mu_0(A)\end{aligned}$$

This means that μ_0 is invariant with respect to P. Since μ_* is the unique invariant probability measure clearly $\mu_0 = c\mu_*$ for some $c > 0$. By the last relation and the definition of μ_0 we have $\mu_*(D) = \mu_0(D) = c\mu_*(D)$. It follows that either $\mu_*(D) = 0$ or $c = 1$. If $\mu_*(D) = 0$ The statement of Theorem 5.7 is obviously satisfied. If $\mu_*(D) > 0$, we have $\mu_*(D) = \mu_0(D) = \mu_0(X) = \mu_*(X) = 1$, which completes the proof. □

6. Semifractals

Given an IFS $\{w_i : i \in I\}$ we define the corresponding *Barnsley-Hutchinson operator* (or *multifunction*) H by formula

$$H(A) = \mathrm{cl} \bigcup_{i\in I} w_i(A). \tag{6.1}$$

A set A_0 such that $H(A_0) = A_0$ is called *invariant* with respect to IFS $\{w_i : i \in I\}$. If there exists a set $A_0 \in \mathcal{F}$ such that $\mathrm{Lt}\, H^n(A) = A_0$ for every bounded set $A \in \mathcal{F}$, then IFS $\{w_i : i \in I\}$ is called *asymptotically stable* and the set A_0 is called *attractor* or *fractal* corresponding to the IFS $\{w_i : i \in I\}$.

If H is defined on the class of compact sets, this definition of fractals coincides with the classical definition used by Barnsley (see [2]). It is well known that if all w_i, $i \in I$, are strictly contractive then IFS $\{w_i : i \in I\}$ is asymptotically stable.

We say that an IFS $\{w_i : i \in I\}$ is *regular* if there is a nonempty subset I_0 of I such that IFS $\{w_i : i \in I_0\}$ is asymptotically stable. The attractor corresponding to the system $\{w_i : i \in I_0\}$ is called *nucleus* of the system $\{w_i : i \in I\}$.

Note that regular IFS in general need not to be asymptotically stable and it may have more that one nucleus.

Proposition 6.1. *Let $\{w_i : i \in I\}$ be a regular IFS and let A_0 be a nucleus of this system. Let H be given by (6.1). Then the set*

$$A_*(A_0) = \mathrm{cl} \bigcup_{n=1}^{\infty} H^n(A_0)$$

has the following properties:

(i) $A_*(A_0) = \mathrm{Lt}\, H^n(A_0)$;
(ii) $H(A_*(A_0)) = A_*(A_0)$;
(iii) $A_*(A_0) \subset A$ *for every* $A \in \mathcal{F}$ *such that* $H(A) \subset A$.

Proof. The condition (i) is obvious because the sequence $(H^n(A_0))$ is increasing. Using relation $\mathrm{cl}(w_i(\mathrm{cl}\, A)) = \mathrm{cl}\, w_i(A)$ we have

$$H(A_*(A_0)) = \mathrm{cl} \bigcup_{i\in I} w_i\Big(\mathrm{cl} \bigcup_{n=1}^{\infty} H^n(A_0)\Big) = \bigcup_{i\in I} \mathrm{cl}\, w_i\Big(\bigcup_{n=1}^{\infty} H^n(A_0)\Big)$$

$$= \mathrm{cl} \bigcup_{i\in I} \bigcup_{n=1}^{\infty} w_i(H^n(A_0)) = \mathrm{cl} \bigcup_{n=1}^{\infty} H\big(H^n(A_0)\big) = A_*(A_0).$$

Finally, let B be a bounded nonempty subset of A and H_0 be the Barnsley-Hutchinson multifunction corresponding to IFS $\{w_i : i \in I_0\}$. From inclusions $H_0^n(B) \subset H^n(B) \subset H^n(A) \subset A$ it follows that $A_0 = \mathrm{Lt}\, H_0^n(B) \subset A$. Consequently $H^n(A_0) \subset A$ for $n \in \mathbb{N}$, whence (iii) follows. □

From Proposition 6.1 it follows that for given regular IFS $\{w_i : i \in I\}$, the set $A_*(A_0)$ does not depend on the choice of the nucleus A_0.

The set

$$A_* = A_*(A_0) = \operatorname{Lt} H^n(A_0), \tag{6.2}$$

where A_0 is an arbitrary nucleus of an regular IFS $\{w_i : i \in I\}$, is called *semifractal* or *semiattractor* corresponding to IFS $\{w_i : i \in I\}$.

Using Proposition 6.1 we can prove the following

Theorem 6.2. *Let $\{w_i : i \in I\}$ be a regular IFS and let A_* be the corresponding semifractal given by* (6.2). *Then*

(i) *A_* is the smallest nonempty closed set such that $\operatorname{cl} H(A_*) = A_*$;*
(ii) *$\operatorname{Lt} H^n(A) = A_*$ for every $A \subset A_*$, $A \neq \emptyset$.*

The semiattractor of an IFS is strictly connected with the invariant measure with respect to this system. Namely we have

Theorem 6.3. *Let X be a Polish space. Assume that IFS $\{(w_i, p_i) : i \in I\}$ is asymptotically stable and that IFS $\{w_i : i \in I\}$ is regular. Then*

$$A_* = \operatorname{supp} \mu_*,$$

where A_* is the semiattractor corresponding to IFS $\{w_i : i \in I\}$ and μ_* is the invariant measure with respect to IFS $\{(w_i, p_i) : i \in I\}$.

Proof. Let $u \in A_*$ and let δ_u be the δ-Dirac measure supported at u. Simple calculation shows that $\operatorname{supp} P^n \delta_u = H^n(u)$, $n \in \mathbb{N}$. Since the sequence $(P^n \delta_u)$ converges weakly to μ_*, by Proposition 5.1 and Theorem 6.2 we have

$$\operatorname{supp} \mu_* \subset \operatorname{Li} H^n(u) = \operatorname{Lt} H^n(u) = A_*. \tag{6.3}$$

Now, let $u \in \operatorname{supp} \mu_*$. By (6.3) we have $u \in A_*$ and from Theorem 6.2 it follows that $\operatorname{Lt} H^n(u) = A_*$. On the other hand, since $H(\operatorname{supp} \mu_*) \subset \operatorname{supp} \mu_*$, hence $H(u) \subset \operatorname{supp} \mu_*$. Consequently $\operatorname{Ls} H^n(u) \subset \operatorname{supp} \mu_*$. From this and the first inclusion in (6.3) we have $\operatorname{Lt} H^n(u) = \operatorname{supp} \mu_*$, which completes the proof. □

The concept of fractal and semifractal defined for Barnsley-Hutchinson multifunction can be introduced in a natural way for a large class of multifunctions.

We say that a multifunction F is *asymptotically stable* if there is $A_0 \in \mathcal{F}$ such that $\operatorname{cl} F(A_0) = A_0$ and $\operatorname{Lt} F^n(A) = A_0$ for every nonempty bounded subset A of X.

Given a multifunction $F : X \to X$ we consider the set

$$C = \bigcap_{x \in X} \operatorname{Li} F^n(x). \tag{6.4}$$

If the set C is nonempty, then the multifunction F is called *asymptotically semistable* and the set C is called *semifractal* or *semiattractor* of F.

Theorem 6.4. *Assume that F is an asymptotically semistable l.s.c. multifunction with the semifractal C. Then the following conditions hold:*

(i) *$C \subset \operatorname{Li} F^n(A)$ for every $A \subset X$, $A \neq \emptyset$;*
(ii) *$\operatorname{cl} F(C) = C$;*

(iii) $\operatorname{Lt} F^n(A) = C$ *for every* $A \subset C$, $A \neq \emptyset$;
(iv) $C \subset A$ *for every nonempty closed subset* A *of* X *such that* $F(A) \subset A$.

Proof. Condition (i) is obvious.

To prove (ii) fix $y \in F(C)$ and choose $x \in C$ such that $y \in F(x)$. Let $z \in X$. Clearly $x \in \operatorname{Li} F^n(z)$. Let $x_n \in F^n(z)$ for $n \in \mathbb{N}$ be such that $x_n \to x$. By Lemma 2.5 we have $F(x) \subset \operatorname{Li} F(x_n)$ and so $y \in \operatorname{Li} F(x_n) \subset \operatorname{Li} F(F^n(z)) = \operatorname{Li} F^n(z)$. Since $z \in X$ was arbitrary, this implies $y \in C$. Thus $\operatorname{cl} F(C) \subset C$. To prove the opposite inclusion observe that $F^n(C) \subset F(C)$, $n \in \mathbb{N}$, which, in turn implies that $\operatorname{Li} F^n(C) \subset \operatorname{cl} F(C)$. Since $C \subset \operatorname{Li} F^n(C)$, this completes the proof of (ii).

To verify (iii) observe that $F(C) \subset C$ implies $\operatorname{Ls} F^n(C) \subset C$. Thus, for an arbitrary nonempty set $A \subset C$ we have

$$C \subset \operatorname{Li} F^n(A) \subset \operatorname{Ls} F^n(A) \subset \operatorname{Ls} F^n(C) \subset C,$$

whence (iii) follows.

Finally, the inclusion $F(A) \subset A$ implies that $F^n(A) \subset A$, $n \in \mathbb{N}$, and consequently $C \subset \operatorname{Li} F^n(A) \subset A$. □

Theorem 6.5. *Let* $F : X \to X$ *be a l.s.c. multifunction. Assume that there exists a l.s.c. and asymptotically semistable multifunction* $F_0 : X \to X$ *such that* $F_0(x) \subset F(x)$, $x \in X$. *Then* F *is asymptotically semistable and its semifractal* C *is given by the formula*

$$C = \operatorname{Lt} F^n(C_0) = \operatorname{cl} \bigcup_{n=1}^{\infty} F^n(C_0), \tag{6.5}$$

where C_0 *is the semifractal of* F_0.

Proof. Let C be a semifractal of F. Since $c_0 \subset C$, the first equality in (6.5) follows from condition (iii) of Theorem 6.4. The condition (ii) of Theorem 6.5 implies that $F^n(C_0) \subset C$ for $n \in \mathbb{N}$ and consequently

$$\operatorname{cl} \bigcup_{n=1}^{\infty} F^n(C_0) \subset C.$$

On the other hand, using the first equality in (6.5) we obtain

$$C = \operatorname{Lt} F^n(C_0) \subset \bigcup_{n=1}^{\infty} F^n(C_0),$$

which completes the proof. □

Theorem 6.6. *Let* P *be an asymptotically stable Markov-Feller operator. Then the corresponding Markov set function* Γ *is asymptotically semistable and*

$$C = \operatorname{supp} \mu_*,$$

where C *is the semifractal of* Γ *and* μ_* *is the measure invariant with respect to* P.

Proof. Fix $x \in X$. Since $(P^n\delta_x)$ converges weakly to μ_*, by Propositions 5.1 and 5.6 we have

$$\operatorname{supp}\mu_* \subset \operatorname{Li}\operatorname{supp} P^n\delta_x = \operatorname{Li}\Gamma^n(x) \subset C.$$

To prove the opposite inclusion, fix a point $z \notin \operatorname{supp}\mu_*$ and choose an $\epsilon > 0$ such that

$$B(z,\epsilon) \cap \operatorname{supp}\mu_* = \emptyset.$$

Let $x \in \operatorname{supp}\mu_*$. By Proposition 5.3 and 5.5 we have

$$\Gamma^n(x) = \operatorname{supp} P^n\delta_x \subset \operatorname{supp} P^n\mu_* = \operatorname{supp}\mu_* \quad \text{for } n \in \mathbb{N}.$$

Thus $\Gamma^n(x) \cap B(z,\epsilon) = \emptyset$. This means that $z \notin \operatorname{Li}\Gamma^n(x)$ and so $z \notin C$. The proof is completed. □

References

[1] M.F. Barnsley, *Fractals Everywhere.* Academic Press, New York 1993.

[2] M.F. Barnsley, S.G. Demko, J.H. Elton and J.S. Geronimo, *Invariant measures for Markov processes arising from iterated function systems with place-dependent probabilities.* Ann.Inst. H. Poincaré Probab. Statist. **24** (1988), 367–394.

[3] R.M. Dudley, *Probabilities and Metrics.* Lecture Notes Series, Aarhus Universitet, Denmark **45**, (1978).

[4] S.N. Ethier, T.G. Kurtz, *Markov Processes.* Wiley Series in Probability and Mathematical Studies, Wiley, New York 1986.

[5] K.J. Falconer, *Techniques in Fractal Geometry.* Wiley, New York, 1997.

[6] S.R. Foguel, *The Ergodic Theory of Markov Processes.* Van Nostrand Mathematical Studies, no. 21, 1969.

[7] J.E. Hutchinson, *Fractals and self-similarity.* Indiana Univ. Math. J. **30** (1981), 713–747.

[8] A. Lasota, M.C. Mackey, *Chaos, Fractals and Noice: Stochastic Aspects of Dynamics.* Applied Mathematical Sciences, Springer, New-York, 1994.

[9] A. Lasota, J. Myjak, *Semifractals.* Bull. Polish Acad. Sci. Math. **44** (1996), 5–21.

[10] A. Lasota, J. Myjak, *Markov operators and fractals.* Bull. Polish Acad. Sci. Math. **45** (1997), 197–210.

[11] A. Lasota, J. Myjak, *Semifractals on Polish spaces.* Bull. Polish Acad. Sci. Math. **46** (1998), 179–196.

[12] A. Lasota, J. Myjak, *Attractors of multifunctions.* Bull. Polish Acad. Sci. Math. **48** (2000), 319–334.

[13] A. Lasota, J. Myjak, T. Szarek, *Markov operators with a unique invariant measure.* J. Math. Anal. Appl. **276** (2002), 343–356.

[14] A. Lasota, J. Yorke, *Lower bounded technique for Markov operators and iterated function systems.* Random Comput. Dynamics **2**, (1994), 41–77.

[15] T. Szarek, *Invariant measures for iterated functions system.* Ann. Polon. Math. **75**, (2001), 87–98.

[16] T. Szarek, *The stability of Markov operators on Polish spaces.* Studia Math. **143** (2000), 145–152.

[17] T. Szarek, *Invariant measures for Markov operators with applications to functions systems.* Studia Math. **154** (2003), 207–222.

[18] T. Szarek, *Invariant measures for nonexpansive Markov operators on Polish spaces.* Dissertationes Math. **415** (2003), 1–62.

Andrzej Lasota
Institute of Mathematics
Silesian University
and
Institute of Mathematics
Polish Academy of Sciences
Bankowa 14, 40-007 Katowice
Poland
e-mail: `lasota@ux2.math.us.edu.pl`

Józef Myjak
Dipartimento di Matematica Pura ed Applicata
Università di L'Aquila
Via Vetoio, 67-100 L'Aquila
Italy
and
WMS AGH
al. Mickiewicza 30, 30-059 Kraków
Poland
e-mail: `myjak@univaq.it`

Tomasz Szarek
Institute of Mathematics
Silesian University
Bankowa 14, 40-007 Katowice
Poland
and
Dipartimento di Matematica Pura ed Applicata
Università di L'Aquila
Via Vetoio, 67-100 L'Aquila
Italy
e-mail: `szarek@itl.pl`

Progress in Probability, Vol. 57, 23–42

On Various Multifractal Spectra

Jacques Lévy Véhel and Claude Tricot

Abstract. We introduce two classes of multifractal spectra, called respectively dimension and continuous spectra. Dimension spectra offer an interesting alternative to the classical Hausdorff spectrum: They are much easier to estimate yet still give relevant information about the geometry of the Hölder function. Continuous spectra are a generalization of the large deviation spectrum that allow to obtain partition free results. Both classes of spectra allow to perform efficient multifractal analysis in an experimental framework.

Mathematics Subject Classification (2000). Primary 28A80; Secondary 28A75, 28A78.

Keywords. Multifractal analysis, large deviation multifractal spectrum.

1. Background and Notation

Multifractal analysis has developed in many directions since its introduction. Progress has been accomplished concerning the domain of validity of the multifractal formalism, both in the deterministic and random frameworks [1, 2, 3, 8]. The analysis has been extended to functions [7] or sequences of capacities [12] in addition to measures. The paper by P. Mörters in this volume gives an account on the multifractal analysis of certain measures related to Brownian paths. More refined spectra and estimation procedures have been defined [4, 9].

The numerical computation of a multifractal spectrum on sampled data remains however a challenging task. We introduce in this work two new classes of spectra, called *dimension spectra* and *continuous spectra*. Our aim is to facilitate the estimation problem, so as to obtain meaningful numerical results even when no assumption on the data structure is made (*i.e.*, in a non-parametric frame).

It is well-known that the Hausdorff multifractal spectrum f_h (see section 2 for definitions) is very hard to calculate in general. Apart from very restricted classes of mathematical models, the exact value of the Hausdorff dimension is difficult to obtain theoretically, and almost impossible to estimate on experimental data. The sets E_α which form a partition of the support are practically inaccessible in a discrete framework. To the contrary, the *box dimension* Δ can be estimated in experimental situations. But Δ is of no use here since in all interesting situations the E_α are either empty, or dense in an interval: In this case $\Delta(E_\alpha) = 1$, and the spectrum f_Δ is trivial. One can try to estimate f_h with the help of the large deviation spectrum f_g, which is most of the times easier to work with. The inequality $f_h \leq f_g$ is always true. However, f_g measures an information which is essentially different from f_h: The Hausdorff spectrum is defined as a dimension function which emphasizes the *geometric* structure of the singularities of a function or measure, whilst f_g yields *statistical* information. We introduce in this paper two spectra, denoted $f_d^{\lim}$ and $f_d^{\limsup}$ where d is any dimension. As is the case for f_h or f_Δ, they are defined by using set dimensions. On the other hand, they share many properties with f_g. For instance, $f_d^{\lim}$ and $f_d^{\limsup}$ are upper semi-continuous functions, and, conversely, every upper semi-continuous function is the $f_d^{\lim}$ or $f_d^{\limsup}$ spectrum for some signal. The major motivation for defining these dimension spectra is that $f_\Delta^{\limsup}$ is both easy to evaluate in practical situations and "more precise" than f_g, i.e. one always has $f_h \leq f_\Delta^{\limsup} \leq f_g$.

Another path consists in focusing on the large deviation spectrum f_g. Although originally introduced as a way to estimate f_h, it has soon been realized that f_g is of independent interest, specially in applications (see for instance [10, 11] for applications in image processing and Internet traffic modelling). A drawback of the large deviation spectrum is that its very definition relies on an arbitrary partitioning of the support of the signal. Different partitions will in general lead to different spectra. We introduce two variants of f_g. Both are *continuous* spectra, and thus allow to get rid of the discretization. As a consequence, these partition-free spectra contain a more intrinsic multifractal information. Moreover, one of them, denoted $\tilde{f}_g^c$, is defined using only one limit ($\eta \to 0$), instead of the two limiting operations ($n \to \infty$ and $\varepsilon \to 0$) used for f_g. We define the corresponding continuous Legendre spectra, and show that they are, under mild conditions, the concave envelopes of the continuous large deviation spectra.

To gain generality, we introduce our spectra for *abstract set functions*, rather than for Hölder exponents of measures or functions of a real variable, as is done classically. More precisely, let $X([0,1])$ be the metric set of all closed sub-intervals (including the singletons) of $[0,1]$ (the extension to $\mathbb{R}$ and $\mathbb{R}^n$ is straightforward). We shall base our multifractal analysis on the study of a function $A : X([0,1]) \longrightarrow \mathbb{R}^+ \cup \{+\infty\}$. The interpretation of A in the classical frame is as follows (the length of an interval u is denoted by $|u|$):

- For the analysis of a Borelian measure μ, take $A(u) = \log \mu(u)/\log|u|$;
- For the analysis of a function z, take $A(u) = \log v_z(u)/\log|u|$, where $v_z(u)$ measures the "variation" of z in u. Common choices are the increment $|z(u_{\max}) - z(u_{\min})|$ (where $u = [u_{\min}, u_{\max}]$), the oscillation $\sup_{t\in u} z(t) - \inf_{t\in u} z(t)$, or, when $u = u_n^k$ is the dyadic interval $[k2^{-n}, (k+1)2^{-n}]$, the wavelet coefficient of z at scale n and location k[1].

[1]This choice requires care, since the resulting spectra will depend on the analyzing wavelet.

Section 2 defines the dimension spectra $f_d^{\lim}$ and $f_d^{\limsup}$. We study their main properties in Section 3: Domain of definition, relationships, maximum. The inverse problem (given an upper semi-continuous function F, find A whose spectrum $f_d^{\lim}$ or $f_d^{\limsup}$ is equal to F) is solved in Section 4 in two cases: d is σ-stable (like the Hausdorff dimension), or $d = \Delta$. Section 5 shows explicit computations of dimension spectra. Finally, we define and study the continuous spectra f_g^c and $\tilde{f}_g^c$ and the corresponding Legendre spectra in Section 6.

2. New dimension spectra

For $x \in [0,1]$, $u_n(x)$ denotes the dyadic interval $u_n^k = [k2^{-n}, (k+1)2^{-n}]$ which contains x (take the right one if there are two such intervals). For any real number α, $N_\alpha(\varepsilon, n)$ denotes the number of 2^{-n}-dyadic intervals such that $|A(u) - \alpha| \le \varepsilon$, and $I_\alpha(\varepsilon, n)$ their union. Recall the definition of the *large deviation spectrum* f_g:

$$f_g(\alpha) = \lim_{\varepsilon \to 0} (\limsup_{n \to \infty} \frac{\log N_\alpha(\varepsilon, n)}{n \log 2}), \tag{1}$$

with the convention that $\log N_\alpha(\varepsilon, n)/n \log 2 = -\infty$ if $N_\alpha(\varepsilon, n) = 0$.
Prior to defining the dimension spectra, let us clarify our notion of a dimension.

Definition 1. *We call* dimension *a function* $d : \mathcal{P}([0,1]) \to \mathbb{R}^+ \cup \{-\infty\}$*, such that*

(i) $E \subset F \Rightarrow d(E) \le d(F)$
(ii) $d(\emptyset) = -\infty$*,* $d(\{x\}) = 0$ *for any* x*, and* $d(E) = 1$ *for any* E *such that* $|E| > 0$.

A dimension may have the following properties:

Definition 2. *The dimension* d *is* stable *if*

$$d(E \cup F) = \max\{d(E), d(F)\}$$

for all sets E*,* F *in* $[0,1]$*. It is* σ-stable *if*

$$d(\bigcup_n E_n) = \sup\{d(E_n)\}$$

for any countable set family (E_n).

Let us now define our new spectra. For any real number x in $[0,1]$, set $\alpha_n(x) = A(u_n(x))$. For any real α, let

$$E_\alpha(\varepsilon, N) = \{x, n \ge N \Rightarrow |\alpha_n(x) - \alpha| \le \varepsilon\}.$$

Note that $E_\alpha(\varepsilon, N)$ increases with N, so that $\bigcup_N E_\alpha(\varepsilon, N)$ may be written as $\sup_N E_\alpha(\varepsilon, N)$. Also, $E_\alpha(\varepsilon, N) = \bigcap_{n \ge N} I_\alpha(\varepsilon, n)$. Let

$$\begin{aligned} E_\alpha(\varepsilon) &= \sup_N E_\alpha(\varepsilon, N) = \{x, \exists N \text{such that } n \ge N \Rightarrow |\alpha_n(x) - \alpha| \le \varepsilon\} \\ &= \liminf_{N \to \infty} I_\alpha(\varepsilon, N). \end{aligned}$$

Since the sets $E_\alpha(\varepsilon)$ decrease with ε, one may define

$$E_\alpha = \lim_{\varepsilon\to 0} E_\alpha(\varepsilon) = \{x, \alpha_n(x) \to_{n\to\infty} \alpha\}.$$

Definition 3. *For any dimension d and any real α, define the following spectra:*

$$f_d(\alpha) = d(E_\alpha) = d(\lim_{\varepsilon\to 0} \sup_N E_\alpha(\varepsilon, N)) \tag{2}$$

$$f_d^{\lim}(\alpha) = \lim_{\varepsilon\to 0} d(E_\alpha(\varepsilon)) = \lim_{\varepsilon\to 0} d(\sup_N E_\alpha(\varepsilon, N)) \tag{3}$$

$$f_d^{\limsup}(\alpha) = \lim_{\varepsilon\to 0} \sup_N d(E_\alpha(\varepsilon, N)). \tag{4}$$

When d is the Hausdorff dimension h, $f_d = f_h$ is the usual Hausdorff spectrum.

3. Properties and relationships

Let us describe a few basic properties verified by dimension spectra.

3.1. Domain of definition

It is clear that the spectra defined above all range in $[0,1]\cup\{-\infty\}$. Let $D = \overline{\mathrm{Im}(A)}$ be the closure of the image of the function A. For every $\alpha \notin D$, there exists ε_0 such that $\varepsilon \le \varepsilon_0 \Longrightarrow E_\alpha(\varepsilon, N) = \emptyset$. Therefore $f_d(\alpha) = f_d^{\lim}(\alpha) = f_d^{\limsup}(\alpha) = -\infty$. Also, $N_\alpha(\varepsilon, n) = 0$, so that $f_g(\alpha) = -\infty$. Thus, while all the spectra are defined on $\mathbb{R}$, their "support" (i.e. the set of α for which the spectrum belongs to $[0,1]$) is included in D.

3.2. Inequalities

Since $E_\alpha \subset E_\alpha(\varepsilon)$, we have

$$f_d(\alpha) \le f_d^{\lim}(\alpha) \tag{5}$$

for all α. Also, $d(E_\alpha(\varepsilon, N)) \le d(\sup_N E_\alpha(\varepsilon, N))$ implies that

$$f_d^{\limsup}(\alpha) \le f_d^{\lim}(\alpha). \tag{6}$$

We will see that there is no relationship in general between f_d and $f_d^{\limsup}$. If d is σ-stable, then $d(\sup_N E_\alpha(\varepsilon, N)) = \sup_N d(E_\alpha(\varepsilon, N))$, so that $f_d^{\lim}$ is identical to $f_d^{\limsup}$. In this particular case, inequalities (5) and (6) reduce to

$$f_d(\alpha) \le f_d^{\lim}(\alpha) = f_d^{\limsup}(\alpha). \tag{7}$$

The spectra f_d and f_g cannot be compared without specifying the dimension d. Let us take for d the *box dimension*, which is defined for any bounded set E as

$$\Delta(E) = \limsup_{n\to\infty} \frac{\log \omega(2^n, E)}{n \log 2}$$

where $\omega(2^n, E)$ denotes the number of 2^n-dyadic intervals covering E. This dimension is stable, but not σ-stable. For all $n \ge N$, $\omega(2^n, E_\alpha(\varepsilon, N)) \le N_\alpha(\varepsilon, n)$, so that for all α, ε, N, $\Delta(E_\alpha(\varepsilon, N)) \le f_g(\alpha)$. Therefore

$$f_\Delta^{\limsup}(\alpha) \le f_g(\alpha). \tag{8}$$

There is no relationship in general between f_g and $f_\Delta^{\lim}$.

Lemma 1. *Let d_1, d_2, be such that for all $E \subset [0,1]$, $d_1(E) \leq d_2(E)$. Then for every A and α:*

$$f_{d_1}(\alpha) \leq f_{d_2}(\alpha)\,,\ f_{d_1}^{\lim}(\alpha) \leq f_{d_2}^{\lim}(\alpha)\,,\ f_{d_1}^{\limsup}(\alpha) \leq f_{d_2}^{\limsup}(\alpha).$$

We leave the proof to the reader.

If h denotes the Hausdorff dimension, it is well known that $h(E) \leq \Delta(E)$ for all E. Gathering previous results, we get the following sequence of inequalities:

Proposition 1. *For any set function A,*

$$f_h(\alpha) \leq f_h^{\limsup}(\alpha) = f_h^{\lim}(\alpha) \leq f_\Delta^{\limsup}(\alpha) \leq \min(f_\Delta^{\lim}(\alpha), f_g(\alpha)). \qquad (9)$$

When the (strong) multifractal formalism holds, $f_h(\alpha) = f_g(\alpha)$ for all α, so that all the above spectra coincide, with the possible exception of $f_\Delta^{\lim}$. Incidentally, this result explains the *a priori* unexpected fact that a naive numerical estimation of f_h on a multinomial measure yields acceptable results: Indeed, estimating simply the box dimension of the sets $E_\alpha(\varepsilon, N)$ gives a correct approximation in this case.

More generally, i.e. without assuming the multifractal formalism, (9) shows that $f_\Delta^{\limsup}(\alpha)$ is always a better approximation to f_h than f_g. In addition, it is not more difficult to estimate.

3.3. Maximum of a spectrum

Since $d([0,1]) = 1$, every spectrum has a maximum not larger than 1. In general, the upper bound depends on A and d. To make this precise, let us introduce new sets. Let

$$S(\varepsilon, N) = \{x, m \geq N, n \geq N \Rightarrow |\alpha_m(x) - \alpha_n(x)| \leq \varepsilon\}.$$

Note that $S(\varepsilon, N)$ increases with N, so that $\bigcup_N S(\varepsilon, N) = \sup_N S(\varepsilon, N)$. Let

$$S(\varepsilon) = \sup_N S(\varepsilon, N) = \{x, \exists N \text{such that } m \geq N, n \geq N \Rightarrow |\alpha_m(x) - \alpha_n(x)| \leq \varepsilon\}.$$

Since the sets $S(\varepsilon)$ decrease as $\varepsilon \to 0$, one may define

$$S = \lim_{\varepsilon \to 0} S(\varepsilon).$$

These constructions are similar to those of $E_\alpha(\varepsilon, N)$, $E_\alpha(\varepsilon)$, E_α, except that they are independent of α. Note that for all $\varepsilon' < \varepsilon$,

$$S(\varepsilon) \subset \{x, \limsup \alpha_n(x) - \liminf \alpha_n(x) \leq \varepsilon\} \subset S(\varepsilon')$$

and

$$S = \{x, \alpha_n(x) \text{ converges}\}.$$

Now define the dimensional indices:

$$d_0 = d(S) = d(\lim_{\varepsilon \to 0} \sup_N S(\varepsilon, N)).$$

$$d_0^{\lim} = \lim_{\varepsilon \to 0} d(S(\varepsilon)) = \lim_{\varepsilon \to 0} d(\sup_N S(\varepsilon, N))$$

$$d_0^{\limsup} = \lim_{\varepsilon\to 0} \sup_N d(S(\varepsilon, N)).$$

Proposition 2. *For every* α,

$$f_d(\alpha) \le d_0\,,\ f_d^{\lim}(\alpha) \le d_0^{\lim}\,,\ f_d^{\limsup}(\alpha) \le d_0^{\limsup}. \tag{10}$$

Proof. Use the inclusions $E_\alpha \subset S$, $E_\alpha(\varepsilon) \subset S(2\varepsilon)$, $E_\alpha(\varepsilon, N) \subset S(2\varepsilon, N)$. □

Proposition 3. *If the set D is bounded, and d is a stable dimension, then $f_d^{\limsup}$ and $f_d^{\lim}$ reach the above upper bounds. In other words,*

$$d_0^{\limsup} \in f_d^{\limsup}(D) \ \ \textit{and} \ \ d_0^{\lim} \in f_d^{\lim}(D).$$

Proof. 1. We first show that there exists $\alpha_1 \in D$ such that $d_0^{\limsup} \le f_d^{\limsup}(\alpha_1)$.

Let $N \in \mathbb{N}$, $\varepsilon > 0$. For any $x \in S(\varepsilon, N)$ and any $n \ge N$, $|\alpha_n(x) - \alpha(x)| \le \varepsilon$, where $\alpha(x) = \frac{1}{2}(\liminf \alpha_n(x) + \limsup_n \alpha_n(x))$. Therefore $x \in E_{\alpha(x)}(\varepsilon, N)$. This implies that $S(\varepsilon, N) \subset \bigcup_{\alpha\in\mathbb{R}} E_\alpha(\varepsilon, N)$.

For every α there exists $k \in \mathbb{Z}$ such that $[\alpha-\varepsilon, \alpha+\varepsilon] \subset [2(k-1)\varepsilon, 2(k+1)\varepsilon]$, so that $\bigcup_{\alpha\in\mathbb{R}} E_\alpha(\varepsilon, N) \subset \bigcup_{k\in\mathbb{Z}} E_{2k\varepsilon}(2\varepsilon, N)$. If D is bounded, then all sets $E_{2k\varepsilon}(2\varepsilon, N)$ are empty but a finite number of them. Using the stability of d, we deduce that

$$d(S(\varepsilon, N)) \le \max_{k\in\mathbb{Z}} d(E_{2k\varepsilon}(2\varepsilon, N)).$$

From this inequality it follows that for all ε, N there exists a real number $\beta(\varepsilon, N)$ such that

$$d(S(\varepsilon, N)) \le d(E_{\beta(\varepsilon,N)}(2\varepsilon, N)).$$

Since the distance from $\beta(\varepsilon, N)$ to D is not more than 2ε, the sequence $(\beta(\varepsilon, N))_N$ has a limiting value $\beta(\varepsilon)$ (for instance $\beta(\varepsilon) = \limsup_N \beta(\varepsilon, N)$). Let N_k be a subsequence such that $\beta(\varepsilon, N_k) \to \beta(\varepsilon)$. If k is large enough, $[\beta(\varepsilon, N_k) - 2\varepsilon, \beta(\varepsilon, N_k) + 2\varepsilon] \subset [\beta(\varepsilon) - 3\varepsilon, \beta(\varepsilon) + 3\varepsilon]$, so that

$$E_{\beta(\varepsilon,N_k)}(2\varepsilon, N_k) \subset E_{\beta(\varepsilon)}(3\varepsilon, N_k).$$

Therefore $d(S(\varepsilon, N_k)) \le d(E_{\beta(\varepsilon)}(3\varepsilon, N_k))$. Since the two sides of this inequality increase as $k \to +\infty$, we obtain

$$\sup_N d(S(\varepsilon, N)) \le \sup_N d(E_{\beta(\varepsilon)}(3\varepsilon, N)).$$

The function $\beta(\varepsilon)$ has a limiting value α_1 in D (for instance $\alpha_1 = \limsup_{\varepsilon\to 0} \beta(\varepsilon)$). Let ε_i be a sequence such that $\beta(\varepsilon_i)$ converges to α_1. Let $\eta > 0$. For i large enough, $\varepsilon_i \le \eta$ and $|\beta(\varepsilon_i) - \alpha_1| \le \eta$. Therefore $[\beta(\varepsilon_i) - 3\varepsilon_i, \beta(\varepsilon_i) + 3\varepsilon_i] \subset [\alpha_1 - 4\eta, \alpha_1 + 4\eta]$. This implies that for all $N \in \mathbb{N}$,

$$E_{\beta(\varepsilon_i)}(3\varepsilon_i, N)) \subset E_{\alpha_1}(4\eta, N),$$

so that

$$\sup_N d(S(\varepsilon_i, N)) \le \sup_N d(E_{\alpha_1}(4\eta, N)).$$

When i tends to ∞, the left hand side tends to $d_0^{\limsup}$. When η tends to 0, the right hand side tends to $f_d^{\limsup}(\alpha_1)$. This proves the required inequality.

2. Let us now show that there exists $\alpha_2 \in D$ such that $d_0^{\text{lim}} \leq f_d^{\text{lim}}(\alpha_2)$. The proof goes along the same lines, but it is somewhat simpler.

First check that $S(\varepsilon) \subset \bigcup_{\alpha \in \mathbb{R}} E_\alpha(\varepsilon) \subset \bigcup_{k \in \mathbb{Z}} E_{2k\varepsilon}(2\varepsilon)$. Deduce that

$$d(S(\varepsilon)) \leq \max_k d(E_{2k\varepsilon}(2\varepsilon)).$$

Now choose $\gamma(\varepsilon)$ such that $d(S(\varepsilon)) \leq d(E_{\gamma(\varepsilon)}(2\varepsilon))$, and a limiting value $\alpha_2 \in D$ of $\gamma(\varepsilon)$. Let $\varepsilon_i \to 0$ be such that $\gamma(\varepsilon_i) \to \alpha_2$. Let $\eta > 0$. If i is large enough, show that $E_{\gamma(\varepsilon_i)}(2\varepsilon_i) \subset E_{\alpha_2}(3\eta)$. This implies that $d(S(\varepsilon_i)) \leq d(E_{\alpha_2}(3\eta))$. Deduce that $d_0^{\text{lim}} \leq f_d^{\text{lim}}(\alpha_2)$. □

Particular cases.
1. If d is σ-stable, then the spectra f_d^{lim} and $f_d^{\text{lim sup}}$ are the same and $d_0^{\text{lim}} = d_0^{\text{lim sup}}$.
2. If $|S| > 0$, then $|S(\varepsilon)| > 0$ for all ε and $S(\varepsilon, N)| > 0$ for N large enough. In this case, $d_0 = d_0^{\text{lim}} = d_0^{\text{lim sup}} = 1$. Both f_d^{lim} and $f_d^{\text{lim sup}}$ reach the value 1. This is also the maximum of f_g.

Remark. Regarding the spectrum f_d, one can show the following: If $\hat{f}_d$ denotes the spectrum

$$\hat{f}_d(\alpha) = \lim_{\varepsilon \to 0} d(\{x / \alpha_n(x) \text{ converges and } |\lim \alpha_n(x) - \alpha| \leq \varepsilon\}),$$

then

$$\sup_\alpha f_d(\alpha) \leq d_0 \leq \sup_\alpha \hat{f}_d(\alpha).$$

This result is less precise than those of Proposition 3. The following example shows that the difference between f_d and the other spectra may be as large as possible.

Example. Consider the generalized Weierstrass function

$$W(x) = \sum_{k=1}^{\infty} \lambda^{-kx} \sin(\lambda^k x),$$

with $\lambda > 1$ and $x \in [0,1]$. It is proved in [13] that $\alpha(x) = x$ for all x. Setting $A(u) = \log v_W(u) / \log |u|$, where $v_W(u)$ is the oscillation of W in u, it is easy to check that the support of all spectra is $[0,1]$. Since E_α contains only one point for every $\alpha \in [0,1]$, $f_d(\alpha) = 0$ identically in $[0,1]$. On the other hand, $\alpha_n(x) \to x$ for all $x \in [0,1]$, thus $d_0 = 1$ and $f_d^{\text{lim sup}}(\alpha) = f_d^{\text{lim}}(\alpha) = f_g(\alpha) = 1$ for all α in $[0,1]$.

3.4. Semi-continuity

We show in this section that $f_d^{\text{lim sup}}$, f_d^{lim} and f_g share a semi-continuity property. Once again f_d does not have this property in general (for a study of the structural properties of f_d with $d = h$, see [12]).

Recall that a function $f : D \subset \mathbb{R} \to \mathbb{R}$ is *upper semi-continuous* if for all $x \in D$, for all sequences (x_n) of D converging to x,

$$\limsup_{n \to \infty} f(x_n) \leq f(x). \qquad (11)$$

Another way to express this property is as follows:

$$\forall x \in D, \forall \varepsilon > 0, \exists \eta(x,\varepsilon) : |x-y| \leq \eta \Rightarrow f(y) \leq f(x) + \varepsilon.$$

Proposition 4. *The functions f_g, $f_d^{\lim}$, $f_d^{\limsup}$ are upper semi-continuous.*

Proof. 1. Let (α_k) be a sequence in D converging to α. Let $\varepsilon > 0$. For any η, $0 < \eta < \varepsilon$, there exists $K(\eta)$ such that $k \geq K(\eta) \Rightarrow [\alpha_k - \eta, \alpha_k + \eta] \subset [\alpha - \varepsilon, \alpha + \varepsilon]$, so that for any n, $N_{\alpha_k}(\eta, n) \leq N_\alpha(\varepsilon, n)$. Therefore,

$$k \geq K(\eta) \Rightarrow \limsup_{n\to\infty} \frac{\log N_{\alpha_k}(\eta,n)}{n \log 2} \leq \limsup_{n\to\infty} \frac{\log N_\alpha(\varepsilon,n)}{n\log 2}.$$

This gives

$$\limsup_{k\to\infty}(\limsup_{n\to\infty} \frac{\log N_{\alpha_k}(n,\eta)}{n\log 2}) \leq \limsup_{n\to\infty} \frac{\log N_\alpha(n,\varepsilon)}{n\log 2}.$$

Since $N_{\alpha_k}(n,\eta)$ decreases as η tends to 0, we deduce that

$$\limsup_{k\to\infty}(\lim_{\eta\to 0}(\limsup_{n\to\infty} \frac{\log N_{\alpha_k}(n,\eta)}{n\log 2})) \leq \limsup_{n\to\infty} \frac{\log N_\alpha(n,\varepsilon)}{n\log 2}.$$

Let ε tend to 0 to get

$$\limsup_{k\to\infty} f_g(\alpha_k) \leq f_g(\alpha). \tag{12}$$

2. Similarly, for any N:

$$k \geq K(\eta) \Rightarrow E_{\alpha_k}(\eta, N) \subset E_\alpha(\varepsilon, N).$$

Letting N tend to ∞, and using the increasing property of d:

$$k \geq K(\eta) \Rightarrow d(E_{\alpha_k}(\eta)) \leq d(E_\alpha(\varepsilon)).$$

Therefore,

$$\limsup_{k\to\infty} d(E_{\alpha_k}(\eta)) \leq d(E_\alpha(\varepsilon)).$$

Since $d(E_{\alpha_k}(\eta))$ decreases as $\eta \to 0$,

$$\limsup_{k\to\infty}(\lim_{\eta\to 0} d(E_{\alpha_k}(\eta))) \leq d(E_\alpha(\varepsilon)).$$

Let ε tend to 0 to get $\limsup_{k\to\infty} f_d^{\lim}(\alpha_k) \leq f_d^{\lim}(\alpha)$. The same type of arguments hold for $f_d^{\limsup}$. □

Notation. For any $f : \mathbb{R} \to \mathbb{R}$, we denote by $\tilde{f}$ the *upper semi-continuous envelope* of f, that is

$$\tilde{f}(\alpha) = \lim_{\varepsilon\to 0}(\sup\{f(\beta), |\beta - \alpha| \leq \varepsilon\}).$$

Inequality (5) and Proposition 4 imply that for all α:

$$\tilde{f}_d(\alpha) \leq f_d^{\lim}(\alpha). \tag{13}$$

In particular, one always has $\tilde{f}_h(\alpha) \leq f_h^{\lim}(\alpha) \leq f_g$.

3.5. Spectrum of a maximum

Proposition 5. *Let A, B be two functions : $X([0,1]) \to \mathbb{R}$, and $C = \max\{A, B\}$. Let $f_g(\alpha, A)$, $f_g(\alpha, B)$, $f_g(\alpha, C)$ be the corresponding spectra. Then, for all α,*

$$f_g(\alpha, C) \leq \max\{f_g(\alpha, A), f_g(\alpha, B)\} \tag{14}$$

Proof. Let $\alpha_n(x) = A(u_n(x))$, $\beta_n(x) = B(u_n(x))$, $\gamma_n(x) = C(u_n(x))$. The relation

$$|\gamma_n(x) - \alpha| \leq \varepsilon \Rightarrow \text{ either } |\alpha_n(x) - \alpha| \leq \varepsilon \text{ or } |\beta_n(x) - \alpha| \leq \varepsilon,$$

valid for all α, n, x, implies the following (with obvious notations):

$$N_\alpha^C(n, \varepsilon) \leq N_\alpha^A(n, \varepsilon) + N_\alpha^B(n, \varepsilon),$$

hence relation (14). □

Proposition 6. *If d is stable, the same result as (14) holds for f_d, $f_d^{\lim}$, $f_d^{\limsup}$.*

Proof. Using similar notations, we observe that

$$E_\alpha^C(\varepsilon, N) \subset E_\alpha^A(\varepsilon, N) \cup E_\alpha^B(\varepsilon, N).$$

The stability of d implies

$$d(E_\alpha^C(\varepsilon, N)) \leq \max\{d(E_\alpha^A(\varepsilon, N)), d(E_\alpha^B(\varepsilon, N))\},$$

so that

$$f_d^{\limsup}(\alpha, C) \leq \max\{f_d^{\limsup}(\alpha, A), f_d^{\limsup}(\alpha, B)\}.$$

The proof for the other spectra uses similar arguments. □

4. The inverse problem for spectra

We have seen that the three spectra $f_d^{\limsup}$, $f_d^{\lim}$ and f_g are upper semi-continuous. It is natural to enquire whether the converse holds, i.e. whether any u.s.c. function F is the spectrum of some function A. The main result of this section answers by the affirmative for $f_d^{\limsup}$ and $f_d^{\lim}$. The case of f_g is treated in [5].

4.1. Generalities on u.s.c. functions

Let us recall some well-known facts. For $E \subset \mathbb{R}$, one has: E is closed $\Leftrightarrow 1_E$ is u.s.c, where 1_E is the characteristic function of E. If f is an upper semi-continuous function defined on a compact set D and ranging in $\mathbb{R}$, then:

- f is bounded from above and reaches its maximum.
- f can be written as $f = \inf\{g \geq f, \text{g continuous on } D\}$.

We shall need the following result:

Lemma 2. *The function $f : D \to \mathbb{R}$ is u.s.c. iff there exists a countable set E dense in D such that*

1. *The restriction $f_{|E}$ is u.s.c.*
2. *In $D - E$, f can be obtained by semi-continuity:*

$$\forall x \in D - E,\ f(x) = \lim_{\varepsilon \to 0} \sup\{f(y) : |x - y| \leq \varepsilon, y \in E\}.$$

Proof. $\Rightarrow$ Assume that f is u.s.c.. Construct E as follows: In every dyadic interval $[k\,2^{-n},(k+1)2^{-n}]$, choose a point $x_{k,n}$ where f reaches its maximum. Take $E=\{x_{k,n}:k\in\mathbb{Z},n\in\mathbb{N}\}$. The restriction $f_{|E}$ is u.s.c.. Let $x\in D-E$. For every n, there exists $x_{k,n}$ in the interval $u_n(x)$ such that $f(x)\le f(x_{k,n})$. Therefore $f(x)\le \sup\{f(y)/y\in E,|x-y|\le 2^{-n}\}$. Letting $n\to+\infty$, we obtain

$$f(x)\le \limsup_{\varepsilon\to 0}\{f(y)/y\in E,|x-y|\le\varepsilon\}.$$

Since f is u.s.c., we obtain an equality.

$\Leftarrow$ It suffices to show that, if (x_n) is a sequence of $D-E$ converging to x (which is necessarily in D), then relation (11) is verified. For all $x'\in D$ and $\varepsilon>0$, let:

$$s(x',\varepsilon)=\sup\{f(y)/|x'-y|\le\varepsilon,y\in E\}.$$

The function $\varepsilon\to s(x',\varepsilon)$ is non decreasing by construction. Using Assumption 2, $s(x_n,\varepsilon)$ converges to $f(x_n)$ as $\varepsilon\to 0$. Thus, for all n, there exists ε small enough to ensure that $f(x_n)\le s(x_n,\varepsilon)\le f(x_n)+1/2n$. There exists $y_n\in E$ such that $|x_n-y_n|\le\min\{\varepsilon,1/n\}$ and $f(y_n)\le s(x_n,\varepsilon)\le f(y_n)+1/2n$. Then $|f(y_n)-f(x_n)|\le 1/n$. The sequence (y_n) tends to x. If $x\in E$, then $f(x)\ge\limsup_n f(y_n)$ from Assumption 1. If $x\in D-E$, the same result stems from Assumption 2. Then $f(x)\ge\limsup_n f(y_n)=\limsup_n f(x_n)$. □

Lemma 2 shows that, as is the case for continuous functions, a u.s.c. function f is fully determined by its values on a countable set E. However, contrarily to continuous functions, the set E here depends on f.

4.2. Construction of A when d is σ-stable

Let a compact set $D\subset\mathbb{R}$, and a semi-continuous function $F:D\longrightarrow[0,1]\cup\{-\infty\}$ such that $\sup_D F=1$, be given. We want to construct a function A, the $f_d^{\lim}$ or $f_d^{\limsup}$ spectrum of which is equal to F. By semi-continuity, the set $F^{-1}([0,1])$ is closed. Then we may assume that $F(D)\subset[0,1]$ without loss of generality. The function F can be extended to $\mathbb{R}$ by defining $F(x)=-\infty$ if $x\notin D$.

Lemma 2 shows that it suffices to match the values of F on a countable set $E=\{\alpha_k\}$ dense in D. On $D-E$, F is obtained by semi-continuity. We can assume that $F(\alpha_0)=1$, and that $\alpha_i\ne\alpha_j$ if $i\ne j$. We will construct a function $A:X([0,1])\longrightarrow\mathbb{R}$ such that

$$f_d(\alpha_k)=f_d^{\lim}(\alpha_k)=f_d^{\limsup}(\alpha_k)=F(\alpha_k)\tag{15}$$

for all $k\ge 0$. This implies that

$$f_d^{\lim}(\alpha)=f_d^{\limsup}(\alpha)=F(\alpha)\tag{16}$$

for all $\alpha\in D$.

Let (I_k) be a family of closed, non degenerate, disjoint intervals in $[0,1]$. For all $k\ge 1$, let $C_k\subset I_k$ be a compact set such that

$$d(C_k)=F(\alpha_k).$$

Let $C_0 = [0,1] - \bigcup_{k=1}^{\infty} C_k$. Since C_0 contains an open set, $d(C_0) = 1$. For any dyadic interval u, define

$$A(u) = \alpha_{l(u)}\ , \text{ where } l(u) = \begin{cases} 0 \text{ if } u \subset C_0 \\ \min\{k \geq 1/u \cap C_k \neq \emptyset\} \text{ otherwise} \end{cases}$$

Since the intervals I_k are disjoint, there exists for every $k \geq 1$ an integer N_k such that for all $n \geq N_k$, for all $x \in C_k$, $u_n(x) \subset C_0 \cup C_k$. Then $A(u_n(x)) = \alpha_k$. Therefore $E_{\alpha_k} = C_k$ and

$$f_d(\alpha_k) = F(\alpha_k).$$

For all $\varepsilon > 0$,

$$E_{\alpha_k}(\varepsilon) = \bigcup\{C_i : |\alpha_i - \alpha_k| \leq \varepsilon\}, \tag{17}$$

so that

$$d(E_{\alpha_k}(\varepsilon)) \geq \sup\{d(C_i) : |\alpha_i - \alpha_k| \leq \varepsilon\}.$$

Since d is σ-stable, this is an equality, so that

$$f_d^{\lim}(\alpha_k) = \lim_{\varepsilon \to 0} \sup\{F(\alpha_i) : |\alpha_i - \alpha_k| \leq \varepsilon\}.$$

The right hand side member is equal to $F(\alpha_k)$. Finally, the set C_k is included in $E_{\alpha_k}(\varepsilon, N)$ for all $N \geq N_k$, so that $F(\alpha_k) \leq f_d^{\lim\sup}(\alpha_k)$. Using (6), we obtain (15) for $k \geq 1$. For $k = 0$, $f_d^{\lim}(\alpha_0) = f_d^{\lim\sup}(\alpha_0) = F(\alpha_0) = 1$.

4.3. Construction of A when $d = \Delta$

We will use the same A as before, with extra conditions on the sequences (I_k) and (C_k).

Lemma 3. *Given a dyadic interval $[0, 2^{-N}]$, and a real number $\delta \in (0,1)$, there exists a compact set C such that $\Delta(C) = \delta$, and for all $n \geq N$:*

$$\omega(2^n, C) \leq 2^{(n-N)\delta}. \tag{18}$$

Proof. For any sequence (ω_n) of integers such that

$$\omega_N = 1\, , \omega_n \leq \omega_{n+1} \leq 2\omega_n, \tag{19}$$

there are infinitely many ways to construct a compact set C as the limit of embedded coverings by ω_n dyadic intervals; such a set verifies $\omega(2^n, C) = \omega_n$. Let us define

$$\omega_n = 2^{E[(n-N)\delta]},$$

where $E[.]$ denotes the integer part. It is clear that $\omega_n \leq \omega_{n+1}$. On the other hand,

$$\delta \leq 1 \Longrightarrow (n - N + 1)\delta \leq (n - N)\delta + 1,$$

so that

$$E[(n - N + 1)\delta] \leq E[(n - N)\delta] + 1,$$

which gives $\omega_{n+1} \leq 2\omega_n$. Then the conditions given in (19) are fulfilled. This proves the existence of the set C. Finally, (18) is also verified, and

$$\Delta(C) = \limsup \frac{\log \omega_n}{n \log 2} = \delta. \qquad \square$$

Lemma 4. *Let $I_k = [2^{-2k}, 2^{-2k+1}]$, $\delta_k \in [0,1]$, and C_k be a compact set included in I_k such that $\Delta(C_k) = \delta_k$ and for all $n \geq 2^{2k}$,*

$$\omega(2^n, C_k) \leq 2^{(n-2k)\delta_k}. \tag{20}$$

Then for any strictly increasing sequence (k_i) of integers:

$$\Delta(\bigcup_i C_{k_i}) = \sup_i \Delta(C_{k_i}). \tag{21}$$

The existence of C_k is proved in Lemma 3.

Proof. Let $\delta = \sup_i \Delta(C_{k_i})$. The inequality $\Delta(\bigcup_i C_{k_i}) \geq \delta$ is trivial. For the reverse inequality, choose for any n the integer i_n such that

$$2k_{i_n} \leq n \leq 2k_{i_n+1}.$$

Then $\cup_{i \geq i_n+1} C_{k_i} \subset [0, 2^{-n}]$, and

$$\omega(2^n, \bigcup_i C_{k_i}) \leq 1 + \sum_{i=1}^{i_n} \omega(2^n, C_{k_i}).$$

Using (20),

$$i \leq i_n \Rightarrow \omega(2^n, C_{k_i}) \leq 2^{(n-2k_i)\delta} \leq 2^{n\delta}.$$

Since $i_n \leq n$,

$$\omega(2^n, \bigcup_i C_{k_i}) \leq 1 + n2^{n\delta}.$$

Thus

$$\Delta(\bigcup_i C_{k_i}) \leq \limsup \frac{\log(1 + n2^{n\delta})}{n \log 2} = \delta. \qquad \square$$

Let us now take (I_k) and (C_k) as in Lemma 4, with $\delta_k = F(\alpha_k)$, and come back to Equation (17). Using (21) we obtain:

$$\Delta(E_{\alpha_k}(\varepsilon)) = \sup\{\Delta(C_i)/|\alpha_i - \alpha_k| \leq \varepsilon\},$$

so that

$$f_\Delta^{\lim}(\alpha_k) = \lim_{\varepsilon \to 0} \sup\{F(\alpha_i)/|\alpha_i - \alpha_k| \leq \varepsilon\} = F(\alpha_k).$$

We conclude as before.

5. Examples of spectra

In this section, we provide various examples of computation of the new spectra. They are meant to show that the inequalities between the spectra may be strict. We choose for d the box dimension, as it is the one most often used in applications.

Example 1. $f_\Delta(\alpha) < f_\Delta^{\lim}(\alpha)$ *for all* α. Let $A([a,b]) = a$ for all $[a,b] \subset [0,1]$. Then $D = [0,1]$, and for all $x \in [0,1]$, $\alpha(x) = x$. For all N,

$$E_\alpha(\varepsilon, N) = E_\alpha(\varepsilon) = [\alpha - \varepsilon, \alpha + \varepsilon] \quad , \quad E_\alpha = \{\alpha\}.$$

Therefore $\Delta(E_\alpha) = 0$, $\Delta(E_\alpha(\varepsilon)) = \Delta(E_\alpha(\varepsilon, N)) = 1$. Finally, $N_\alpha(\varepsilon, n) \simeq 2\varepsilon/2^{-n}$. For all $\alpha \in D$, we obtain

$$f_\Delta(\alpha) = 0 \, , \; f_\Delta^{\lim}(\alpha) = f_\Delta^{\limsup}(\alpha) = 1 \, , \; f_g(\alpha) = 1. \tag{22}$$

It is not possible to find a function or a measure whose Hölder regularity is exactly A above. However, it is easily checked that the set function defined on intervals by $C([a,b]) = |b-a|^a$ extends to a Choquet capacity on the Borel subsets of $[0,1]$, to which a multifractal analysis may be applied (see [12]). Alternatively, one may relax the condition $A([a,b]) = a$. Indeed, the computations above still apply when, for all $u \subset [0,1]$,

$$A(u) + O(\frac{1}{\log |u|}) \in u \tag{23}$$

uniformly with respect to $|u|$. This situation is illustrated by the generalized Weierstrass function W (see Section 3.3), with $A(u) = \log v_W(u)/\log|u|$ and $v_W(u)$ the oscillation of W in u. There exists constants $C_1 \neq 0$ and C_2 such that:

$$\forall\, u = [a,b], \quad C_1|u|^b \le v_W(u) \le C_2|u|^a.$$

Then (23) is verified, and the spectra take the values shown in (22).

Example 2. $f_\Delta^{\limsup}(\alpha) < f_\Delta^{\lim}(\alpha)$ *for some* α. This example shows that $f_d^{\limsup}$ may be different from $f_d^{\lim}$ when d is not σ-stable.

Let $\gamma > 0$ and $F = \{k^{-\gamma} : k \ge 1\}$. Then $\Delta(F) = 1/(\gamma+1)$. The left extremity of u is denoted by $u_{\min}$. Let $p : [0,1] \longrightarrow \mathbb{R}$ be a strictly increasing, continuous function such that $p(0) = 0$. For all $u \in X([0,1])$, let

$$A(u) = \begin{cases} 0 & \text{if } u \cap F \neq \emptyset \text{ and } p(|u|) < u_{\min} \\ 1 & \text{otherwise} \end{cases}$$

If $x \in F$, $u_n(x) \cap F \neq \emptyset$. There exists a smallest integer $N(x)$ such that $n \ge N(x) \Rightarrow p(2^{-n}) \le u_n(x)_{\min}$, so that $\alpha_n(x) = 0$. For all $\varepsilon = 2^{-n}$, the set $E_0(\varepsilon, N) = \{k^{-\gamma}/N(k^{-\gamma}) \le N\}$ is finite, and $E_0(\varepsilon) = F$. If $x \notin F$, and n is large enough, $\alpha_n(x) = 1$. Let us take two real numbers $c < d$ and an integer K such that $2^{-\gamma} < c - 2^{-K} \le d + 2^{-K} < 1$. For all $x \in [c,d]$ and $n \ge K$, then $u_n(x) \cap F = \emptyset$, so that $\alpha_n(x) = 1$. Therefore $[c,d] \subset E_1(2^{-K}, N)$ for all $N \ge K$. We deduce that for all $\varepsilon < 2^{-K}$, $\sup_N \Delta(E_1(\varepsilon, N)) = 1$. Therefore,

$$f_\Delta^{\limsup}(\alpha) = \begin{cases} 0 & \text{if } \alpha = 0 \\ 1 & \text{if } \alpha = 1 \end{cases}$$

and

$$f_\Delta^{\lim}(\alpha) = f_\Delta(\alpha) = \begin{cases} \Delta(F) & \text{if } \alpha = 0 \\ 1 & \text{if } \alpha = 1 \, . \end{cases}$$

These results do not depend on the function p.

To illustrate this case in the frame of classical multifractal analysis, we deal with a slightly more complex but similar situation, and consider the measure defined as

$$\mu = \mathcal{L} + \sum_{k=1}^{\infty} a_k \delta_{k^{-\gamma}}$$

where $\mathcal{L}$ is the Lebesgue measure on $[0,1]$, δ_x is the Dirac mass at x and (a_k) is a sequence of real numbers decreasing to 0 and such that $\sum_{k=1}^{\infty} a_k$ converges. Let $A(u) = \log\mu(u)/\log|u|$. If $x = k^{-\gamma}$, then $\mu(u_n(x)) = 2^{-n} + a_k$ if n is large enough. Since $a_k > 0$, $\alpha(u_n(k^{-\gamma})) \to 0$. But this convergence is not uniform with respect to k, so that for every N the set $E_0(\varepsilon, N)$ is finite as before. If $x \notin F$, and $x \neq 0$, then $\mu(u_n(x)) = 2^{-n}$ when n large, so that $\alpha(u_n(x)) \to 1$. The results on the spectra are the same.

Example 3. $f_\Delta^{\lim}(\alpha) \neq f_g(\alpha)$ *for some value of* α. We need to show that there is no general relationship between these two spectra. Let β, γ, δ, ω be such that $0 < \delta < \min\{\beta, 1\}$ and $0 < \omega < \gamma/(\gamma+1)$. Let $F = \{k^{-\gamma}\}$ as in Example 2. For all $u \in X([0,1])$, let

$$A(u) = \begin{cases} 0 & \text{if } |u|^\beta \le u_{\min} \le |u|^\delta, \\ & \text{or } |u|^\omega < u_{\min} \text{ and } u \cap F \neq \emptyset \\ 1 & \text{otherwise.} \end{cases}$$

The function $p(x)$ of Example 2 is replaced by x^ω. Results on the spectra f_Δ, $f_\Delta^{\lim}$, $f_\Delta^{\lim\sup}$ are unchanged. Let us compute f_g.

The number of dyadic intervals which do not meet F is of the order of 2^n. Therefore $f_g(1) = 1$.

Let us evaluate the number K_n of dyadic intervals u_n such that $u_n \cap F \neq \emptyset$ and $(u_n)_{\min} > 2^{-n\omega}$. Note that the minimum distance between two points of F in $[x, 1]$ is $\rho(x) \simeq k^{-\gamma} - (k+1)^{-\gamma} \simeq k^{-\gamma-1}$, where $k^{-\gamma} \simeq x$. Therefore $\rho(x) \simeq x^{(\gamma+1)/\gamma}$. Letting $x = 2^{-n\omega}$, $\rho(2^{-n\omega})$ is equivalent to $2^{-n\omega(\gamma+1)/\gamma}$ which is larger than 2^{-n} since $\omega < \gamma/(\gamma+1)$. We deduce that $K_n \simeq \mathrm{Card}(F \cap [2^{-n\omega}, 1])$. Since $K_n^{-\gamma} \simeq 2^{-n\omega}$, then $K_n \simeq 2^{n\omega/\gamma}$. Since $\delta < 1$, the number of dyadic intervals u_n such that $2^{-n\beta} \le (u_n)_{\min} \le 2^{-n\delta}$ is $K'_n \simeq 2^n(2^{-n\delta} - 2^{-n\beta}) \simeq 2^{n(1-\delta)}$.

The number of u_n verifying $A(u_n) = 0$ is equivalent to $K_n + K'_n$. Therefore

$$f_g(\alpha) = \begin{cases} \max\{\frac{\omega}{\gamma}, 1-\delta\} & \text{if } \alpha = 0 \\ 1 & \text{if } \alpha = 1\,. \end{cases}$$

This result does not depend on β.

For a numerical application, take $\gamma = 1$, $\omega = 1/3$, $\delta = 2/3$: Then $f_g(0) = 1/3$, $f_\Delta^{\lim}(0) = 1/2$. Let now $\gamma = 1$, $\omega = 1/3$, $\delta = 1/3$: Then $f_g(0) = 2/3$, $f_\Delta^{\lim}(0) = 1/2$. The spectrum $f_\Delta^{\lim}$ can thus be larger or smaller than f_g for some values of α.

To exhibit a function whose Hölder regularity is similar to the above function A, we shall use a technique based on wavelets. Fix a wavelet ψ in the Schwartz class such that the functions $t \to \psi_{j,k}(t) = 2^{j/2}\psi(2^j t - k)$, $j, k \in \mathbb{Z}$ form an orthonormal basis of L^2, and define a function z by its wavelet coefficients in this basis, $c_{j,k} = 2^j \int \psi_{j,k}(t)z(t)dt$ (note that we use here an L^∞ normalization rather than an L^2 one). We set:

$$c_{j,k} = \begin{cases} 1 & \text{if } 2^{-n\beta} \leq k2^{-n} \leq 2^{-n\delta}, \\ & \text{or } 2^{-n\omega} < k2^{-n} \text{ and } u_n^k \cap F \neq \emptyset \\ 2^{-n} & \text{otherwise} \end{cases}$$

where $u_n^k = [k2^{-n}, (k+1)2^{-n}]$. One immediately checks that z is in L^2 if $\beta < \gamma$. Choose now as is usual $A(u_n^k) = |\log(c_{n,k})|/n$. This yields the desired behaviour. Note that z depends on the wavelet ψ.

6. Continuous spectra

The definition of the large deviation spectrum f_g has two drawbacks: First it depends on the choice of the interval partitions (usually the dyadic intervals), and second it uses two limiting operations (n tends to $+\infty$, then ε tends to 0) which makes it difficult to evaluate from a given set of data (see [9] for more on this topic). This section introduce variants of f_g, denoted by f_g^c and $\tilde{f}_g^c$, and called the continuous large deviation spectra. They are independent of any interval partition. Moreover $\tilde{f}_g^c$ uses only one limiting operation. These spectra are helpful in numerical applications of multifractal analysis.

As before, A is a function $X([0,1]) \longrightarrow \mathbb{R}^+ \cup \{+\infty\}$. Recall that for any set family $\mathcal{F}$, $\cup\mathcal{F}$ denotes the union of all sets in $\mathcal{F}$. For any measurable set E of the line, $|E|$ is its Lebesgue measure. Let $\eta \in (0,1)$. We introduce families of intervals:

$$\mathcal{R}_\eta = \{u \in X([0,1]) \text{ such that } |u| = \eta\}$$

$$\mathcal{R}_\eta(\alpha) = \{u \in X([0,1]) \text{ such that } |u| = \eta \text{ and } A(u) = \alpha\}$$

$$\mathcal{R}_\eta^\varepsilon(\alpha) = \{u \in X([0,1]) \text{ such that } |u| = \eta \text{ and } |A(u) - \alpha| \leq \varepsilon\}.$$

By convention $\log 0/\log\eta = +\infty$.

Definition 4. *The continuous large deviation spectra are*

$$f_g^c(\alpha) = \lim_{\varepsilon\to 0}\limsup_{\eta\to 0} \frac{\log\left(|\bigcup\mathcal{R}_\eta^\varepsilon(\alpha)|/\eta\right)}{|\log\eta|} = \lim_{\varepsilon\to 0}\limsup_{\eta\to 0}\left(1 - \frac{\log|\bigcup\mathcal{R}_\eta^\varepsilon(\alpha)|}{\log\eta}\right)$$

$$\tilde{f}_g^c(\alpha) = \limsup_{\eta\to 0} \frac{\log\left(|\bigcup\mathcal{R}_\eta(\alpha)|/\eta\right)}{|\log\eta|} = \limsup_{\eta\to 0}\left(1 - \frac{\log|\bigcup\mathcal{R}_\eta(\alpha)|}{\log\eta}\right)$$

The integer $N_\alpha(\varepsilon, n)$ used in the definition of f_g has been replaced by a mean number $|\bigcup\mathcal{R}_\eta^\varepsilon(\alpha)|/\eta$ in the definition of f_g^c. In many applications, A is continuous and $\mathcal{R}_\eta(\alpha)$ is non empty for a whole range of values of α: This allows to define the spectrum $\tilde{f}_g^c$, which avoids the use of the "ε-tolerance".

Recall that, when the so-called weak multifractal formalism holds, f_g can be obtained as the Legendre transform of the following function:

$$\tau(q) = \liminf_{n\to\infty} \frac{\log S_n(q)}{-n}$$

where, for all $q \in \mathbb{R}$, $S_n(q)$ is defined with dyadic intervals as follows:

$$S_n(q) = \sum_{k=0}^{2^n-1} 2^{-nqA(u_n^k)} \tag{24}$$

with the convention $0^q = 0$. One denotes by f_l the Legendre transform of τ:

$$f_l(\alpha) := \tau^*(\alpha) = \inf_{q\in\mathbb{R}} (q\alpha - \tau(q)).$$

The function f_l is called the *Legendre multifractal spectrum*, and the equality $f_l = f_g$, when it holds, is essentially a consequence of Ellis theorem (see [6]).

Let us introduce similar notions in a continuous framework. An interval family is a *packing* if all of its intervals are disjoint. For any $q \in \mathbb{R}$ and for any interval family $\mathcal{R}$, let

$$H^q(\mathcal{R}) = \sup\{\sum_{u\in\mathcal{R}'} |u|^{qA(u)}, \mathcal{R}' \subset \mathcal{R}, \mathcal{R}' \text{ is a packing}\}.$$

By convention $H^q(\emptyset) = 0$. Here are some basic properties of $H^q(\mathcal{R})$:

Lemma 5. *Let $\mathcal{R}_1, \mathcal{R}_2$ be two families of interval.*

1. $\mathcal{R}_1 \subset \mathcal{R}_2 \Longrightarrow H^q(\mathcal{R}_1) \leq H^q(\mathcal{R}_2)$.
2. $H^q(\mathcal{R}_1 \cup \mathcal{R}_2) \leq H^q(\mathcal{R}_1) + H^q(\mathcal{R}_2)$.
3. *If $|u| = \eta$ and $\alpha \leq A(u) \leq \beta$ for all $u \in \mathcal{R}$, then*

$$\forall q \geq 0, \qquad \frac{1}{2}|\bigcup \mathcal{R}|\,\eta^{\beta q-1} \leq H^q(\mathcal{R}) \leq |\bigcup \mathcal{R}|\,\eta^{\alpha q-1}. \tag{25}$$

Proof. Property 1 comes from the fact that for any packing $\mathcal{R}_3 \subset \mathcal{R}_1$ we have $\mathcal{R}_3 \subset \mathcal{R}_2$. For Property 2, take a packing $\mathcal{R}'$ in $\mathcal{R}_1 \cup \mathcal{R}_2$. Let $\mathcal{R}'_1 = \mathcal{R}' \cap \mathcal{R}_1$ and $\mathcal{R}'_2 = \mathcal{R}' \cap \mathcal{R}_2$. Then

$$\sum_{u\in\mathcal{R}'} |u|^{qA(u)} \leq \sum_{u\in\mathcal{R}'_1} |u|^{qA(u)} + \sum_{u\in\mathcal{R}'_2} |u|^{qA(u)} \leq H^q(\mathcal{R}_1) + H^q(\mathcal{R}_2).$$

For Property 3, it suffices to consider the case where $\bigcup \mathcal{R}$ is an interval. Let M_η be the maximum number of intervals of $\mathcal{R}$ covering $\bigcup \mathcal{R}$. Show that

$$\eta M_\eta \leq |\bigcup \mathcal{R}| \leq 2\eta M_\eta.$$

Then use the inequalities $\eta^{\beta q} \leq |u|^{qA(u)} \leq \eta^{\alpha q}$ for all $u \in \mathcal{R}$ and $q > 0$. □

To define a continuous counterpart to $S_n(q)$, we simply take $H^q_\eta := H^q(\mathcal{R}_\eta)$. It can be written as

$$H^q_\eta = \sup\{\sum_{u\in\mathcal{R}'} \eta^{qA(u)}, \mathcal{R}' \text{ is a packing of } [0,1] \text{ by intervals of length } \eta\}.$$

The relevant quantity that corresponds to $\tilde{f}_g^c$ is:

$$\begin{aligned} J_\eta^q &= \sup_\alpha H^q(\mathcal{R}_\eta(\alpha)) \\ &= \sup_\alpha \sup\{\sum_{u\in\mathcal{R}'} \eta^{\alpha q},\ \mathcal{R}' \text{ is a packing such that } |u|=\eta \text{ and } A(u)=\alpha\}. \end{aligned}$$

Definition 5. *Let*

$$\tau^c(q) = \liminf_{\eta\to 0} \frac{\log H_\eta^q}{\log \eta}$$

and

$$\tilde{\tau}^c(q) = \liminf_{\eta\to 0} \frac{\log J_\eta^q}{\log \eta}.$$

The continuous Legendre spectra *are defined as* $f_l^c = (\tau^c)^*$ *and* $\tilde{f}_l^c = (\tilde{\tau}^c)^*$.

Here are some obvious properties of $f_g^c, \tilde{f}_g^c, f_l^c, \tilde{f}_l^c$.

Proposition 7.

1. f_l^c *and* $\tilde{f}_l^c$ *are concave.*
2. $\forall\ \alpha, \tilde{f}_g^c(\alpha) \le f_g^c(\alpha)$ *and* $f_g(\alpha) \le f_g^c(\alpha)$.
3. *If* μ *is a multinomial measure,* $f_g^c(\alpha) = \tilde{f}_g^c(\alpha) = f_l^c(\alpha) = \tilde{f}_l^c(\alpha) = f_g(\alpha)$.

A fundamental property of the continuous spectrum is that, under a mild restriction, f_l^c is the concave envelope of f_g^c:

Proposition 8.

1. *For all* $\alpha, f_l^c(\alpha) \ge (f_g^c(\alpha))^{**}$ *and* $\tilde{f}_l^c(\alpha) \ge (\tilde{f}_g^c(\alpha))^{**}$.
2. *Assume that* $f_g^c = -\infty$ *outside a compact interval. Then:*

$$\forall \alpha, \quad f_l^c(\alpha) = (f_g^c(\alpha))^{**}.$$

For part 2, we will need a corollary of the next Lemma, which is of independent interest:

Lemma 6. *Let* $\mathcal{X}(\mathbb{R})$ *be the family of all closed intervals in* $\mathbb{R}$. *Let* $F : \mathcal{X}(\mathbb{R}) \to \mathbb{R} \cup \{-\infty\}$ *be such that*

$$I \subset J \Rightarrow F(I) \le F(J).$$

For any $\alpha \in \mathbb{R}$, *define* $f(\alpha) = \lim_{\varepsilon\to 0} F([\alpha-\varepsilon, \alpha+\varepsilon])$. *Then, for any compact interval* J *and for any real* q,

$$\lim_{\varepsilon\to 0} \inf_{\alpha\in J} \{\alpha q - F([\alpha-\varepsilon,\alpha+\varepsilon])\} = \inf_{\alpha\in J}\{\alpha q - f(\alpha)\}. \tag{26}$$

Proof. Fix $q \in \mathbb{R}$. Let $l_\varepsilon = \inf_{\alpha\in J}\{q\alpha - F([\alpha-\varepsilon,\alpha+\varepsilon])\}$, $l = \inf_{\alpha\in J}\{q\alpha - f(\alpha)\}$. Since $F([\alpha-\varepsilon,\alpha+\varepsilon])$ decreases to $f(\alpha)$ as $\varepsilon \to 0$, l_ε increases and is bounded by l. Therefore we must show that $l \le \lim_\varepsilon l_\varepsilon$. For all $\varepsilon > 0$, there exists $\alpha(\varepsilon) \in J$ such that

$$q\,\alpha(\varepsilon) - F([\alpha(\varepsilon)-\varepsilon, \alpha(\varepsilon)+\varepsilon]) \le l_\varepsilon + \varepsilon. \tag{27}$$

Let β be a limit value of the sequence $(\alpha(1/n))$ in the compact J, and $r > 0$. Let $N \geq 2/r$. There exists $n \geq N$ such that $|\alpha(1/n) - \beta| \leq r/2$. Since $1/n \leq r/2$,

$$[\alpha(\frac{1}{n}) - \frac{1}{n}, \alpha(\frac{1}{n}) + \frac{1}{n}] \subset [\beta - \varepsilon, \beta + \varepsilon]$$

so that $F([\alpha(1/n) - 1/n, \alpha(1/n) + 1/n]) \leq F([\beta - r, \beta + r])$. We obtain

$$q\,\beta - F([\beta - r, \beta + r]) \leq q(\alpha(\frac{1}{n}) + \frac{r}{2}) - F([\alpha(\frac{1}{n}) - \frac{1}{n}, \alpha(\frac{1}{n}) + \frac{1}{n}]) \quad (28)$$

$$\leq l_{1/n} + \frac{1}{n} + q\frac{r}{2}. \quad (29)$$

Since n can be taken to be arbitrarily large, inequality (29) implies

$$q\,\beta - F([\beta - r, \beta + r]) \leq \lim_{\varepsilon \to 0} l_\varepsilon + q\frac{r}{2}.$$

When $r \to 0$, $F([\beta - r, \beta + r])$ tend to $f(\beta)$, so that $q\,\beta - f(\beta) \leq \lim_{\varepsilon \to 0} l_\varepsilon$. □

This result may be applied directly to $f_g(\alpha, \varepsilon)$ or to $f_g^c(\alpha, \varepsilon)$, when these functions have compact support. Therefore:

Corollary 1. *If there exists a compact interval J such that $f_g(\alpha, \varepsilon) = -\infty$ for all $\alpha \notin J$ and $\varepsilon \leq 1$, then*

$$\lim_{\varepsilon \to 0} \inf_\alpha \{q\alpha - f_g(\alpha, \varepsilon)\} = \inf_\alpha \{q\alpha - f_g(\alpha)\}.$$

If there exists a compact interval J such that $f_g^c(\alpha, \varepsilon) = -\infty$ for $\alpha \notin J$ and $\varepsilon \leq 1$:

$$\lim_{\varepsilon \to 0} \inf_\alpha \{q\alpha - f_g^c(\alpha, \varepsilon)\} = \inf_\alpha \{q\alpha - f_g^c(\alpha)\}.$$

Remark. The corollary remains true even when $f_g^c(\alpha)$ (resp. $f_g(\alpha)$) does not have a compact support (see [5]).

Proof of Proposition 8. We treat the case $q \geq 0$.
1. We shall prove the equivalent statements:

$$\tau^c(q) \leq \inf_\alpha (q\alpha - f_g^c(\alpha)) \text{ and } \tilde{\tau}^c(q) \leq \inf_\alpha (q\alpha - \tilde{f}_g^c(\alpha)).$$

For all α and $\varepsilon > 0$, $\mathcal{R}_\eta^\varepsilon(\alpha) \subset \mathcal{R}_\eta$, so that

$$H^q(\mathcal{R}_\eta^\varepsilon(\alpha)) \leq H_\eta^q. \quad (30)$$

Relation (25) implies that

$$H^q(\mathcal{R}_\eta^\varepsilon(\alpha)) \geq \frac{1}{2}|\bigcup \mathcal{R}_\eta^\varepsilon(\alpha)|\,\eta^{q(\alpha+\varepsilon)-1}. \quad (31)$$

For all $\delta > 0$, there exists $\varepsilon_0 > 0$ such that for all $\varepsilon \leq \varepsilon_0$,

$$f_g^c(\alpha) \leq \limsup_{\eta \to 0} \frac{\log(|\bigcup \mathcal{R}_\eta^\varepsilon(\alpha)|/\eta)}{|\log \eta|} + \frac{\delta}{2}.$$

For all $\eta_0 > 0$, there exists $\eta \leq \eta_0$ such that

$$f_g^c(\alpha) \leq \frac{\log(|\bigcup \mathcal{R}_\eta^\varepsilon(\alpha)|/\eta)}{|\log \eta|} + \delta,$$

so that

$$|\bigcup \mathcal{R}_\eta^\varepsilon(\alpha)| \geq \eta^{-f_g^c(\alpha)+\delta+1}. \tag{32}$$

Gathering (30), (31), (32), we get: $H_\eta^q \geq \frac{1}{2}\eta^{q(\alpha+\varepsilon)-f_g^c(\alpha)+\delta}$. Taking the logarithm on both sides and using the definition of τ^c as a lim inf:

$$\tau^c(q) \leq q\alpha - f_g^c(\alpha) + q\varepsilon + \delta.$$

Let $\varepsilon \to 0$, then $\delta \to 0$ to obtain $\tau^c(q) \leq \inf_\alpha(q\alpha - f_g^c(\alpha))$.

Let us now consider $\tilde{f}_g^c(\alpha)$. For all α, $J_\eta^q \geq \frac{1}{2}H^q(\mathcal{R}_\eta(\alpha)) \geq \frac{1}{2}|\bigcup \mathcal{R}_\eta(\alpha)|\,\eta^{q\alpha-1}$. We deduce that for all $\delta > 0$, $\eta_0 > 0$, there exists $\eta \leq \eta_0$ such that

$$J_\eta^q \geq \frac{1}{2}\eta^{q\alpha-\tilde{f}_g^c(\alpha)+\delta}.$$

Therefore $\tilde{\tau}^c(q) \leq q\alpha - \tilde{f}_g^c(\alpha) + \delta$.

2. Let $\varepsilon > 0$. Let $L_\varepsilon(q) = \inf_\alpha\{q\alpha - f_g^c(\alpha,\varepsilon)\}$. For every α there exists $\eta(\alpha,\varepsilon) < 1$ such that

$$\eta \leq \eta(\alpha,\varepsilon) \Rightarrow \frac{\log(|\bigcup \mathcal{R}_\eta^\varepsilon(\alpha)|/\eta)}{|\log \eta|} \leq f_g^c(\alpha,\varepsilon) + \varepsilon.$$

Using (25):

$$\begin{aligned} H^q(\mathcal{R}_\eta^\varepsilon(\alpha)) &\leq |\bigcup \mathcal{R}_\eta^\varepsilon(\alpha)|\eta^{q\alpha-|q|\varepsilon-1} \\ &\leq \eta^{q\alpha-f_g^c(\alpha,\varepsilon)-\varepsilon(1+|q|)} \\ &\leq \eta^{L_\varepsilon(q)-\varepsilon(1+|q|)}. \end{aligned}$$

Let α_1, ..., α_K be a finite sequence such that $J \subset \bigcup_i[\alpha_i - \varepsilon, \alpha_i + \varepsilon]$. For $\eta \leq \min_i\{\eta(\alpha_i,\varepsilon)\}$:

$$H_\eta^q \leq K\,\eta^{L_\varepsilon(q)-\varepsilon(1+|q|)}.$$

Taking logarithms:

$$\frac{\log H_\eta^q}{\log \eta} \geq L_\varepsilon(q) - \varepsilon(1+|q|) + \frac{\log K}{\log \eta}.$$

Now let η and ε tend to 0 to get $\tau^c(q) \geq \lim_{\varepsilon\to 0} L_\varepsilon(q)$. Use corollary 1 to conclude.

The case $q < 0$ goes along the same lines. □

Remark. The same technique shows that $f_l = (f_g)^{**}$ when f_g has compact support.

We end this section with some easily proved properties of τ^c and $\tilde{\tau}^c$.

Proposition 9.

1. *τ^c and $\tilde{\tau}^c$ are increasing and concave.*
2. *$\tau^c(0) = \tilde{\tau}^c(0) = -\Delta(\mathrm{Supp}(A))$, where* $\mathrm{Supp}(A) = \mathbb{R} - \cup\{u, |u| > 0, A(u) = +\infty\}$.

3. *If* $A(u) = \log \mu(u)/\log |u|$ *with* μ *a probability measure, then* $\tau^c(1) = \tilde{\tau}^c(1) = 0$.
4. *For any sequence* (η_n) *tending to zero such that* $\lim_{n\to\infty} \log \eta_n / \log \eta_{n+1} = 1$, $\tau^c(q) = \liminf_{n\to\infty} \log H^q_{\eta_n} / \log \eta_n$ *and* $\tilde{\tau}^c(q) = \liminf_{n\to\infty} \log J^q_{\eta_n} / \log \eta_n$.

The last property is important for numerical applications, since it allows to evaluate τ^c and $\tilde{\tau}^c$ using discrete sequences such as $\eta_n = 2^{-n}$.

References

[1] M. Arbeiter and N. Patzschke. Random self-similar multifractals. *Math. Nachr.*, 181:5–42, 1996.

[2] J. Barral. Continuity of the multifractal spectrum of a random statistically self-similar measure. *J. Theo. Prob*, 13-4:1027–1060, 2000.

[3] G. Brown, G. Michon, and J. Peyrière. On the multifractal analysis of measures. *J. Statist. Phys.*, 66(3-4):775–790, 1992.

[4] A. Chhabra and R.V. Jensen. Direct determination of the $f(\alpha)$ singularity spectrum. *Phys. Rev. Letters*, 62:1327–1330, 1989.

[5] A. Echelard, J. Lévy Véhel and C. Tricot. Some properties of the 2-microlocal and large deviation multifractal spectra. *Preprint*, 2004.

[6] R.S. Ellis. Large Deviations for a General Class of Random Vectors. *Ann. Prob.*, 12(1):1–12, 1984.

[7] S. Jaffard. Multifractal formalism for functions II: selfsimilar functions. *SIAM J. Math. Anal*, 28(4):971–998, 1997.

[8] K.S. Lau and S.M. Ngai. L^q spectrum of the Bernouilli convolution associated with the golden ration. *Studia Math.*, 131(3):225–251, 1998.

[9] J. Lévy Véhel. Numerical computation of the large deviation multifractal spectrum. In *CFIC*, Rome, 1996.

[10] J. Lévy Véhel. Introduction to the multifractal analysis of images. *Fractal Images Encoding and Analysis, Y. Fisher, Ed.*, Springer Verlag, 1998.

[11] J. Lévy Véhel and R. Riedi. Fractional Brownian motion and data traffic modeling: The other end of the spectrum. *Fractals in Engineering, Lévy Véhel, J. Lutton, E. and Tricot, C., Eds.*, Springer Verlag, 1997.

[12] J. Lévy Véhel and R. Vojak. Multifractal analysis of Choquet capacities. *Adv. in Appl. Math.*, 20(1):1–43, 1998.

[13] C. Tricot. *Courbes et dimension fractale.* Springer Verlag, 2d edition, 1999.

Jacques Lévy Véhel
Projet Fractales, INRIA, BP 105, 78153 Le Chesnay Cedex, France
e-mail: jacques.levy-vehel@inria.fr

Claude Tricot
Lab. de Mathématiques Pures, Université Blaise Pascal, 63177 Aubière Cedex, France
e-mail: ctricot@wanadoo.fr

Progress in Probability, Vol. 57, 43–56

One-Dimensional Moran Sets and the Spectrum of Schrödinger Operators

Zhi-Ying Wen

Abstract. We study the structure and dimension of the spectrum of one-dimensional Schrödinger operators with Sturmian potentials. For this purpose we introduce the class of recurrent Moran sets which are generalizations of the well-known self-similar fractals.

Mathematics Subject Classification (2000). Primary 28A78; Secondary 81Q10.

Keywords. Moran set, spectrum, Schrödinger operator.

1. Structure of Moran sets

In this part we consider some general classes of geometrical constructions and discuss the dimension properties of the resulting sets. First we recall some properties of the classical self similar sets, and then consider some possible generalizations.

Let $F \subset \mathbb{R}^d$ be a closed set. A finite *family of similarity contractions* consists of a family of similarity contractions $\{S_1, \dots, S_m\}$ on $\mathbb{R}^d$, that is,

$$|S_i(x) - S_i(y)| = c_i|x - y| \text{ where } 0 < c_i < 1.$$

Denote by $K(\mathbb{R}^d)$ the class of non-empty compact subsets of $\mathbb{R}^d$. Let $S(A) = \bigcup_{i=1}^m S_i(A)$ for $A \in K(\mathbb{R}^d)$. Then S is a contraction mapping on $K(\mathbb{R}^d)$ [13].

Theorem A. [13] *Let $\{S_1, \dots, S_m\}$ be a family of similarity contractions on $\mathbb{R}^d$. Then there exists a unique, non-empty compact set $E \subset \mathbb{R}^d$ such that*

$$E = \bigcup_{i=1}^m S_i(E).$$

Moreover, for any $A \in K(\mathbb{R}^d)$ such that $S_i(A) \subset A$ for all i,

$$E = \bigcap_{k=0}^{\infty} S^k(A),$$

where S^k is the k-th iterate of S.

This set E is called the *self-similar set (SSS)*, or *attractor*, or *invariant set* of the family of similarity contractions $\{S_1, \ldots, S_m\}$. The sets $S_{i_1} \circ \cdots \circ S_{i_k}(A), 1 \leq i_1, \ldots, i_k \leq m$ are called the basic elements of A of order k.

We say that the family of similarity contractions $\{S_1, \ldots, S_m\}$ satisfies the *open set condition (OSC)* if there exists a non-empty open set $U \subset \mathbb{R}^d$ such that $\bigcup_{i=1}^m S_i(U) \subset U$ where the union is disjoint.

For $s \geq 0$ and $E \subset \mathbb{R}^d$, we use $\mathcal{H}^s(E)$, $\mathcal{P}^s(E)$, $\dim_H E$, $\dim_P E$, $\underline{\dim}_B E$ and $\overline{\dim}_B E$ to denote respectively the s-dimensional Hausdorff measure, packing measure, Hausdorff dimension, packing dimension, lower and upper box-counting dimension of the set E.

Theorem B. [13] *Let E be the attractor of a family of similarity contractions $\{S_1, \ldots, S_m\}$ where S_i has similarity ratio c_i. If the open set condition is satisfied then*

$$\dim_H E = \dim_P E = \underline{\dim}_B E = \overline{\dim}_B E = s$$

and

$$0 < \mathcal{H}^s(E), \mathcal{P}^s(E) < \infty,$$

where s is the unique positive solution of

$$\sum_{i=1}^m c_i^s = 1.$$

So far we assumed that

1° All $S_i (1 \leq i \leq m)$' are similarity contractions.
2° The family of similarities is finite.
3° At each step, the position of basic elements are completely determined.
4° At each step, the contraction ratios keep unchanged.
5° The open set condition holds so that Theorem B is true.

There are many possible ways to generalize the SSS by modifying some of the above conditions. Replacing similarities by conformal mappings but keeping the other conditions unchanged, we get a *Cookie-cutter set*, which is the nonlinear analogue of SSS. For Cookie-cutter sets the conclusions of Theorem B are still valid , but the dimension is presented by a variational principle, see Bedford [1] and the references therein. If we consider self affine mappings instead of similarities, we will get a *self-affine set*. In this case the conclusions of Theorem B are false in general, and it is extremely difficult to calculate the dimensions. For a typical case see McMullen [20]. The extension of SSS to the case of infinitely many similarity contractions has been studied by Feng De-Jun [7] and Mauldin and Urbanski [24]. Without the OSC, the dimensions also coincide but the other parts of Theorem B will be false in general. A typical case has been studied by Rao and Wen [26]. As a generalization of SSS, Pesin and Weiss [25] considered a rather large class of *Cantor-like sets*. For the geometric constructions of this class, the placements at each step can be arbitrary, and they need not be disjoint in some sense, but the contraction ratios are not changed. Using uniform and non-uniform mass distribution principles, they searched for the conditions guaranteeing the coincidence of the Hausdorff and Box-counting dimensions. Feng, Wen and Wu [8], introduced *one-dimensional homogeneous Moran sets.* At each step, both the placement of basic elements and the contraction ratios may change. They studied how these changes influence the dimensions of the resulting set. Hua [10], Hua and Li [11], and Marion [21] studied a kind of generalization of SSS which they called *generalized self-similar set* and *ensemble à similitude interne,* respectively. They considered a sequence of families of similarity contractions, and in each step of their construction, they applied different contraction ratios, but the placements of the basic elements are determined completely by the contraction schedules. Moreover they both required the lower limit of contraction ratios to be strictly positive. Marion discussed the Hausdorff dimension and Hua Su discussed both Hausdorff and packing dimension.

We shall present the dimensions and measures of the following generalization of SSS:

1) The placements of the basic sets at each step of the constructions can be arbitrary;

2) The contraction ratios may be different at each step;

3) The lower limit of the contraction ratios permits zero.

As we will show, the requirements 1) and 2) can lead to differences between Hausdorff dimension and packing dimension. If the lower limit of the contraction ratios is positive, then the packing and upper Box-counting dimensions coincide. Otherwise the situation becomes rather complicated.

We introduce now one-dimensional Moran classes and Moran sets, the higher-dimensional cases can be treated similarly. Let $J \subset \mathbb{R}$ be a closed interval; $\{n_k\}_{k\geq 1}$ a sequence of positive integers; and $\Phi = \{\Phi_k\}_{k\geq 1}$ a sequence of positive real vectors, where $\Phi_k = (c_{k,1}, \ldots, c_{k,n_k})$, $0 < c_{k,j} < 1$, $k \in \mathbb{N}$, $1 \leq j \leq n_k$. For any $k \in \mathbb{N}$, set $D_k = \{(i_1, \ldots, i_k);\ 1 \leq i_j \leq n_k,\ 1 \leq j \leq k\}$, and $D = \bigcup_{k\geq 0} D_k$. By convention, take $D_0 = \emptyset$.

Let $\sigma = (\sigma_1, \ldots, \sigma_k)$, $\tau = (\tau_1, \ldots, \tau_m) \in D$, denote $\sigma * \tau = (\sigma_1, \ldots, \sigma_k, \tau_1, \ldots, \tau_m)$.

Let $\mathcal{J} := \{J_\sigma \subset J; \sigma \in D\}$ be a family of closed subintervals of J, $\mathcal{J}$ is called to form a **Moran structure** if

1° $J_\emptyset = J$.

2° For any $k \geq 0$ and any $\sigma \in D_k$, $J_{\sigma*1}, \ldots, J_{\sigma*n_{k+1}} \subset J_\sigma$, and for any $i \neq j$, $int(J_{\sigma*i}) \cap int(J_{\sigma*j}) = \emptyset$, i.e., the structure satisfies the open set condition.

3° For any $k \geq 1$, $\sigma \in D_{k-1}$ and $1 \leq j \leq n_k$,

$$\frac{|\ J_{\sigma*j}\ |}{|\ J_\sigma\ |} = c_{k,j}.$$

Let $\mathcal{F}_k := \{J_\sigma; \sigma \in D_k\}$, then $\mathcal{F} := \cup_{k\geq 0}\mathcal{F}_k = \{J_\sigma;\ \sigma \in D\} = \mathcal{J}$, any set in $\mathcal{F}_k$ is called a basic interval of order k of E.

Now let $E_k = \bigcup_{\sigma \in D_k} J_\sigma$ and $E = \bigcap_{k\geq 0} E_k$. Then $E := E(\mathcal{J}, J, \{n_k\}, \{\Phi_k\})$ is called the Moran set corresponding to the structure $(J, \{n_k\}, \{\Phi_k\})$. Let $\mathcal{M} := \mathcal{M}(J, \{n_k\}, \{\Phi_k\})$ be the class of all Moran sets corresponding to $(J, \{n_k\}, \{\Phi_k\})$. Then $\mathcal{M}$ is called the Moran class associated with the structure $(J, \{n_k\}, \{\Phi_k\})$.

The sequences $\{n_k\}_{k\geq 1}$ and $\{\Phi_k\}_{k\geq 1}$, and the relative positions of basic intervals play an important role in the study of fractal properties of the Moran classes and Moran sets.

Suppose $E, E' \in \mathcal{M}$ with $\mathcal{J}$ and $\mathcal{J}'$ their generating families respectively, then the relative positions of the basic intervals of $\mathcal{J}$ and that of $\mathcal{J}'$ may be different, although but both $\mathcal{J}$ and $\mathcal{J}'$ satisfy the same conditions of Moran structure $1°, 2°, 3°$.

To discuss the dimension properties of the Moran class $\mathcal{M}(J, \{n_k\}, \{\Phi_k\})$, we define the pre-dimensional sequence $\{s_k\}_{k\geq 1}$, where s_k is the value satisfying

$$\prod_{i=1}^{k} \sum_{j=1}^{n_i} c_{i_j}^{s_k} = 1, \quad k \geq 1.$$

Let $s_* = \liminf_{k\to\infty} s_k$ and $s^* = \limsup_{k\to\infty} s_k$.

The simplest case is $\inf\{c_{ij}\} > 0$. In this case, we have

Theorem 1.1. [10, 11, 12, 21] *If $c_* = \inf\{c_{ij}\} > 0$, then any $E \in \mathcal{M}(J, \{n_k\}, \{\Phi_k\})$ satisfies*

$$\dim_H E = \underline{\dim}_B(E) = s_*, \quad \dim_P E = \overline{\dim}_B(E) = s^*. \qquad (*)$$

Moreover, $0 < \mathcal{H}^{s_*}(E) < \infty$ and $0 < \mathcal{P}^{s^*}(E) < \infty$, respectively, if and only if

$$0 < \liminf_{k\to\infty} \prod_{i=1}^{k} \sum_{j=1}^{n_i} c_{i_j}^{s_k} < \infty \text{ and } 0 < \limsup_{k\to\infty} \prod_{i=1}^{k} \sum_{j=1}^{n_i} c_{i_j}^{s_k} < \infty.$$

The above theorem, in particular the formula $(*)$, can be seen as the generalization of Theorem B for the SSS, but in the case $c_* = 0$, the formula $(*)$ doesn't hold in general. Additional conditions are needed.

Theorem 1.2. [8] *Let $\mathcal{M} = \mathcal{M}(J_0, \{n_k\}, \{\Phi_k\})$ be a Moran class.*

Condition A.

1) $\sup_k n_k := \lambda < \infty$;

2) $0 < \inf_i \max_j\{c_{ij}\} \leq c* := \sup_i \max_j\{c_{ij}\} < 1$.

Condition B.

Let $d_k = \min_{1\leq j\leq n_k} \{c_{k,j}\}$, $M_k = \max_{\sigma\in D_k} |J_\sigma|$. Suppose $\lim_{k\to\infty} \frac{\log d_k}{\log M_k} = 0$.

Then under condition A or B, for any $E \in \mathcal{M}$,

$$\dim_H E = \underline{\dim}_B(E) = s_*, \quad \dim_P E = \overline{\dim}_B(E) = s^*.$$

For some special structure, we can get some further properties even for $c_* = 0$. For any $k \geq 1$, let $\Phi_k = c_k(1, 1, ...1)$, i.e. $c_{k,1} = c_{k,2} = \cdots = c_{k,n_k} = c_k$. The resulting Moran class $\mathcal{M}(J_0, \{n_k\}, \{c_k\})$ is called a Homogeneous Moran class. Put

$$t_* = \liminf_{k\to\infty} \frac{\log n_1 \cdots n_k}{-\log c_1 \cdots c_{k+1} n_{k+1}}, \quad t^* = \limsup_{k\to\infty} \frac{\log n_1 \cdots n_{k+1}}{-\log c_1 \cdots c_k + \log n_{k+1}},$$

we have

Theorem 1.3. [8] *For any* $E \in \mathcal{M}(J_0, \{n_k\}, \{c_k\})$,

$$t_* \leq \dim_H E \leq s_* \leq s^* \leq \dim_P E \leq \overline{\dim}_B E \leq t^*.$$

The following theorem shows a continuity property of the dimensions of the sets in $\mathcal{M}(J_0, \{n_k\}, \{c_k\})$.

Theorem 1.4. [8] *Let* $\mathcal{M}(J_0, \{n_k\}, \{c_k\})$ *be a Homogeneous Moran class.*

1) *Suppose* $t_* < s_*$ *and* $t_* < \alpha < s_*$, *then there exists* $E \in \mathcal{M}$, *such that* $\dim_H E = \alpha$;
2) *Suppose* $s^* < t^*$ *and* $s^* < \beta < t^*$, *then there exists* $E \in \mathcal{M}$, *such that* $\dim_P E = \beta$.

Above four parameters t_*, s_*, s^*, t^* can be attained by the following Moran sets:

Homogeneous uniform Cantor set $\mathcal{C}$**:** $\mathcal{C} \in \mathcal{M}(J_0, \{n_k\}, \{c_k\})$, and for any $\sigma \in D_{k-1}$, the subintervals $J_{\sigma*1}, \ldots, J_{\sigma*n_k}$ of J_σ satisfy

- The left end point of the subinterval $J_{\sigma*1}$ and the right end point of the subinterval $J_{\sigma*n_k}$ coincide with the respective points of the interval J_σ;
- Any two adjacent subintervals have the same distance, i.e., the distance of $J_{\sigma*i}, J_{\sigma*i+1}$ is equal to $c_1 \cdots c_{k-1}(1 - n_k c_k)/(n_k - 1)$, $1 \leq i \leq n_k - 1$.

Homogeneous partial Cantor set $\mathcal{C}^*$**:** $\mathcal{C}^* \in \mathcal{M}(J_0, \{n_k\}, \{c_k\})$, and for any $\sigma \in D_{k-1}$, the subintervals $J_{\sigma*1}, \ldots, J_{\sigma*n_k}$ of J_σ satisfy

- The left end point of the subinterval $J_{\sigma*1}$ coincides with that of the interval J_σ;
- the distance of any two adjacent subintervals is 0.

Proposition 1.5. ([8]) *With the above notation, we have*

$$\dim_H \mathcal{C}^* = t_*, \ \dim_H \mathcal{C} = s_*, \ \dim_P \mathcal{C}^* = \overline{\dim}_B \mathcal{C}^* = s^*, \ \dim_P \mathcal{C} = \overline{\dim}_B \mathcal{C} = t^*.$$

2. Spectrum of quasi-crystals

The studies of the spectrum of general one-dimensional Schrödinger operators has a long history. For a survey, we refer to Sütö [29]. In the 1980's many people studied difference and differential Schrödinger operator with random potential.

In 1984, Shechtman et. al. found a new kind of alloy phase in sharp cooling Al-Mn alloy, it's crystallogram is distinct, and moreover, the crystallogram have not only order 2,3 rotational symmetry, but also order 5 rotational symmetry, which had been excluded by classical crystal diffraction theory. On one hand, the existence of distinct crystallogram implies that the lattice of sharp cooling Al-Mn alloy is orderly, on the other hand, the existence of order 5 rotational symmetry implies that the lattice of the alloy it is not periodic. This new kind of crystal is called quasi-crystal now. After the discovery of quasi-crystal, people began to study Schrödinger operator with quasi-periodic potential. One-dimensional Schrödinger operators with Sturmian potential or substitution potential are two classes of closely studied operators. In this article we consider operators with Sturmian potential.

Let $\beta \in (0,1)$ be irrational. We define

$$s_n(\beta) := [(n+1)\beta] - [n\beta] \text{ for } n \in \mathbb{Z}.$$

Then the 0-1-sequence $\{s_n\}_{n\in\mathbb{Z}}$ is called a Sturmian sequence with slope β. Let H_β be the self-adjoint operator on l^2 defined by

$$(H_\beta\psi)_n = \psi_{n-1} + \psi_{n+1} + v_n\psi_n, \ \forall n \in \mathbb{Z}, \ \psi = \{\psi_n\}_{n\in\mathbb{Z}} \in l^2,$$

where the sequence $v_n = Vs_n(\beta)$ is called the potential of H_β and $V > 0$ is called the potential strength. H_β is usually called a Schrödinger operator with Sturmian potential $\{s_n\}_{n\in\mathbb{Z}}$. We shall discuss the structure of the spectrum of H_β:

$$\sigma(H_\beta) = \{x \in \mathbb{R};\ x - H_\beta \text{ has bounded inverse}\}.$$

Casdagli([3],1986) proved that the pseudo-spectrum B_∞ of Schrödinger operator for H_β is a Cantor set, where $\beta = (\sqrt{5}-1)/2$ is the golden mean, and potential strength $V > 24$. The role of B_∞ is to construct a bridge to apply trace map theory. By establishing a relation between B_∞ and $\sigma(H_\beta)$ and using a method of periodical approach, Sütö [28] obtained the same result for the typical case $V \geq 4$. By using some techniques on random Schrödinger operator [16, 23], Bellissard, Iochum, Scoppola and Testart [2] proved 1989 that $\sigma(H_\beta)$ is a Cantor set of zero Lebesgue measure for any irrational β and $V > 0$. In 2002, the same result for the substitution potential was independently obtained by Lenz [17] and by Liu, Tan, Wen and Wu [18] with different methods. The question whether the Hausdorff dimension of $\sigma(H_\beta)$ for irrational β is strictly less than 1 and strictly greater than 0 has absorbed a lot of attention. The first step was done by Raymond [27] in 1997, using the continued fraction expansion of β. He exhibited a key idea: the recurrent structure of the spectrum. For $V > 4$ and the golden mean, he gave an upper bound of the Hausdorff dimension of the corresponding spectrum, which is strictly less than 1. He also gave a conjecture on the lower bound of Hausdorff dimension. On the other hand, Damanik, Killip and Lenz [4], by applying the theory developed by del Rio, Jitomirskaya, Last and Simon ([5],[14],[15], 1995-99), proved that if β has bounded density (this means if $[a_1, a_2, \ldots]$ is the continued fraction expansion of β, then $\limsup_{k\to\infty} \frac{1}{k}\sum_{i=1}^{k} a_i < \infty$), then the Hausdorff dimension of the spectral measure of H_β is strictly greater than 0. Since the spectral measure is supported by the spectrum $\sigma(H_\beta)$, the Hausdorff dimension of the spectrum is bounded away from zero. Note that the set of irrational numbers with bounded density has Lebesgue measure zero, so their result is only valid for a set of β's of Lebesgue measure zero.

Based on the observation of Raymond [27], we will show a recurrent structure of Moran type of the spectrum $\sigma(H_\beta)$, and develop some techniques for Moran sets presented in §1 (see also [8, 12]).

2.1. Spectral structure

It is well known that if the potential sequence of the operator H is periodic, then the spectrum of H is a finite union of some intervals. To get the spectrum of the operator with non-periodic potential $\{v_n\}_{n\in\mathbb{Z}}$, one adopts often the truncating technique, consider a sequence of positive integers $\{n_k\}_{k\geq 1}$, for each k, take subsequence $\{v_i\}_{|i|\leq n_k}$, then we extend it to be a periodic sequence $\mathcal{S}_k$. Let H_k be the operator with $\mathcal{S}_k$ as its potential, then its spectrum $\sigma(H_k)$, we call the approximate spectrum of order k of $\sigma(H_k)$, will be a finite union of some intervals, the spectrum of the original operator will be the limit in some sense of the family of $\{\sigma(H_k)\}_{k\geq 1}$. A way for determining the spectrum $\sigma(H_k)$ is the method of transfer matrix, see for example [29]. The advantage of this method is that the structure of the $\sigma(H_k)$ is much simpler than $\sigma(H)$. But there are some difficulties: (1) How to choose the sequence $\{n_k\}_{k\geq 1}$? (2) For each k, we wish to know the explicit structure of $\sigma(H_k)$, in particular, the main difficulty is to estimate the length of the intervals in the construction of $\sigma(H_k)$; (3) The relations between the approximate spectra of different orders.

By this way, we are going to analyze the spectrum of the operator H_β.

Let $\beta = [a_1, a_2, \ldots, a_i, \ldots] \in (0,1)$ be an irrational, and let $p_k/q_k (k > 0)$ be the k-th asymptotic fraction of β:

$$\begin{aligned} p_{-1} &= 1, \quad p_0 = 0, \quad p_{k+1} = a_{k+1}p_k + p_{k-1}, \ k \geq 0, \\ q_{-1} &= 0, \quad q_0 = 1, \quad q_{k+1} = a_{k+1}q_k + q_{k-1}, \ k \geq 0. \end{aligned}$$

Let $k \geq 1$ and $x \in \mathbb{R}$, we define the transfer matrix $M_k(x)$ over q_k sites by

$$\mathbf{M}_k(x) = \begin{bmatrix} x - v_{q_k} & -1 \\ 1 & 0 \end{bmatrix} \begin{bmatrix} x - v_{q_k - 1} & -1 \\ 1 & 0 \end{bmatrix} \cdots \begin{bmatrix} x - v_2 & -1 \\ 1 & 0 \end{bmatrix} \begin{bmatrix} x - v_1 & -1 \\ 1 & 0 \end{bmatrix}.$$

By convention, take

$$\mathbf{M}_{-1}(x) = \begin{bmatrix} 1 & -V \\ 0 & 1 \end{bmatrix}, \quad \mathbf{M}_0(x) = \begin{bmatrix} x & -1 \\ 1 & 0 \end{bmatrix}.$$

Let $t_{(k,p)}(x) = \operatorname{tr} \mathbf{M}_{k-1}(x)\mathbf{M}_k^p(x)$ and $\sigma_{(k,p)} = \{x \in \mathbb{R} : |t_{(k,p)}(x)| \leq 2\}$, where trM stands for the trace of the matrix M. With these notations, the following facts have been found in [2, 27, 29, 30] which can be used to estimate the length of the intervals of approximating spectra.

(A) Renormalization relation: for any $k \geq 0$,

$$\mathbf{M}_{k+1}(x) = \mathbf{M}_{k-1}(x)(\mathbf{M}_k(x))^{a_{k+1}}, \quad \text{so, } t_{(k+2,0)}(x) = t_{(k,a_{k+1})}(x). \tag{2.1}$$

(B) Trace relation: for any $k \geq 0, p \geq -1$,

$$t^2_{(k+1,0)}(x) + t^2_{(k,p)}(x) + t^2_{(k,p+1)}(x) - t_{(k+1,0)}(x)t_{(k,p)}(x)t_{(k,p+1)}(x) = V^2 + 4. \quad (2.2)$$

(C) Let $\deg t_{(k,p)}(x)$ denote the degree of the polynomial $t_{(k,p)}(x)$, then for $V > 4$, $\sigma_{(k,p)}$ is composed of $\deg t_{(k,p)}(x)$ pairwise disjoint closed intervals which we call the *bands* of $\sigma_{(k,p)}$, we note $B_{(k,p)}$ these bands. Moreover $t_{(k,p)}(x)$ is monotone on each band $B_{(k,p)}$.

The following proposition shows the relations between different approximating spectra and the original spectrum.

Proposition 2.1. [2, 27] *With the notation above, we have*

(1) $\sigma_{(k+2,0)} \cup \sigma_{(k+1,0)} \subset \sigma_{(k+1,0)} \cup \sigma_{(k,0)}$;
(2) $\sigma(H_\beta) = \bigcap_{k \geq 0}(\sigma_{(k+1,0)} \cup \sigma_{(k,0)})$;
(3) *For any* $k \in \mathbb{N}$ *and* $p \geq -1$, $\sigma_{(k,p+1)} \subset \sigma_{(k+1,0)} \cup \sigma_{(k,p)}$;
(4) *For any* $k \in \mathbb{N}$, $p \geq 0$ *and* $V > 4$, $\sigma_{(k+1,0)} \cap \sigma_{(k,p)} \cap \sigma_{(k,p-1)} = \emptyset$.

By Proposition 2.1.(3), 2.1.(4) and the fact that $\sigma_{(k,p)}$ is a finite union of bands, we have that

Corollary 2.2. *For any* $k \in \mathbb{N}$ *and* $p \geq -1$, *any band of* $\sigma_{(k,p+1)}$ *is contained entirely either in a band of* $\sigma_{(k+1,0)}$ *or in a band of* $\sigma_{(k,p)}$.

Raymond found that different bands of the same order may play some different roles. He distinguished these bands by introducing the notion of types.

Definition 2.3. [27] Let $k \in \mathbb{N}$, we define three types of bands as follows:
(k, I)-type band: a band of $\sigma_{(k,1)}$ contained in a band of $\sigma_{(k,0)}$;
(k, II)-type band: a band of $\sigma_{(k+1,0)}$ contained in a band of $\sigma_{(k,-1)}$;
(k, III)-type band: a band of $\sigma_{(k+1,0)}$ contained in a band of $\sigma_{(k,0)}$.

These three kinds of types of bands are well defined [27] because of Proposition 2.1, (3) and (4). We call these bands *spectral generating bands of order k*. Denoted by $\mathcal{G}_k$ the set of all spectral generating bands of order k. Note that for $k = 0$, there is only one $(0, \mathrm{I})$-type band $\sigma_{(0,1)} = [V-2, V+2]$ and only one $(0, \mathrm{III})$ type band $\sigma_{(1,0)} = [-2, 2]$. Notice that all bands of $\sigma_{(k+1,0)}$ is either (k, II) or (k, III)-type (see Lemma 2.5), but there exist some bands of $\sigma_{(k,1)}$ which are not generating bands.

The following lemma 2.4 shows us some recurrent relation between spectral generating bands of order k and order $k+1$, which concludes any spectral generating bands of order k contains certainly some spectral generating bands of order $k+1$; and the lemma 2.5 determines completely the recurrent relation of successive spectral generating bands and establishes the relation with the spectrum.

Lemma 2.4. [27] *Let* $k \in \mathbb{N}$,

(1) *A* (k, I)*-type band contains a unique band of* $\sigma_{(k+2,0)}$ *which is a* $(k+1, \mathrm{II})$*-type band;*

(2) *A (k, II)-type band contains $a_{k+1}+1$ bands of $\sigma_{(k+1,1)}$ which are of $(k+1, \mathrm{I})$-type bands; and contains a_{k+1} bands of $\sigma_{(k+2,0)}$ which are of $(k+1, \mathrm{III})$-type bands;*

(3) *A (k, III) type band contains a_{k+1} bands of $\sigma_{(k+1,1)}$ which are of $(k+1, \mathrm{I})$-type bands, and contains $a_{k+1}-1$ bands of $\sigma_{(k+2,0)}$ which are of $(k+1, \mathrm{III})$-type bands.*

Lemma 2.5. [19]

(a) *Given $k \geq 0$, let $G_k = \bigcup_{B \in \mathcal{G}_k} B$ be the union of all the spectral generating bands of order k, then $\sigma(H_\beta) = \bigcap_{k \geq 0} G_k$.*

(b) *Let $k > 0$, then any spectral generating band of order $k+1$ is contained in a spectral generating band of order k.*

We find that for $V > 4$ the structure is more complex but still a little analogous to Moran structure. So on one hand we define a new kind of set class–recurrent Moran class, on the other hand we apply a lot of techniques used in study of Moran class.

2.2. Recurrent Moran structure of the spectrum

Last section shows some information on the recurrent structure of the Schrödinger operator with Sturmian potential. It seems to have some similarities with the structure of Graph-directed sets and Moran sets, as we will see that, it is much more complex than both classes above, for further discussion, we recall first the definition and properties of Graph-directed sets which we copy from [6]. Let V be a set of 'vertices' which we label $\{1, 2, \ldots, q\}$, and let E be a set of 'directed edges' with each edge starting and ending at a vertex so that (V, E) is a directed-graph. We write $E_{i,j}$ for the set of edges from vertex i to vertex j, and $E_{i,j}^k$ for the set of sequence of k edges $(e_1, \ldots, e_k)$ which form a directed path from vertex i to vertex j. For each $e \in E$, let $F_e : \mathbb{R}^n \mapsto \mathbb{R}^n$ be a contracting similarity of ratio r_e with $0 < r_e < 1$. Then there exists unique $G_1, \ldots, G_q$, such that

$$G_i = \bigcup_{j=1}^{q} \bigcup_{e \in E_{i,j}} F_e(G_j).$$

The sets $\{G_1, \ldots, G_q\}$ are called a family of graph-directed sets. We assume that the unions above are disjoint for all i.

For any $s > 0$, we associate a $q \times q$ matrix $A^{(s)}$ with (i,j)-th entry $A_{i,j}^{(s)} = \sum_{e \in E_{i,j}} r_e^s$.

If $A^{(s)}$ is primitive(that is , there exists an integer $p \geq 1$ such that all entries of the matrix $(A^{(s)})^p$ are strictly positive), and if $s \geq 0$ such that the largest eigenvalue of $A^{(s)}$ is 1, then

$$\dim_H G_i = \underline{\dim}_B G_i = \overline{\dim}_B G_i = s, \qquad \text{for } i \in V.$$

By comparing the definitions of Moran sets, graph-directed sets and the structure of the spectrum described above, we introduce the following recurrent Moran structure for the spectrum.

Keeping the notations above, let $\mathcal{G} = \bigcup_{k\geq 0} \mathcal{G}_k$, then $\mathcal{G}$ is the set of all spectral generating bands. We introduce now a kind of matrices for describing the structure of $\mathcal{G}$.

Let $\mathcal{A} = \{\text{I,II,III}\}$ be an alphabet. Let $\mathcal{T} = \{\mathbf{T}_n\}_{n\geq 1}$ be a sequence of non-negative 3×3 matrices

$$\mathbf{T}_n := (t_{ij}(n))_{i,j\in\mathcal{A}} := \begin{bmatrix} 0 & 1 & 0 \\ a_n+1 & 0 & a_n \\ a_n & 0 & a_n - 1 \end{bmatrix}, \quad n \geq 1,$$

where a_n is the n-th partial quotient of the continued fraction expansion of β.

Intuitively, $t_{ij}(n)$ represents a $(n-1, i)$-type band which generates $t_{ij}(n)$ bands of (n, j)-type, where $i, j \in \mathcal{A}$.

Let $k \geq 0$ be an integer. Then from Lemma 2.5, for any spectral generating band $B \in \mathcal{G}_k$, there exists a unique word $i_0 i_1 i_2 \cdots i_k \in \mathcal{A}^{k+1}$, such that B is a (k, i_k)-type band which is contained in a (l, i_l)-type band for any $0 \leq l < k$, we call $i_0 i_1 i_2 \cdots i_k$ the type index of B. With these notations, we see that

$$D_k := \{i_0 i_1 i_2 \cdots i_k \mid i_0, i_1, \ldots, i_k \in \mathcal{A},\ i_0 = \text{I or III},\ \mathrm{t}_{\mathrm{i}_{\mathrm{j}-1}\mathrm{i}_{\mathrm{j}}}(\mathrm{j}) > 0, 1 \leq \mathrm{j} \leq \mathrm{k}\}$$

is the set of all possible type indices of bands in $\mathcal{G}_k$. Set $D = \cup_{k\geq 0} D_k$, by convention we assume $D_0 = \{\text{I, III}\}$. Note that two different bands of order k may have the same type index. In fact, for any $k > 0$ and $\tau = i_0 i_1 \cdots i_k \in D$, let

$$K_\tau = t_{i_0 i_1}(1) t_{i_1 i_2}(2) \cdots t_{i_{k-1} i_k}(k),$$

then there are just K_τ bands in $\mathcal{G}_k$ with type index τ.

$\mathcal{G}$ has the following recurrent Moran structure: for any $B \in \mathcal{G}$, there exists $k \geq 0$ such that $B \in \mathcal{G}_k$, let $\tau = i_0 i_1 \cdots i_k$ be the type index of B, then by Lemma 2.4, for any $s \in \mathcal{A}$, B contains $t_{i_k s}(k+1)$ bands in $\mathcal{G}_{k+1}$ with type index $\tau * s \in D_{k+1}$ (where '$*$' stands for the concatenation of words), and all the spectral generating bands of order $k+1$ contained in B are mutually disjoint.

For estimating the Hausdorff dimension of the recurrent Moran structure, we need to estimate the length of bands in $\mathcal{G}$ (this is difficult because even the bands in $\mathcal{G}_k$ with the same type index may have different length), so we introduce two auxiliary sequences of matrices $\mathcal{P} = \{\mathbf{P}_n\}_{n\geq 1}$ and $\mathcal{Q} = \{\mathbf{Q}_n\}_{n\geq 1}$ as follows:

$$\mathbf{P}_n := (p_{ij}(n))_{i,j\in\mathcal{A}} := \begin{bmatrix} 0 & u_1^{a_n - 1} & 0 \\ u_1/a_n & 0 & u_1/a_n \\ u_1/a_n & 0 & u_1/a_n \end{bmatrix},$$

$$\mathbf{Q}_n := (q_{ij})_{i,j\in\mathcal{A}} := \begin{bmatrix} 0 & u_3^{a_n - 1} & 0 \\ u_3(a_n+2)^{-3} & 0 & u_3(a_n+2)^{-3} \\ u_3(a_n+2)^{-3} & 0 & u_3(a_n+2)^{-3} \end{bmatrix},$$

where $u_1 = \frac{3}{V-8}, u_3 = \frac{1}{V+5}$.

The following two propositions will be used to estimate the length of the bands, which play an important role in the estimating of the dimensions of the spectrum.

Let $\mathcal{W} = \{\mathbf{W}_n = (w_{ij}(n))_{i,j\in\mathcal{A}}\}_{n\geq 1}$ be a sequence of matrices with $w_{ij}(n) \geq 0 (i, j \in \mathcal{A}, n \geq 1)$, then for any $\tau = i_0 i_1 \cdots i_k$, define

$$L_\tau(\mathcal{W}) = w_{i_0 i_1}(1) w_{i_1 i_2}(2) \cdots w_{\tau_{k-1}\tau_k}(k).$$

Proposition 2.6. [19] *Let $\beta = [a_1, a_2, \ldots] \in (0,1)$ be an irrational and $V > 20$. Then for any spectral generating band $B \in \mathcal{G}$ with type index τ,*

$$4L_\tau(\mathcal{Q}) \leq |B| \leq 4L_\tau(\mathcal{P}).$$

Now we introduce two other auxiliary sequences of matrices $\mathcal{T}' = \{\mathbf{T}'_n\}_{n\geq 1} = (t'_{ij}(n))_{i,j\in\mathcal{A}}$ and $\mathcal{Q}' = \{\mathbf{Q}'_n\}_{n\geq 1} = \{(q'_{ij}(n))_{i,j\in\mathcal{A}}\}_{n\geq 1}$ as follows.

If $a_n < 3$, define

$$\mathbf{T}'_n = \mathbf{T}_n, \quad \mathbf{Q}'_n = \begin{bmatrix} 0 & u_2(a_n+2)^{-1} & 0 \\ u_2(a_n+2)^{-1} & 0 & u_2(a_n+2)^{-1} \\ u_2(a_n+2)^{-1} & 0 & u_2(a_n+2)^{-1} \end{bmatrix};$$

and if $a_n \geq 3$, define

$$\mathbf{T}'_n = \begin{bmatrix} 0 & 1 & 0 \\ b_n & 0 & b_n \\ b_n & 0 & b_n \end{bmatrix}, \quad \mathbf{Q}'_n = \begin{bmatrix} 0 & u_2^{a_n-1} & 0 \\ u_2(a_n+2)^{-1} & 0 & u_2(a_n+2)^{-1} \\ u_2(a_n+2)^{-1} & 0 & u_2(a_n+2)^{-1} \end{bmatrix},$$

where $u_2 = \frac{1}{4(V+8)}$ and $b_n = \max\{2, [(a_n - 1)/2]\}$.

Proposition 2.7. [19] *With the notations and the hypotheses above, there exists $\mathcal{G}' \subset \mathcal{G}$ such that $\mathcal{G}' \bigcap \mathcal{G}_0 = \mathcal{G}_0$, for each band $B \in \mathcal{G}'$ of type index $\tau = i_0 i_1 \cdots i_k (k \geq 1)$ contains at least $t'_{i_k s}(k+1)$ bands in $\mathcal{G}' \cap \mathcal{G}_{k+1}$ of type index $\tau * s$, and*

$$|B| \geq 4L_\tau(\mathcal{Q}').$$

2.3. Dimensions of the spectrum

Keep the definitions and the notations introduced in the previous sections, by using Propositions 2.6 and 2.7, we can estimate the dimensions of the spectrum of one-dimensional Schrödinger operator with Sturmian potential, we summarize some results here, for more details, we offer to [19].

Theorem 2.8. [19] *Let $[a_1, a_2, \ldots]$ be the continued fraction expansion of the irrational number $\beta \in (0,1)$ and let $M(\beta) := \liminf_{k\to\infty}(a_1 a_2 \cdots a_k)^{1/k}$. Let $V > 20$ and $u_1 = \frac{3}{V-8}$, $u_2 = \frac{1}{4(V+8)}$, then*

(1) *If $M(\beta) < \infty$, then*

$$\max\{\frac{\ln 2}{10\ln 2 - 3\ln u_2}, \frac{\ln M(\beta) - \ln 3}{\ln M(\beta) - \ln(u_2/3)}\} \leq \dim_H(\sigma(H_\beta)) \leq \frac{2\ln M(\beta) + \ln 3}{2\ln M(\beta) - \ln u_1}.$$

(2) *If $M(\beta) = \infty$, then*

$$\dim_H(\sigma(H_\beta)) = 1.$$

From the above theorem, we get immediately

Corollary 2.9. *If $V > 20$, then for any irrational $\beta \in (0,1)$:*

(i) $\dim_H(\sigma(H_\beta)) > 0$.

(ii) *Let $[a_1, a_2, \ldots]$ be the continued fraction expansion of β, then*

$$\dim_H(\sigma(H_\beta)) < 1 \text{ if and only if } \liminf_{k\to\infty}(a_1a_2\cdots a_k)^{1/k} < \infty.$$

Remark 2.10. Let

$$\Gamma := \{\beta = [a_1, a_2, \ldots] \in (0,1);\ \liminf_{k\to\infty}(a_1a_2\cdots a_k)^{1/k} < \infty,\}$$

then by the metrical theory of continued fractions, the set Γ has Lebesgue measure one ([16]). Consequently, if $V > 20$ then $\dim_H(\sigma(H_\beta)) > 0$ for all $\beta \in (0,1)$, and $1 > \dim_H(\sigma(H_\beta)) > 0$ for almost all $\beta \in (0,1)$.

Remark 2.11. Here is an example such that the Hausdorff dimension of the spectrum is 1: Take $\beta = 1/(e-1)$, it is known that $\beta = [1,1,2,1,1,4,\ldots,1,1,2k,\ldots]$, thus $\dim_H \sigma(H_\beta) = 1$ for $V > 20$ by Theorem 2.8.(2).

Remark 2.12. Let $M(\beta) < \infty$, then for any $\epsilon > 0$, there exist V large enough, such that $\dim_H \sigma(H_\beta) < \epsilon$.

Acknowledgement. The author thanks Dr. Qin-Hui Liu for the helpful discussions and support in the preparation of the article.

References

[1] Bedford T, Applications of dynamical systems to fractals - a study of cookie-cutter Cantor sets, In *Fractal geometry and analysis* (eds J Bélaire and S Dubuc), Canada: Kluwer, 1991, 1–44.

[2] J. Bellissard, B. Iochum, E. Scoppola and d. Testart, Spectral properties of one-dimensional quasi-crystals, *Commun. Math. Phys.* **125** (1989), 527–543.

[3] M. Casdagli, Symbolic dynamics for the renormalization map of quasiperiodic Schrödinger equation, *Commun. Math. Phys.* **107** (1986), 295–318.

[4] D. Damanik, R. Killip, D. Lenz, Uniform spectral properties of one-dimensional quasicrystals, III. α-continuity, *Commun. Math. Phys.* **212** (2000), 191–204.

[5] R. del Rio, S. Jitomirskaya, Y. Last, and B. Simon, Phys. Rev. Lett. **75** (1995), 117–119; *Operators with singular continuous spectrum, IV. Hausdorff dimensions, rank one perturbations, and localization, J. D'Anal. Math.* **69** (1996), 153–200.

[6] K.J. Falconer, *Techniques in fractal geometry,* Wiley 1997.

[7] D.-J. Feng, Ph.D. Thesis, Wuhan University, 1997.

[8] D.-J. Feng, Z.-Y. Wen and J. Wu, Some dimensional results for homogeneous Moran sets, *Science of China* (Series A), **40:5** (1997), 475–482.

[9] R. A. Horn and C. R. Johnson, *Matrix analysis,* Cambridge University Press, 1985.

[10] S. Hua, Hausdorff dimensions of generalized Cantor sets, *Acta Math. Appl. Sinica* **17** (1994), 551–558.

[11] S. Hua, W.-X. Li, Packing dimensions of generalized Moran sets, *Progress in Natural Science* **6** (1996), 148–152.

[12] S. Hua, H. Rao, Z.-Y. Wen and J. Wu, On the structures and dimensions of Moran sets, *Science of China*(Series A) **43:8** (2000), 836–852.

[13] Hutchinson J E, Fractals and self-similarity. *Indiana Univ. Math. J.* **30** (1981), 713–747.

[14] S. Jitomirskaya and Y. Last, Power law subordinacy and singular spectra, I. Half-line operators, *Acta Math.* **183** (1999) ,171–189.

[15] S. Jitomirskaya and Y. Last, Power law subordinacy and singular spectra, II. Line operators, *Comm. Math. Phys.* **211** (2000), 643–658.

[16] S. Kotani, Jacobi matrices with random potentials taking finitely many values, *Rev. Math. Phys.* **1** (1990), 129–133

[17] D. Lenz, Singular Spectrum of Lebesgue Measure Zero for One-Dimensional Quasicrystals, *Comm. Math. Phys.* **227** (2002), 119–130.

[18] Liu Qing-Hui, Tan Bo, Wen Zhi-Xiong and Wu Jun, Measure Zero Spectrum of a class of Schrodinger operators, *J. Statist. Phys.* **106** (2002), 681–691.

[19] Q.-H. Liu, Z.-Y. Wen, Hausdorff dimension of spectrum of one-dimensional Schrödinger operator with Sturmian potentials, *Potential analysis* **19** (2003).

[20] C. McMullen, The Hausdorff dimension of general Siepinski carpets. *Nogaya Math. J.* **96** (1984), 1–9.

[21] J. Marion, Mesures de Hausdorff d'un fractals à similitude interne. *Ann. Sci. Math. Quebec* **10** (1986), 111–132.

[22] P.A.P. Moran, Additive functions of intervals and Hausdorff measure. *Proc. Camb. Phil. Soc.* **12** (1946), 15–23.

[23] F. Martinelli, E. Scoppola, Introduction to the mathematical theory of Anderson localization, *Rivista del Nuovo Cimento* **10** (1987).

[24] R.D. Mauldin, M. Urbanski, Dimensions and measures in infinite iterated function systems. *Proc. London Math. Soc.* **73** (1996), 105–154.

[25] Y. Pesin, H. Weiss, On the dimension of deterministic and random Cantor-like sets symbolic dynamics, and the Eckmann-Ruelle conjecture. *Commun. Math. Phys.* **182** (1996), 105–153.

[26] Rao Hui, Wen Zhi-Ying, Some studies of a class of self-similar fractals with overlap structures. *Advances in Applied Math.* **20** (1998), 50–72.

[27] L. Raymond, *A constructive gap labelling for the discrete schrödinger operater on a quasiperiodic chain.* Preprint.

[28] A. Süto, The spectrum of a quasipeoriodic Schödinger operator, *Commun. Math. Phys.* **111** (1987), 409–415.

[29] A. Sütö, Schrödinger difference equation with deterministic ergodic potentials, *Beyond Quasicrystals*, Eds. F. Alex and D. Gratias, Springer-Verlag, 1994.

[30] M. Toda, *Theory of Nonlinear Lattices*, Number 20 in Solid-State Sciences, Springer-Verlag, second enlarged edition, 1989. Chap. 4.

Zhi-Ying Wen
Department of Mathematics
Tsinghua University
P.O. Box 100084
Beijing
China
e-mail: wenzy@tsinghua.edu.cn

Part 2

Fractals and Dynamical Systems

Progress in Probability, Vol. 57, 59–78

Small-scale Structure via Flows

Albert M. Fisher

Abstract. We study the small scale of geometric objects embedded in a Euclidean space by means of the flow defined by zooming toward a point in the object. For Fuchsian and Kleinian limit sets this "scenery flow" corresponds naturally to the geodesic flow. We explore this analogy in a variety of examples, coming from dynamics, geometry and probability theory.

Mathematics Subject Classification (2000). 37D40, 37F50, 60J65, 60G18, 37A40, 37A50, 60F17.

Keywords. Geodesic flow, horocycle flow, scenery flow, Brownian local time, Kleinian limit set, Julia set, scaling function, Hausdorff dimension, average density.

1. Introduction

Suppose we center a mathematical microscope at some point of a fractal set, and turn the knob continuously; as we zoom down toward smaller scales, ever-changing scenes go past us, as if we were riding on a train taking us deeper and deeper into the heart of the fractal landscape. Let us try to model such a fractal excursion mathematically. The continuously changing nature of the process suggests that a precise description will involve a continuous–time dynamical system, in other words a *flow*. In this article, we shall sketch how such a flow (the *scenery flow* of the fractal set) can be defined, and indeed, constructed rigorously for a variety of examples, and we shall see how the scenery flow can be usefully applied in studying the fractal geometry. We shall, moreover, see that this flow of magnification, and a related translation flow, provide close analogues of two familiar flows: the geodesic and horocycle flows of a Riemann surface.

To begin our story, we shall recall some basic properties of these classical flows.

2. Geodesic and horocycle flows

In the complex plane $\mathbb{C}$, let $\mathbb{H} = \{z = x + iy : \ y \geq 0\}$ denote the upper half plane. The interior of $\mathbb{H}$, those points with $y > 0$, is given the *hyperbolic metric*, defined by $ds^2 = (dx^2 + dy^2)/y^2$, which makes it isometric to the Poincaré disk Δ. The orientation-preserving isometries for this metric are the real Möbius transformations Möb($\mathbb{R}$), with

$$f_A(z) = (az + b)/(cz + d)$$

for $A = \begin{bmatrix} a & b \\ c & d \end{bmatrix}$ where $a, b, c, d \in \mathbb{R}$ and $\det(A) = ad - bc = 1$; that is, $A \in SL(2, \mathbb{R})$, the *special linear group*.

The matrices A and γA for $\gamma \neq 0$ give the same Möbius transformation, and this is the only such identification, so Möb($\mathbb{R}$) $\cong PSL(2, \mathbb{R})$, the *projective* special linear group. Let $\Gamma \subseteq$ Möb($\mathbb{R}$) be a discrete subgroup. Then the identification space $\Gamma \backslash \mathbb{H}$ is a Riemann surface; this may be compact, or be noncompact with either finite or infinite area. The unit tangent bundle of $\mathbb{H}$ can be identified with $PSL(2, \mathbb{R})$. This correspondence is easily described. We will associate (arbitrarily) the identity matrix I to the vector i_i which is located at the point $i \in \mathbb{H}$ and points in the vertical direction; then, given $A \in SL(2, \mathbb{R})$, let $f_A^*(i_i)$ be the image of this vector by the derivative map of f_A, that is, it is the vector located at the point $f_A(i)$ which has been rotated appropriately by the argument of the complex derivative. This image vector also has hyperbolic length one, as Möbius transformations are isometries for the hyperbolic metric; so this defines a map from $PSL(2, \mathbb{R})$ to the unit tangent bundle $T_1(\mathbb{H})$. The group Γ acts on $PSL(2, \mathbb{R})$ by left multiplication and one sees that $\Gamma \backslash PSL(2, \mathbb{R})$ is the unit tangent bundle of the surface $\Gamma \backslash \mathbb{H}$.

The geodesic flow on the surface is by definition the flow on this unit tangent bundle which moves a vector along its tangent geodesic at unit speed. Algebraically, this is given by right multiplication by $E_t \equiv \begin{bmatrix} e^{\frac{t}{2}} & 0 \\ 0 & e^{-\frac{t}{2}} \end{bmatrix}$ on $\Gamma \backslash PSL(2, \mathbb{R})$. To understand this, note that this matrix is equivalent as a Möbius transformation to $\begin{bmatrix} e^t & 0 \\ 0 & 1 \end{bmatrix}$ which dilates the plane by the factor e^t, and hence moves the vector i_i up the imaginary axis at unit speed in the hyperbolic metric. The action on a general unit vector is then given by the conjugation by f_A^* which is a hyperbolic isometry, so this is indeed the geodesic flow. The unstable horocycle flow h_t^u is given by the right action of $H_t^u \equiv \begin{bmatrix} 1 & 0 \\ t & 1 \end{bmatrix}$; the stable flow acts by its transpose. As the names suggest, these preserve the unstable and stable horocycles. Recall here that the *stable set* of a point x in a flow τ_t is the set of all y which are forward asymptotic to x, i.e. such that $d(\tau_t(x), \tau_t(y)) \to 0$ as $t \to +\infty$, while for the unstable set we trade $+$ for $-\infty$. The stable set of the point i_i for the geodesic flow is the set of all vertical unit vectors based on the line $y = 1$, see the right side of Fig. 2. This line is a "circle" tangent to ∞. Therefore by conjugation, stable sets of the geodesic

FIGURE 1. Covering space for punctured torus; the geodesic flow

flow are one dimensional, and are exactly the orbits of the stable horocycle flow, which are carried by those circles classically known as *horocycles*: those tangent to the boundary $\mathbb{R}$ of $\mathbb{H}$.

For the simplest example of a noncompact, finite area surface, see Fig.1; here (depicted in the disk model) Γ is a free group on two generators, these being two hyperbolic Möbius transformations, one which shoves the interior of the disk to the right and one which moves everything up; the curved quadrilateral in the center is a fundamental domain for this action. The left side is glued to the right, and the bottom to the top, so the resulting surface is a torus, just like for the usual gluings of a square, to get the quotient space $\mathbb{R}^2/\mathbb{Z}^2$, except that now the corner point gives a cusp, as it goes out to ∞ in the hyperbolic metric: this is a *punctured torus* (Fig. 1). Classical results are:

Theorem 2.1. *The geodesic and horocycle flows g_t, h_t^u, h_t^s preserve Riemannian volume of the unit tangent bundle of the surface M. This measure is finite iff the surface area is finite. For this case, if M is compact (equivalently has no cusps) then:*

(i) *g_t is ergodic, indeed is measure-theoretically isomorphic to a Bernoulli flow of finite entropy (= 1);*

(ii) *h_t^u, h_t^s are uniquely ergodic, with entropy zero.*

In the finitely generated, finite area case with cusps, all this is true except that h_t^u, h_t^s are only nearly uniquely ergodic; normalized Riemannian volume is the only non-atomic invariant probability measure if we disallow measures which give mass to horocycle orbits tangent to cusps.

More interesting for us will be the infinite area case, where the cusp opens up to flare out in a hyperbolic trumpet, Fig. 3, we return to this below.

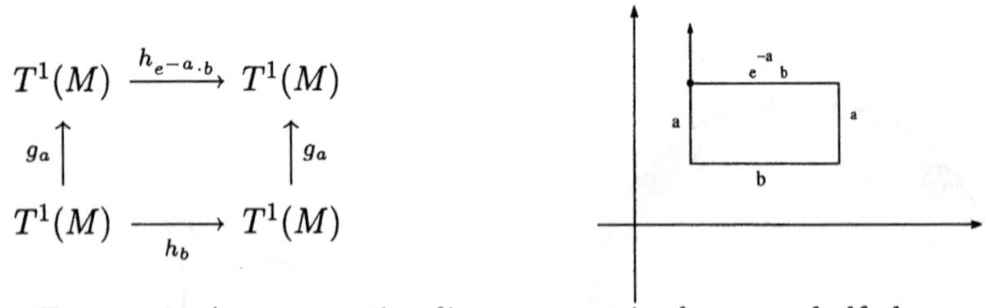

FIGURE 2. A commutative diagram, seen in the upper half plane; hyperbolic distances = flow times.

FIGURE 3. After opening the cusp; gluing, we now have infinite area.

The flows g_t and $h_t = h_t^s$ do not commute, but do satisfy the following commutation relation:

$$h_b g_a = g_a h_{e^{-a}b}.$$

In other words, the diagram in the left part of Fig. 2 commutes. One can prove this algebraically, or see it geometrically in the upper half plane, as illustrated in the right-hand part of the figure.

Remark 2.1. Because of the commutation relation, the pair (geodesic flow, horocycle flow) gives an action of the $(ax+b)$-group (the real affine group) on $T^1(M)$. This already hints that there might be a relation with fractal geometry, as fractal sets generally exhibit symmetries with respect to both dilation and translation.

Observation. The commutation relation tells us that $h_t = h_t^s$ is isomorphic to a speeded-up version of itself. An ergodic theorist immediately will recognise that this is very special, as the entropy of a sped-up transformation or flow multiplies by that factor, so in this case:

$$\text{entropy}(h_t) = e^{-a} \cdot \text{entropy}(h_t).$$

There are, thus, only two possibilities for the entropy of the flow h^s (or h^u): 0, or ∞!

We have already seen an example of zero entropy (the finite area Riemann surface case); next we shall see a situation where infinite entropy occurs, and this example will lead us into the fractal realm.

3. Brownian motion and stable processes

Let Ω be the space of continuous functions from $\mathbb{R}$ to $\mathbb{R}$, with the topology of uniform convergence on compact sets. This makes Ω into a Polish space, that is, a complete separable metric space, which is ideal from the point of view of measure theory.

Let μ denote Wiener measure on Ω, conditioned to be 0 at time 0, and defined both for future and past times. Then the *scaling property* of Brownian motion says: for $B \in \Omega$ and $a > 0$,

$$B(at)/\sqrt{a}$$

is "distributed like" $B(t)$. What this probability language means to an analyst is: the transformation $B \mapsto \Delta_a(B)$ defined by

$$(\Delta_a(B))(t) = B(at)/\sqrt{a}$$

preserves Wiener measure. To an ergodic theorist this suggests the following: defining

$$g_t = \Delta_{e^{-t}},$$

the action $g_t : \Omega \to \Omega$ is a measure- preserving flow! Next question: what flow is it? Answer: up to measure theoretic isomorphism, it is (the) Bernoulli flow of infinite entropy. (See [Fis87], and see [Fis04] for related work.) Call this the **scaling flow** of Brownian motion; geometrically it dilates time and rescales space appropriately to give another Brownian path.

Next, consider the **increment flow**,

$$((h^u)_a(B))(t) = B(a+t) - B(a).$$

This simply shifts the origin point $(0,0)$ along the graph of $B(t)$, and since the process $B(t)$ has *stationary increments*, again preserves the measure μ.

Now comes the magical part: these two flows satisfy the same commutation relation as the geodesic and horocycle flows of our surface! But what is $(h^u)_t$, measure-theoretically? Answer: it is now infinite entropy. (One way to see this is to note that the increment flow is naturally isomorphic to the shift flow on white noise, which is an infinite entropy Bernoulli flow; the isomorphism is given by integration). So we have our "infinite entropy horocycle flow" example.

It is clear that the same commutation relation holds for any self-similar process with stationary increments; examples are the stable processes, see [FL02]; see [Zä91] for closely related work.

4. Brownian zero sets

This study of Brownian motion gives us almost for free another example, which will make the link to fractal subsets of the line. As is well known, the zero set of a Brownian path, $Z_B \subseteq \mathbb{R}$, has Hausdorff dimension $1/2$, and for the gauge function $\phi = (2t \log\log(1/t))^{\frac{1}{2}}$, has positive locally finite Hausdorff measure H_ϕ. Now define the map from Ω to itself by $B(t) \mapsto L_B(t)$ where $L_B(t) = H_\phi(Z_B \cap [0,t])$, the total measure up to time t (and similarly for $t < 0$, but with negative sign). Write μ_Z for the image measure on path space Ω; this is Paul Lévy's *local time.* (Note: to visualize the local time $L_B(t)$ of a Brownian path B, it helps to know – by a theorem of Lévy – that local time paths are exactly maximum paths $M_{\widehat{B}}(t) = \sup\{B(s) : s \in [0,t]\}$ for a different Brownian path $\widehat{B}$.) The flow g_t preserves these correspondences, hence the scaling flow on (Ω, μ_Z) is also a Bernoulli flow of infinite entropy (being a factor of the flow on (Ω, μ)) [BF92]. But what about the increment flow? Here things change: we slide into the gaps of the local time; the appropriate measure has become *infinite.* We shall return to examine the consequences of this in §11.

5. The extended Cantor function (or Devil's Staircase)

The graph of Brownian local time is a continuous, nondecreasing function with a dense set of flat spots, reminiscent of the Cantor function– which suggests to us that we study that non-random example in a similar way. In [Fis92] we see an extended version of the usual Cantor function as depicted, say, in Mandelbrot's book [Man]. Note the upper and lower envelopes of the form ct^d for $d = \log 2/\log 3$, the Hausdorff dimension of the Cantor set. Now this extended function $L(t)$, defined to be identically 0 for $t \leq 0$, satisfies

$$L(3t)/2 = L(3t)/3^d = L(t);$$

this means that for the scaling flow g_t of exponent d, $g_{t_0}(L) = L$, where $t_0 = \log 3$; in other words, the scaling flow of this graph is a single periodic orbit!

Rescaling this path represents zooming down toward the point 0 in the Cantor set. But though the set C is, as we all know, "the same everywhere", on closer inspection this isn't quite so true. What do we see, for instance, if we slide the graph of f over to the point $1/4$ (which happens to be in C) by the increment flow, and then begin rescaling? In this case, the orbit is no longer periodic, but will converge asymptotically to a different periodic orbit, one with twice the previous length: $2\log 3 = \log 9$. But now differently from the point 0, there is asymptotic scenery on both sides of the point $1/4$ [Fis92].

And this is still very special: what happens when a general point of C replaces the number $1/4$? Answer: after taking the forward orbit closure, we get a mixing ergodic flow! This will be naturally isomorphic to the scenery flow of the fractal set C, with which we began the article. So now it is time to give some precise definitions.

6. The scenery flow

Let $\Omega = \Omega(\mathbb{R}^n)$ be the collection of all closed subsets of $\mathbb{R}^n$; topologise this by taking first the Hausdorff metric on closed subsets of the one-point compactification and then taking the relative topology on this subset Ω. This topology (the *Attouch-Wetts* or *geometric* topology, see §2 of [BF96]) makes Ω compact; for an example consider $F_n = \{n\}$; this sequence of subsets of $\mathbb{R}$ converges in $\Omega(\mathbb{R})$ to the empty set.

Define the **magnification flow** g_t on Ω by $A \mapsto e^t \cdot A$. Choose a closed subset $F \subseteq \mathbb{R}^n$, choose a point $z \in F$, and define $\Omega_{(F,z)}$ to be the omega-limit set of $(F - z)$, the set translated so as to place z at the origin. Thus,

$$\Omega_{F,z} = \bigcap_{T \geq 0} \text{closure}\{g_t(F - z) : t \geq T\}.$$

The flow $(\Omega_{F,z}; g_t)$ is the **scenery flow** of the set F at the point z.

Example. Let C be the Cantor set $C \subseteq [0, 1]$; define $\widehat{C} = \cup_{k=0}^{\infty} 3^k C$, so $3 \cdot \widehat{C} = \widehat{C}$. Then

$$g_{\log 3}(\widehat{C}) = e^{\log 3} \cdot \widehat{C} = \widehat{C},$$

so the scenery flow of C at 0 is $\{e^t \widehat{C} : t \in [0, \log 3]\}$, and this is a single periodic orbit of length $\log 3$. The total Hausdorff measure $L(t) = H_d(\widehat{C} \cap [0, t])$ gives a non-decreasing continuous function, the extended Cantor function, and this produces a flow isomorphism to the scaling flow of the Cantor function, described before.

Definition 6.1. The scenery flow of the set F is the flow of magnification by e^t on Ω_F, the union of the scenery flow spaces of each point $z \in F$ (so $\Omega_F = \cup \Omega_{F,z}$).

One can show [Fis92] that the scenery flow of the set C is naturally isomorphic to the following special flow: the height $\log 3$ suspension of the natural extension of the map $3x(\mod 1)$ on C. As a first consequence, since this flow is recurrent, a.e. orbit is dense and so the landscape encountered while zooming down toward a general point passes through all possible limiting scenes; the orbit closure is the whole space. Moreover, the flow is ergodic, so, by the Birkhoff ergodic theorem, the scenes will be encountered with the right frequency. So in particular this proves that the order-two (average) density exists and is a.s. constant. See [BF92] for a slightly different proof.

But what else can one say about this flow? The base map is isomorphic to the Bernoulli shift σ with weights $(1/2, 1/2)$ on $\Sigma = \Pi_{-\infty}^{\infty}\{0, 1\}$ via ternary expansion, and this map has entropy $\log 2$. Now we see something interesting. By Abramov's formula, flow entropy $=$ (base entropy)/(expected return time), so we conclude: entropy$(g_t) = \log 2/\log 3 = d = \dim(C)$! This gives us a formula for dimension:

The Hausdorff dimension of C equals the entropy of the scenery flow.

This formula is not always valid: a counterexample is a Brownian zero set; the dimension is $1/2$ but the scenery flow is isomorphic to the scaling flow on local time, which, as we have seen, has infinite entropy. Nevertheless, we wonder: is it possible that this formula is valid elsewhere?

7. The Fuchsian case

In fact we shall find just such an example in Fuchsian limit sets. Recall our modified punctured torus, where we opened up the cusp into a trumpet. The **limit set** Λ of the group Γ is now a Cantor subset of the boundary (Fig. 3; all the open intervals which are unions of the closed interval where the fundamental domain meets the boundary are removed, leaving a topological Cantor set). So, what is its scenery flow? Well, first of all, the Cantor set is a subset of the circle, and zooming down toward this we see a tangent line: the infinitesimal scenery of a circle is a line. That's rather boring! But if we consider the scenery flow of the limit set, things become much more interesting. What we will see asymptotically is a collection of fractal subsets of the tangent space to the circle (the real line) which is invariant under dilation. But what are these sets? The answer is satisfyingly simple: they are the limit set Λ moved to the boundary $\mathbb{R}$ of $\mathbb{H}$, the upper-half space model of the hyperbolic disk. Here we have to allow all possible correspondences via Möbius transformations from Δ to $\mathbb{H}$ such that a point of the limit set occurs both at 0 and ∞. Next, what can we say about the dynamics or ergodic theory of this scenery flow? We find:

Theorem 7.1. *Let Δ be the unit disk with Poincaré metric, with boundary $\partial\Delta = S^1$. Let Γ be a finitely generated Fuchsian group of second type (that is, the limit set Λ is not all of the boundary). Then the scenery flow of Λ is a finite-to-one factor of the geodesic flow of the surface $M = \Gamma\backslash\Delta$, and the limiting scenes are images of Λ in $\mathbb{R}$ with respect to Möbius transformations from Δ to $\mathbb{H}$.*

Proof. We construct the factor map directly. Let v_p be a unit tangent vector based at the point $p \in \Delta$. Consider the geodesic tangent to v_p; this is a circle which meets $\partial\Delta$ orthogonally in two points η, ξ in the past, future directions respectively. Since a complex Möbius transformation is determined uniquely by where it sends three points, there exists a unique Möbius transformation $\alpha = \alpha_{v_p}$ such that α takes v_p to the unit vector $-i_i$ which points in the direction $-i$ at the location $i \in \mathbb{H}$, taking p to i, η, ξ to $\infty, 0$ and the geodesic to the imaginary axis.

More precisely, $\alpha(p) = i$ and $\alpha^*(v_p) = -i_i$, where α^* is the derivative of α. Now define a map $\hat{\pi} : T^1(\Delta) \to \Omega = \{$ closed subsets of $\mathbb{R}\}$ by $v_p \mapsto \alpha_{v_p}(\Lambda)$.

Claim 1. For $\gamma \in \Gamma$, $\hat{\pi} \circ \gamma = \hat{\pi}$.

This holds since $\gamma(\Lambda) = \Lambda$.

Now let $\hat{g}_t$ be the geodesic flow on $T^1(\Delta)$.

Claim 2. $\hat{\pi}(\hat{g}_t v_p) = e^t \cdot \hat{\pi}(v_p)$.

The reason is that in $\mathbb{H}$, moving along the geodesic tangent to $-i_i$ (the imaginary axis) toward the point $0 \in \mathbb{H}$ is isomorphic by conjugacy to keeping the vector fixed at location i and dilating $\mathbb{H}$ by the factor e^t.

Claim 1 tells us that $\hat{\pi}$ induces a well-defined map $\pi : T^1(M) \to \Omega$ on the factor space $M = \Gamma \backslash \Delta$.

Claim 3. π is finite-to-one.

Proof: Suppose $\hat{\pi}(v_p) = \hat{\pi}(w_q)$ for some other vector w_q tangent to Δ. Then $(\alpha_{w_q})^{-1} \circ \alpha_{v_p}$ is a Möbius transformation of Δ which preserves the limit set Λ. Let $\widetilde{\Gamma}$ be the subgroup of Möb(Γ) which has been extended from Γ by adjoining all such elements. Then the limit set of $\widetilde{\Gamma}$ is also Λ. Now Margulis' Lemma implies that if $\Gamma_1 \subseteq \Gamma_2$ are groups of hyperbolic isomorphisms of $\mathbb{H}^n$ and have the same limit set, then Γ_1 is of finite index in Γ_2. Hence in the factor space $T^1(M)$there are at most finitely many such vectors w_q for a given v_p. □

We thank Bernie Maskit and Peter Waterman for their help with this part of the proof.

It remains to show:

Claim 4. The asymptotic limiting sceneries of the limit set in the circle S^1 are the images by the stereographic projections.

Proof: Choose a point z in the limit set and place the circle so this point is at the origin, in the upper half space and tangent to the real axis. Zooming toward z for time t is equivalent to dilating this picture by the factor e^t; let us at the same time consider the downward-pointing unit vector at Euclidean height 1 in each picture. The Poincaré disks get larger and larger, approximating $\mathbb{H}$, and our tangent vector is moving via the geodesic flow. Now superimpose on this picture the stereographic projection determined by that vector, that is, by its image in the original Poincaré disk. The Möbius transformations converge to stereographic projections, proving the Claim. □

We mention that the above proof extends immediately to limit sets of geometrically finite groups of isometries of hyperbolic n- space $\mathbb{H}^n$. There is one main change: "geodesic flow" is replaced by "geodesic *frame* flow". Consider for instance $n = 3$. Then Δ^3 is the unit ball in $\mathbb{R}^3$ with the Poincaré metric; a *frame* is an orthonormal set of three vectors, one tangent to the geodesic flow, and the frame flow by definition simply moves this triple according to parallel transport. The choice of a frame defines a unique "stereographic projection" to the upper half-space model $\mathbb{H}^3$, sending the frame f_p to the standard basis frame based at the point $(0,0,1) \in \mathbb{H}^3 \subseteq \mathbb{R}^3$. The reason we need frames is that pictures of the scenery flow are different when rotated, and the frame flow will include this information.

8. Ergodic theory and Sullivan's formula

For the case of a Fuchsian group of second type, the geodesic flow of the surface $M = \Gamma \backslash \Delta$ at first glance is not nice from the point of view of ergodic theory: not only does $T^1(M)$ have infinite volume (this is not a priori a problem, as infinite measure ergodic theory can be brought in) but, much worse than that, a.e. vector v_p is non-recurrent, i.e. eventually leaves any compact region. Sullivan's insight is that interesting dynamics can be recovered if we restrict attention precisely to those vectors which are recurrent for positive and negative times; these are exactly those vectors v_p for which the endpoints η, ξ at $-\infty, +\infty$ belong to the limit set Λ.

Next question: is there a natural measure to put on this recurrent set, replacing Riemannian volume? Sullivan's answer is a modification of Patterson's measure μ on Λ. (For nice cases, μ is the Hausdorff covering measure on Λ; for some other cases, as Sullivan showed, it is the packing measure.) Sullivan's measure on $T^1(M)$ is described as follows. It is defined first on $T^1(\Delta)$, in a Γ-equivariant way; hence it projects to the factor space $T^1(M)$. A unit tangent vector v_p is parametrized by the two endpoints $\eta \neq \xi$, plus a real number (where it is along the geodesic). Hence $T^1(\Delta)$ can be parameterized by $(S^1 \times S^1 - \text{diagonal}) \times \mathbb{R}$. The recurrent set is represented by $(\Lambda \times \Lambda - \text{diagonal}) \times \mathbb{R}$. The measure is equivalent (shares the same sets of measure zero) with $(\mu \times \mu) \times$ Lebesgue measure; the Radon-Nikodym derivative with respect to this product measure is $1/|\eta - \xi|^{2d}$, with distance measured in the Euclidean metric on the disk and d the Hausdorff dimension of Λ.

This guarantees that the resulting measure (known as the **Patterson-Sullivan measure**) is Γ-equivariant and invariant for the geodesic flow. Sullivan then proves the following theorem (we state the n-dimensional version); some background references are [Pat76],[Pat87], [Sul84], [Sul70]:

Theorem 8.1. *Let Γ be a geometrically finite subgroup of Möb(Δ^n). Then the geodesic flow is ergodic for the Patterson-Sullivan measure $\hat{\mu}$, and $\hat{\mu}$ is the unique measure of maximal entropy, with entropy equal to the Hausdorff dimension dim$(\Lambda) = d$.*

We can hence conclude:

Corollary 8.2. *The topological entropy of the scenery flow of the limit set of a geometrically finite Kleinian group is equal to the Hausdorff dimension of that set.*

Proof. The scenery flow is a finite index factor of the frame flow. The entropy of the frame flow equals that of the geodesic flow, since it is an isometric extension. Since it is a finite-to-one factor, entropy is preserved. □

9. The scenery flow of a Julia set and hyperbolic Cantor set

The definition of the scenery flow (as an omega-limit set of the magnification flow acting on closed subsets of $\mathbb{R}^n$) makes sense for general fractal sets; sometimes we can identify this flow, constructing it concretely. We have already discussed the example of the middle-1/3 Cantor set; this is a linear Cantor set, and many other such linear flows (such as those generated by linear conformal IFS's with the open set condition) can be studied by a straightforward modification of this. The case of Kleinian limit sets was also easy, as this is also a "linear" case; the maps are linear fractional transformations, and, as we have seen above, the limiting sceneries are just given by stereographic projection.

We describe next the case of hyperbolic Julia sets; conformal mixing repellors and hyperbolic $\mathbb{C}^{1+\alpha}$ Cantor sets are dealt with in a similar way.

There is an analogy to the Kleinian case. There we had a homomorphism π from the hyperbolic manifold $T^1(M) = T^1(\Gamma\backslash\Delta)$ to the space Ω of scenes, conjugating the geodesic frame flow to the magnification flow on sets.

Here we construct a "model scenery flow" which will play the role of $T^1(M)$.

Let $f : \mathbb{C} \to \mathbb{C}$ be a rational map which is hyperbolic on its Julia set $\mathcal{J}$. Write Df for its derivative, and f_* for the action of the derivative on the unit tangent bundle of $\mathcal{J}$ (by which we mean the unit tangent bundle of $\mathbb{C}$, restricted to $\mathcal{J}$). Let $\tilde{f} : \widehat{\mathcal{J}} \to \widehat{\mathcal{J}}$ denote the natural extension of this map. The space $\widehat{\mathcal{J}}$ is an inverse limit and can be identified with the space of sequences $\underline{\hat{z}} = (\ldots \hat{z}_{-1}\hat{z}_0\hat{z}_1 \ldots)$ where $\hat{z}_i = (z_i, \theta_i) \in \mathbb{C} \times [0, 2\pi]$ represents a unit tangent vector and the sequence satisfies $\hat{z}_{i+1} = f_*(\hat{z}_i)$. Write $\widehat{\Omega}$ for this collection of sequences, with shift map $\hat{\sigma}$; let $\widetilde{\Omega}$ denote the suspension flow over $(\widehat{\Omega}, \hat{\sigma})$ with height (= return time) $\log|Df(z_0)|$. This is our model flow.

The flow homomorphism π from $\widetilde{\Omega}$ to Ω is defined as follows: on the base, $\pi(\underline{\hat{z}}) = L_{\underline{\hat{z}}} = \lim_{n\to\infty}(Df^n(z_{-n}) \cdot (\mathcal{J} - z_{-n}))$. The limit is in the Hausdorff metric on closed subsets of $\widehat{\mathbb{C}} \equiv \mathbb{C} \cap \{\infty\}$; thus the Julia set has been centered at the nth preimage z_{-n} and expanded and rotated by the derivative map. That this always converges is a consequence of bounded distortion, see [BFU]. We have:

Theorem 9.1. *([BFU], [FU]). For the hyperbolic Julia set of a rational map, there is, up to rotation, a unique measure of maximal entropy for the model scenery flow. Its entropy satisfies the formula "model flow entropy equals dim($\mathcal{J}$)". When we include rotations, then for all but a few exceptional cases, there is a unique measure of maximal entropy and the model scenery flow is ergodic with respect to this measure.*

Remark on proof. We refer the reader to [BFU], but mention two interesting points. First, ergodicity for the model flow implies a rotational symmetry for the scenery flow; that is, when zooming down toward a.e. point we not only eventually see all possible scenes, but they occur at all angles (and with the expected frequency). This question can, by a lemma of Furstenberg [Fur61] see [BFU] be formulated in terms of the non-existence of a certain *circle-valued cocycle* or equivalently as a problem about the non-existence of *Hausdorff measure invariant line fields* (or cross fields); so, the hard work in proving ergodicity has been transferred to that setting: see [FU] and [May].

Second, we sketch here the proof of the entropy formula because it is so instructive and works out so nicely. Bowen's formula for dimension [Bo79] reads: there exists a unique d such that $P(-\log|Df|^d) = 0$; this is dim($\mathcal{J}$). Here $P(\phi)$ is the pressure of a function ϕ, which can be defined as $P(\phi) = \sup(h(\mu) + \int \phi \, d\mu)$ where the supremum is taken over invariant probability measures and $h(\mu)$ is the entropy of the transformation $(\mathcal{J}, f, \mu)$. We know from the theory of Sinai-Ruelle-Bowen that there exists a unique invariant measure μ such that the sup is attained. So we have for this invariant measure μ:

$$0 = h(\mu) + \int -\log|Df|^d \mathrm{d}\mu$$

hence

$$h(\mu) = \int \log|Df|^d \mathrm{d}\mu = d \int \log|Df| \mathrm{d}\mu$$

and so

$$\frac{h(\mu)}{\int \log|Df| \mathrm{d}\mu} = d.$$

Now the formula on the left is (base entropy)/(expected return time), hence (by Abramov's formula) equals the special flow entropy.

The scenery flow is, a priori, a magnification flow defined on *sets*; however, it is natural to carry along more information, given (for the previous examples) by a labelling inherited from the Kleinian group or map f respectively. This information is provided by the model flows; we call this the *labelled* or *marked* scenery flow. We conclude with a dimension formula which unites Sullivan's formula for Kleinian limit sets, with Bowen's for Julia sets and "cookie cutter" Cantor sets:

Theorem 9.2. *For geometrically finite Kleinian groups in dimension n, hyperbolic rational maps, conformal mixing repellors, and hyperbolic $\mathrm{C}^{1+\alpha}$ Cantor sets, we have the formula:*

The topological entropy of the marked scenery flow of the limit set is equal to the Hausdorff dimension of the limit set.

10. Doubling maps and the Riemann surface lamination

The simplest example of a hyperbolic Julia set is for the map $f(z) = z^2 + c$ with $c = 0$; then $\mathcal{J}$ is the circle S^1 and f restricted to $\mathcal{J}$ is the usual doubling map of S^1, isomorphic to $x \mapsto 2x(\mod 1)$ on the unit interval. In this case, the natural extension of f is the hyperbolic map on the solenoid $\widehat{f} : \widehat{S} \to \widehat{S}$, which in turn is an a.s. one-to-one factor of the Bernoulli shift $\sigma : \Sigma \to \Sigma$ for $\Sigma \equiv \Pi_{-\infty}^{\infty}$. Here $|Df| = 2$ so the model flow is the special flow of height $\log 2$ over the solenoid. The base entropy is $\log 2$, so the flow entropy is (by Abramov) $\log 2/\log 2 = 1$, which is, indeed, the Hausdorff dimension of the Julia set S^1. In this case the model flow space is identical to Sullivan's *Riemann surface lamination* [Sul92], [Sul91]; we thank especially D. Sullivan and J. Kahn for conversations on this point. See also Chapter VI of [dMvS] regarding these papers of Sullivan.

This suggests that the model flow for a general hyperbolic Julia set is, on the one hand, a generalization of the Riemann surface lamination and on the other, an analogue for rational maps of the recurrent part of the frame bundle of the hyperbolic n-manifold $\Gamma\backslash\Delta^n$. Minsky and Lyubich, building partly on our construction of sceneries, and extending that to general rational maps, showed the following remarkable result: that a rigidity theorem of Thurston for rational maps can be proved analogously to Mostow's rigidity theorem for Kleinian groups, replacing the 3-manifold by the "hyperbolic 3-manifold lamination". See [ML], and [Lyu] for related work. All of this, also [BFU] and [FU], fits the philosophy of the Sullivan-Thurston "dictionary" between Kleinian groups and rational maps.

The Riemann surface lamination is a *double suspension*: the solenoid is a suspension of the adic transformation (odometer), giving a flow $(h^u)_t$, and the scenery flow space is a suspension flow g_t over that (now suspending the hyperbolic map). This space carries two flows: the flow which spins around the solenoid direction, which is just the lift of the rotation flow on the circle, and which preserves each level, and the vertical flow. This pair of flows h^u, g satisfy the same commutation relation as before–which is, indeed, exactly what should happen, since the solenoid leaves at a given height are the unstable leaves of the vertical flow.

11. The horocycle flow: infinite measures, return times and average density

Let us consider the case of a Fuchsian group of first type (the limit set is the circle); here the appropriate measure for the geodesic flow and for the horocycle flows h^u, h^s is the same: Riemannian volume on $T^1(M)$. For a group of second type, the situation is radically different: our recurrent measure is Patterson-Sullivan measure $\hat{\mu}$, but this is no longer invariant for h^u or h^s. The reason is that any given horocycle tangent to the limit set Λ meets the recurrent set in a fractal subset of the horocycle. The natural measure, therefore, is a modification of this: it is now equivalent to (Lebesgue measure $\times$ μ) $\times$ Lebesgue measure. That is, e.g. for h^s, the boundary point ξ at $+\infty$ of a vector v_p in the support of this measure is required to be in the limit set, while the infinite past boundary point η is free to wander along the real line; these points have conditional measure given by Patterson measure μ and by Lebesgue measure on $\mathbb{R}$ respectively.

In nice cases, this measure (which we call **Kenny measure**), is the unique invariant Radon measure up to multiplication by a constant (since it is infinite there is always this choice of normalization). See [Ken] for the general no-cusp case, [Bur] for the no-cusp case with dimension $> 1/2$. With M. Burger we have a proof different from Kenny's which includes the case with cusps allowed (where we rule out atomic measures and measures supported on horocycles around the cusps in the statement of unique ergodicity); manuscript in preparation.

It is natural to wonder what happens for other fractal sets. For instance for hyperbolic $\mathbb{C}^{1+\alpha}$ Cantor sets, one can also prove infinite-measure unique ergodicity, using techniques of [BM], combined with [BF96] and [BF97], see also [Fis03a].

The philosophy suggested by these examples is:

For some infinite measure preserving transformations,
the returns to a finite measure subset are a fractal-like subset of times.

This idea of a "fractal-like" subset of the integers can be made precise by use of the scaling flow (as $t \to -\infty$ for the way we have defined that flow here). The notion of average (or order-two) density extends to integer fractal sets, playing the role of a finitely additive Hausdorff measure, see [BF92]. There is another interpretation: since the average density can be thought of as a parameter which measures the *lacunarity* of a fractal set, we have here an analogue of Mandelbrot's concept [Man], for subsets of the integers. The integer average density leads to a new type of ergodic theorem for certain infinite measure transformations, given by normalization by the "dimension" followed by a log average [Fis92]. Then the average density reappears in a different guise: the limiting value of the time average is the expected value of the observable times the average density of the fractal integer set. The fractal point of view leads to two new isomorphism invariants for infinite measure-preserving transformations: the dimension of return times, and the average density (providing the defining limits exist). See [Fis92] and [Fis03b].

Further insight is given by examples coming from probability theory. Certain countable state Markov chains [Fel] exhibit this type of behavior. This led us to an extension of a beautiful but little-known result of Chung and Erdös [CE], see [ADF].

Now since infinite measure-preserving transformations can have a geometrical significance, related to fractal sets, one might wonder whether there might be examples of a *transition* from finite to infinite measure, based on this point of view; that might be regarded as analogous to a *change of phase* in physics.

Just such a phenomenon occurs for certain maps of the interval with an indifferent (or *neutral* or *parabolic*) fixed point. Here there is a 1- parameter family of maps (see [FL01], [FL02], [FL04a], [FL04b]), related to the Markov chain examples just discussed and also to the *Manneville-Pomeau maps* (see also [Lop]) as well as to the interesting counterexamples of Hofbauer [Hof], which exhibits three "phases" of behavior, as the parameter α ranges from 0 to ∞.

We consider the distribution of returns to the right half of the interval. For $\alpha \in (2,\infty)$, the mean and variance are finite; for $\alpha \in (1,2)$, variance is infinite but mean is finite, while for $\alpha \in (0,1)$, both are infinite. For $(1,2]$ and $(2,\infty)$ the unique absolutely continuous invariant measure for the map is finite; for $(0,1]$ it is infinite. The asymptotic return-time behavior for $[2,\infty)$ is Gaussian; for $(1,2]$ it is stable, passing through all the completely asymmetric stable laws; for $(0,1)$ it continues on, through all the Mittag-Leffler processes.

This last region (infinite measure) is the realm of fractal-like return times. The Mittag-Leffler paths are similar to Cantor functions. The increment flow along these processes has infinite measure (for background see [Aar], [Zwe]), and is a horocycle flow for the corresponding scaling flow (which is the "geodesic flow"). For all parameters $\alpha \neq 1$ the scaling flow on paths is infinite entropy Bernoulli (the case $\alpha = 1$ is handled in a special way as it has an extra drift parameter). This example completes a circle of ideas begun in [Fis92], [ADF].

12. Spaces of tilings

Fractal sets are but one of the geometric forms with an interesting small-scale structure. Other examples are tilings of $\mathbb{R}^n$ which have some sort of self-similar nesting character.

The simplest example is the binary tiling of the unit interval; at level $n \geq 0$, there are 2^{n+1} tiles given by the intervals of the form $[k \cdot 2^{-(n+1)}, (k+1) \cdot 2^{-(n+1)}]$. The tiling structure is generated by the dynamics of the map $f : x \mapsto 2x (\mod 1)$ and can be thought of as a "cookie cutter Cantor set without the gaps". Spaces of nested tilings of $\mathbb{R}^m$ are topologized in a natural way related to the Attouch-Wetts topology discussed above: the hierarchy of tilings is indexed by $\mathbb{Z}$; two nested tilings are close if on a large ball in $\mathbb{R}^m$, the boundaries are ε-close in the Hausdorff metric, for each index l between $-k$ and k. The scenery flow for this space of tilings is modelled by the height $\log 2$ suspension flow over the natural extension of f, discussed before, i.e. by Sullivan's Riemann surface lamination.

Now the binary tilings are the joins of pullbacks of the standard Markov partition $\mathcal{P} = \{P_0, P_1\} = \{[0,1/2],[1/2,1]\}$ for this map. This suggests that Markov partitions give interesting candidates for a scenery flow. A next example to consider is that of an Anosov toral diffeomorphism, such as $\begin{bmatrix} 2 & 1 \\ 1 & 1 \end{bmatrix}$. Here we will have two scenery flows, one for the stable and one for the unstable foliation. The resulting model scenery flow (for the unstable case) is the height $\log |Df^u|$ suspension of the Anosov diffeomorphism; the horocycle flows are the translations of the tilings, and this is modelled by the unstable flow of the suspension.

Many fractal tilings associated to "complex number systems" can be treated in a similar way, with more complicated Markov partitions. A theorem of Bowen [Bo78] says that a Markov partition for an $n-$toral automorphism for $n \geq 3$, or toral endomorphism for $n \geq 2$, cannot have a smooth boundary. Examples with fractal Markov partition boundaries were studied by Bedford [Bed], see also [Dek], [Pra], [KV], [Ma02]. The scenery flow of the hierarchy of nested tilings generated by such Markov partition is constructed in a similar way as for the doubling map of the circle, see §10.

Another class of interesting one-dimensional tilings are those given by the **renormalization hierarchy** of an interval exchange transformation. See [AF03] for a development of the basic theory and an in-depth study of the simplest example (exchanges of two intervals), and [AF01], [Fis03b] for more information.

If all exchanges of k intervals are considered as *parameter space,* the construction places parameter space and dynamical space (the interval exchanges themselves) in a single unified picture. Following this example, we can return to the geodesic and horocycle flows of a Riemann surface, and reinterpret them in this new light. The commutation relation now says: *The horocycle flow is a fixed point of renormalization, as the geodesic flow renormalizes the horocycle flow to itself.* This may make us think of the scenery flow of the attracting Cantor set C of the Coullet-Tresser-Feigenbaum map f, also a fixed point for a renormalization operator. There is a second map g, conjugate to the shift σ, which has an analytic extension, and for which C is a *repellor,* see [Fal85], [Ran], [Sul87]. Using this hyperbolic map g we construct its scenery as in §9. The maps g and f satisfy the commutation relation $g \circ f^2 = f \circ g$, a discrete-time version of that for the geodesic and horocycle (dilation and translation) scenery flows.

13. Nonlinear doubling maps and smooth classification

Now we return to the basic example of the doubling map on the circle, but with a twist which makes things even more interesting: we take a nonlinear version, given by a degree 2 hyperbolic $\mathbb{C}^{1+\alpha}$ map $f : S^1 \to S^1$. Using the same machinery discussed above for Julia sets, and similar also to the cookie cutter Cantor set case of [BF97], one constructs the scenery flow for the space of tilings. This is a complete invariant of smooth conjugacy. A closely related invariant is Sullivan's *scaling function* [Sul87]. We construct a solenoid much as before, but with relative lengths of subtiles now given by this scaling function. As Sullivan noted [Sul92] there is a correspondence between conformal structures on the Riemann surface lamination and scaling functions. See Theorem VI.6.1 of [dMvS] for a proof in the analytic case.

The scenery flow picture helps to make the conformal structure more easily understood: for it is just the natural conformal structure on the height $\log |Df|$ suspension flow over the solenoid. It turns out that a smooth equivalence class of hyperbolic $C^{1+\alpha}$ maps is determined by the topological data (degree) plus any one of these: a Gibbs measure class, a Hölder scaling function, a scenery flow with Hölder return time, and a conformal structure on the Riemann surface lamination.

Here is one way of viewing all this. Given a hyperbolic doubling map, Cantor set or Julia set, the construction of the scenery is a sort of linearization procedure – and the space of scenes is a tangent object, the analogue of a tangent space. As such, it is acted on (linearly) by the derivative of the map. This gives the return map to a cross-section of the scenery flow. The flow itself is simply dilation, and is linear as well; the nonlinearity of the original map has been coded into the flow space, by means of the identifications made when defining that space; and that information is, in turn, remembered by the conformal structure of the Riemann surface lamination.

14. Further directions of research; Acknowledgements

Some classes of examples not mentioned so far are self-affine fractals (as for example the graphs $(t, B(t))$ of Brownian motion paths, see U. Zähle [Zä91]), space-filling curves (on which the author is currently working with P. Arnoux), and spatially inhomogeneous fractals such as the "theater curtain" pictures in [Man], which interpolate middle- interval Cantor sets from the whole unit interval to the empty set, see also p. 107 of [Fal97] and [PZ]. The holy grail of non-homogeneous examples must be the Mandelbrot set boundary M. As Tan Lei showed [Tan], at certain points of M the scenery flow can be precisely analysed: for a Misiurewicz point c, the scenery flow of M at $c \in M$ is that of the Julia set $\mathcal{J}_c$ at c for the map $z \mapsto z^2 + c$. However what happens at this set of points, even though it is a countable dense subset of M, says next to nothing about the nearby points. The limiting scenery there is anybody's guess (though see [McM] regarding "deep points") and seems to be a deep and difficult problem.

Lastly we mention two closely related but quite complementary approaches to the study of small-scale structure of fractal sets, each of which is much like what we have called the scenery. Furstenberg's *microsets* were motivated especially by the study of intersections, sums and products of arithmetically derived fractal sets see [Fur70]. Preiss' *tangent measures* were introduced to study geometric measures at small scales [Pre], [Mat]. For certain self-similar measures, the tangent measures at almost every point form the same tangent measure distribution [Gra]. Mörters and Preiss [MP] have shown that tangent measure distributions are so-called Palm measures, and so have connected the geometric study of measures to an important series of papers by Ulrich Zähle [Zä88], [Zä90], [Zä91] who studied self-similarity of random measures. See Patzschke and M. Zähle [PZ], Moerters (who generalized the average density introduced in [BF92]) [Mö97], [Mö98], [KM], Falconer and Xiao [Fal97], [FX], [Xia] for some related developments.

Perhaps the future will see a general theory which integrates all these approaches. My own sense is that our understanding is still at a very beginning stage, and that there is a world of exciting exploring yet to be done at the small scales of these remarkable landscapes.

I wish especially to thank coworkers on these and related projects: Jon Aaronson, Pierre Arnoux, Tim Bedford, Marc Burger, Herold Dehling, Manfred Denker, Sebastien Ferenczi, Pascal Hubert, Artur Lopes, Jerome Los, Yair Minsky, Luis Fernando Carvalho da Rocha and Mariusz Urbański. Figures 1 and 3 were made using McMullen's *lim* program; Tony Phillips helped with the other graphics. There are many more people who should be acknowledged here, see [Fis03b]. I wish to add that without the encouragement and participation of these friends, teachers, and colleagues, most of this would never have been worked out and in any case the process would not have been nearly so much fun.

References

[Aar] J. Aaronson. *An Introduction to Infinite Ergodic Theory.* Math. Surveys and Monographs. AMS, 1997.

[ADF] J. Aaronson, M. Denker, and A. M. Fisher. Second order ergodic theorems for ergodic transformations of infinite measure spaces. *Proc. Amer. Math. Soc.* 114(1):115–128, 1992.

[AF01] P. Arnoux and A.M. Fisher. The scenery flow for geometric structures on the torus: the linear setting. *Chinese Ann. of Math.*, 4:427–470, 2001.

[AF03] P. Arnoux and A. M. Fisher. Anosov families, renormalization and nonstationary subshifts. Technical report, LMD, Univ. de Marseilles, 2003.

[Bed] T. Bedford. Generating special Markov partitions for hyperbolic toral automorphisms using fractals. *Ergod. Th. Dynam. Sys.*, 6:325–333, 1986.

[BF92] T. Bedford and A.M. Fisher. Analogues of the Lebesgue density theorem for fractal sets of reals and integers. *Proc. London Math. Soc.*, 64:95–124, 1992.

[BF96] T. Bedford and A.M. Fisher. On the magnification of Cantor sets and their limit models. *Monatsh. Math.*, 121:11–40, 1996.

[BF97] T. Bedford and A.M. Fisher. Ratio geometry, rigidity and the scenery process for hyperbolic Cantor sets. *Erg. Th. Dyn. Sys.*, 17:531–564, 1997.

[BFU] T. Bedford, A.M. Fisher, and M. Urbański. The scenery flow for hyperbolic Julia sets. *Proc. London Math. Soc.*, 3(85):467–492, 2002.

[BM] R. Bowen and B. Marcus. Unique ergodicity for horocycle foliations. *Israel Jour. Math.*, 26(1):43–67, 1977.

[Bo78] R. Bowen. Markov partitions are not smooth. *Proceedings of the American Mathematical Society*, 71:130–132, 1978.

[Bo79] R. Bowen. Hausdorff dimension of quasi-circles. *Publications Mathématiques (Institut des Hautes Etudes Scientifiques, Paris)*, 50:11–25, 1979.

[Bur] M. Burger. Horocycle flow on geometrically finite surfaces. *Duke Math. J.*, 61(3):779–803, 1990.

[CE] K.L. Chung and P. Erdös. Probability limit theorems assuming only the first moment I. *Memoirs Amer. Math. Soc.*, 6, 1951.

[Dek] M. Dekking. Recurrent sets. *Adv. in Math.*, 44:78–104, 1982.

[dMvS] Welington de Melo and Sebastien van Strien. *One-Dimensional Dynamics*. Springer, 1993.

[Fal85] K.J. Falconer. *The Geometry of Fractal Sets*. Cambridge U. Press, New York/Cambridge, 1985.

[Fal97] K. Falconer. *Techniques in Fractal Geometry*. John Wiley and Sons, 1997.

[Fel] W. Feller. Fluctuation theory of recurrent events. *Trans. AMS*, 67:98–119, 1949.

[Fis87] A.M. Fisher. Convex-invariant means and a pathwise central limit theorem. *Adv. Math.*, 63:213–246, 1987.

[Fis92] A.M. Fisher. Integer Cantor sets and an order-two ergodic theorem. *Ergod. Th. Dynam. Sys.*, 13:45–64, 1992.

[Fis03a] A.M. Fisher. Notes on Dynamics: the ergodic theory of adic transformations, 2003.

[Fis03b] A.M. Fisher. Small-scale Structure and Randomness: the Scenery Flow in Dynamics, Geometry and Probability. http://ime.usp.br/~afisher, March 2003. thesis for livre docencia, Univ. of São Paulo.

[Fis04] A.M. Fisher. A Pathwise Central Limit Theorem for random walks. *Adv. Prob. (to appear)*, 2004.

[FL01] A.M. Fisher and A. Lopes. Exact bounds for the polynomial decay of correlation, 1/f noise and the CLT for the equilibrium state of a non-Hölder potential. *Nonlinearity*, 14:1071–1104, 2001.

[FL02] A.M. Fisher and A. Lopes. Invariance principles in log density and convergence to stable flows. Technical report, USP, 2002.

[FL04a] A.M. Fisher and A. Lopes. Invariance principles in log density for renewal processes. Technical report, USP, 2004.

[FL04b] A.M. Fisher and A. Lopes. Self-similar return sets and stable laws for some maps with an indifferent fixed point. Technical report, USP, 2004.

[FU] A.M. Fisher and M. Urbański. On invariant line fields. *Bulletin London Math. Soc.*, 32:555–570, 2000.

[Fur61] H. Furstenberg. Strict ergodicity and transformation of the torus. *Amer. J. Math.*, 83:573–601, 1961.

[Fur70] H. Furstenberg. Intersections of Cantor sets and transversality of semigroups. In R. C. Gunning, editor, *Problems in Analysis, a symposium in honor of S. Bochner*. Princeton Univ. Press, 1970.

[FX] K.J. Falconer and Yimin Xiao. Average densities of the image and zero set of stable processes. *Stochastic Processes Appl.*, 55:271–283, 1995.

[Gra] S. Graf. On Bandt's tangential distribution for self-similar measures. *Monatsh. Math.*, 120:223–246, 1995.

[Hof] F. Hofbauer. Examples for the non-uniqueness of the equilibrium states. *Trans. Amer. Math. Soc.*, 228:133-141, 1977.

[Ken] P. Kenny. *Ergodic measures for a class of horocycle flows*. PhD thesis, McGill University, 1983.

[KM] D. Krieg and P. Mörters. Tangent measure distributions of hyperbolic Cantor sets. *Monatshefte Math.*, 312:53–93, 1998.

[KV] R. Kenyon and A. Vershik. Arithmetic construction of sofic partitions of hyperbolic toral automorphisms. *Erg. Th. Dyn. Syst.*, 18:357–372, 1998.

[Lop] A. Lopes. The Zeta function, non-differentiabilty of pressure and the critical exponent of transition. *Adv. in Math.*, 101(2):133–165, 1993.

[Lyu] M. Lyubich. Laminations and holomorphic dynamics. Technical report, Lecture Notes of the mini-course given at the Conference "New Directions in Dynamical Systems in Kyoto, August, 2002.

[Man] B. Mandelbrot. *The Fractal Geometry of Nature*. W. H. Freeman, 1982.

[Ma02] A. Manning. A Markov partition that reflects the geometry of a hyperbolic toral automorphism. *Trans. Amer. Math. Soc.*, to appear.

[Mat] P. Mattila. *Geometry of Sets amd Measures in Euclidean Spaces*, volume 44. Cambridge Advanced Studies in Mathematics, 1995.

[May] V. Mayer. Comparing measures and invariant line fields. *Ergod. Th. Dynam. Sys.*, 22:555–570, 2002.

[McM] C.T. McMullen. *Renormalization and 3-manifolds which fiber over the circle*, volume 142 of *Studies in Math.* Princeton Univ. Press, Princeton, New Jersey, 1996.

[ML] Y. Minsky and M. Lyubich. Laminations in holomorphic dynamics. *J. Diff. Geometry*, 47:17–94, 1997.

[Mö97] P. Mörters. Average densities, tangent measures and rectifiability. *Periodica Mathematica Hungarica*, 37:313–324, 1997.

[Mö98] P. Mörters. Average densities and linear rectifiability of measures. *Mathematika*, 37:65–79, 1998.

[MP] P. Mörters and D. Preiss. Tangent measure distributions of fractal measures. *Math. Annalen*, 126:313–328, 1998.

[Pat76] S. J. Patterson. The limit set of a Fuchsian group. *Acta Math*, 136:241–273, 1976.

[Pat87] S. J. Patterson. Measures on limit sets of Kleinian groups. In D. B. A. Epstein, editor, *Analytical and Geometric Aspects of Hyperbolic Space*, volume 111 of *LMS Lecture Notes Series*, pages 281–323. Cambridge Univ. Press, 1987.

[Pra] B. Praggastis. Numeration systems and Markov partitions from self similar tilings. *Trans. Amer. Math. Soc.*, 351:3315–3349, 1999.

[Pre] D. Preiss. Geometry of measures in $\mathbb{R}^n$: Distribution, rectifiability, and densities. *Annals Math.*, 125:537–643, 1987.

[PZ] N. Patzschke and M. Zähle. Fractional differentiation in the self-affine case III: The density of the Cantor set. *Proc. AMS*, 117:132–144, 1993.

[Ran] D.A. Rand. Global phase space universality, smooth conjugacies and renormalization: I. the $C^{1+\alpha}$ case. *Nonlinearity*, 1:181–202, 1988.

[Sul70] D.P. Sullivan. The density at infinity of a discrete group of hyperbolic motions. *Publ. Matématiques*, 50:171–202, 1970.

[Sul84] D.P. Sullivan. Entropy, Hausdorff measures old and new, and limit sets of geometrically finite Kleinian groups. *Acta Math.*, 153:259–277, 1984.

[Sul87] D.P. Sullivan. Differentiable structures on fractal-like sets, determined by intrinsic scaling functions on dual Cantor sets. *AMS Proc. Symp. Pure Math.*, 48:15–23, 1987.

[Sul91] D.P. Sullivan. Bounds, quadratic differentials, and renormalization conjectures. *AMS Centennial Publications*, 2, 1991.

[Sul92] D.P. Sullivan. Linking the universalities of Milnor-Thurston, Feigenbaum and Ahlfors-Bers. In *Topological Methods in Modern Mathematics*, pages 543–564. Publish or Perish, 1992.

[Tan] Lei Tan. Similarity between the Mandelbrot set and Julia sets. *Comm. Math. Phys.*, 134:587–617, 1990.

[Xia] Yimin Xiao. Random fractals and Markov processes. In *Proceedings of the Conference on Fractal Geometry and Applications: A Jubilee of Benoit Mandelbrot*, 2003. (to appear).

[Zä88] U. Zähle. Self-similar random measures I: Notion, carrying Hausdorff dimension, and hyperbolic distribution. *Prob. Theor. Rel. Fields*, 80:79–100, 1988.

[Zä90] U. Zähle. Self-similar random measures II: a generalization to self-affine measures. *Math. Nachr.*, 146:85–98, 1990.

[Zä91] U. Zähle. Self-similar random measures III: Self-similar random processes. *Math. Nachr.*, 151:121–148, 1991.

[Zwe] R. Zweimueller. Probabilistic properties of dynamical systems with infinite invariant measures. Diplomarbeit, Univ. Salzburg 1995.

Albert M. Fisher
Dept Mat IME-USP, Caixa Postal 66281, CEP 05315-970 São Paulo, Brazil
URL: `http://ime.usp.br/~afisher`
e-mail: `afisher@ime.usp.br`

Progress in Probability, Vol. 57, 79–92

Hausdorff Dimension of Hyperbolic Attractors in $\mathbb{R}^3$

Károly Simon

To Erzsébet

Abstract. One of the most exciting open problems in the field of dimension theory of dynamical systems is to find the Hausdorff dimension of hyperbolic attractors in space. After an introduction we will give an account of the known results.

Mathematics Subject Classification (2000). Primary 37C45; Secondary 28A80, 28A78.

Keywords. Hyperbolic attractor, non-conformal, Hausdorff dimension.

1. Introduction

In this paper we are always working in $\mathbb{R}^d$ where d is usually not bigger than 3. The simplest dynamically defined sets are the self-similar ones. It has been known since the papers Hutchinson [13] or even Moran [15] that for a self-similar iterated function system (IFS) $\mathcal{F} = \{f_1, \dots, f_m\}$, where

$$|f_i(x) - f_i(y)| = \lambda_i |x - y|, \quad 0 < \lambda_i < 1, \quad x, y \in \mathbb{R}^d,$$

we can compute the Hausdorff dimension of the attractor Λ, which is the unique nonempty compact set satisfying

$$\Lambda = \bigcup_{i=1}^{m} f_i(\Lambda), \tag{1.1}$$

as the solution s of the equation

$$\lambda_1^s + \dots + \lambda_m^s = 1. \tag{1.2}$$

More precisely,

$$\dim_H(\Lambda) = \min\{s, d\} \tag{1.3}$$

The research in this paper was supported by OTKA foundation #T32022 .

holds if the cylinders $\Lambda_i := f_i(\Lambda)$, $i = 1, \dots, m$, are pairwise disjoint or, more generally, the so-called open set condition holds. However, if we assume only that the maps f_i above are affine maps (in this case the set defined in (1.1) is called self-affine), then the problem is much more difficult. On the one hand, we can expect only 'almost all' type results, on the other hand, even 'almost all' type results hold only in the case when $\|f_i\| < \frac{1}{2}$, see [11] and [38]. In this note we are interested in the following problem:

Problem. How can we express the dimension of uniformly hyperbolic (axiom A) invariant sets?

On the line and on a surface this problem has similar flavor to the computation of dimension of self-similar sets with disjoint cylinders, and that is why on the line and on a surface the answer has been known for more than 20 years. However, in space (and higher dimension) the problem has similar flavor to the computation of the dimension of self-affine sets, which is much more difficult. That is why in space the problem has been open for more than twenty years. In this note we would like to provide a gentle introduction to this field for everybody interested in fractals but not necessarily having any background in the field of dynamical systems. There are at least three other survey papers [28], [2] and [31] of much wider scope dealing with (among some other things) the problem above.

2. The dimension of self-affine sets

First we study how to compute the dimension of the attractor for a typical self-affine IFS if all contractions are 'stronger' than $\frac{1}{2}$.

2.1. Falconer's Theorem

The most important result in this field is due to Falconer [11]. This is an 'almost all' type result in the following sense:

Let Λ be the attractor of a contractive self-affine IFS $\mathcal{F} = \{f_1, \dots, f_m\}$ on $\mathbb{R}^d$, that is, Λ satisfies (1.1). For every $\mathbf{t} = \{\mathbf{t}_1, \dots, \mathbf{t}_m\} \in \mathbb{R}^d \times \dots \times \mathbb{R}^d$ we define $\Lambda^{\mathbf{t}}$ as the attractor of the 'shifted' IFS $\mathcal{F}^{\mathbf{t}} = \{f_1 + \mathbf{t}_1, \dots, f_m + \mathbf{t}_m\}$.

Theorem 2.1 (Falconer [11], **Solomyak** [38]**).** *If $\|f_i\| < \frac{1}{2}$ holds*[1] *for all $i = 1, \dots, m$, then for Lebesgue almost all $\mathbf{t} \in \mathbb{R}^{m \cdot d}$ we have*

$$\dim_H \Lambda^{\mathbf{t}} = \dim_B \Lambda^{\mathbf{t}} = \min\{d, \alpha(f_1, \dots, f_m)\}, \tag{2.1}$$

where $\alpha(f_1, \dots, f_m)$ is defined via the singular value function $s \mapsto \varphi^s$ in the following way:

$$\alpha(f_1, \dots, f_m) := \inf\left\{ s : \sum_{k=1}^{\infty} \sum_{i_1 \dots i_k} \varphi^s(f_{i_1} \circ \dots \circ f_{i_k}) < \infty \right\}. \tag{2.2}$$

[1] In Falconer's paper this bound was 1/3. Solomyak improved it to 1/2.

We recall that the *singular values* $1 \geq \alpha_1 \geq \cdots \geq \alpha_m > 0$ of a contracting linear map $T : \mathbb{R}^d \to \mathbb{R}^d$ are defined as the square roots of the eigenvalues of the symmetric transformation T^*T. Equivalently, we obtain the singular values of T as the semi-axes of the ellipsoid $T(B)$, where B is the unit ball. Then the singular value function $s \mapsto \varphi^s(T)$ for $0 \leq s \leq d$ is defined as

$$\varphi^s(T) = \alpha_1 \cdots \alpha_{[s]} \alpha_{[s]+1}^{s-[s]}. \tag{2.3}$$

2.2. The importance of the singular value function in computation of dimension We give a very **heuristic** explanation in the special case when $d = 2$ and $1 < s < 2$. Recall that the s-dimensional Hausdorff measure of Λ is defined as:

$$\mathcal{H}^s(\Lambda) := \lim_{\delta \to 0} \inf \left\{ \sum_{i=1}^{\infty} \operatorname{diam}(U_i)^s : \text{ where } \Lambda \subset \bigcup_{i=1}^{\infty} U_i \text{ and } \operatorname{diam}(U_i) \leq \delta \right\}. \tag{2.4}$$

For simplicity we assume that the unit ball B contains the attractor Λ . For an arbitrary n the cylinder $B_{i_1\dots,i_n} := f_{i_1} \circ \cdots \circ f_{i_n}(B)$ is an ellipse of semi-axis α_1, α_2, the singular values of $T := f_{i_1} \circ \cdots \circ f_{i_n}$.

Let us call $\mathcal{C}_{i_1\dots i_n}$ a cover of $B_{i_1\dots,i_n}$ with the minimal number of squares of side $2\alpha_2$. Then $\mathcal{C}_{i_1\dots i_n}$ consists of α_1/α_2 squares of side $2\alpha_2$. Further, $\bigcup_{i_1,\dots,i_n} \mathcal{C}_{i_1\dots i_n}$ is a cover of the attractor Λ. In (2.4) we take the sum of the diameter of the sets in a cover, raised to the power s and take infimum for all covers with sets of small diameters. Compute this sum for the cover: $\bigcup_{i_1,\dots,i_n} \mathcal{C}_{i_1\dots i_n}$. The sum in (2.4) is $2^s \sum\limits_{i_1\dots i_k} \varphi^s(f_{i_1} \circ \cdots \circ f_{i_k})$ since $\varphi^s(T) = \frac{\alpha_1}{\alpha_2}\alpha_2^s$ for $T = f_{i_1} \circ \cdots \circ f_{i_k}$. So, if $s > \alpha(f_1, \dots, f_m)$ then the sum $\sum\limits_{i_1\dots i_k} \varphi^s(f_{i_1} \circ \cdots \circ f_{i_k})$ tends to zero as $n \to \infty$. Therefore $\alpha(f_1, \dots, f_m)$ is an upper bound of the Hausdorff dimension.

However, there are very natural examples (see e.g. [16] and [25]) showing that formula (2.1) does not always hold. Edgar proved that for $d \geq 2$ the upper bound $\frac{1}{2}$ on the norm of the contraction in Falconer Theorem is sharp. In [34] we proved with B. Solomyak that one cannot improve this upper bound even if all the maps in Falconer's Theorem are similarities (assuming again that $d \geq 2$).

3. Hyperbolic attractors on the line

Let $\{f_i\}_{i=1}^m$ be a contracting IFS on the line. We assume that $f_i \in C^2$ (for simplicity) and

$$0 < \inf_{i,x} |f_i'(x)| < \sup_{i,x} |f_i'(x)| < 1. \tag{3.1}$$

The Hausdorff dimension of the attractor Λ (defined by (1.1)) can be computed in a very similar way to the self similar case, if the cylinders are disjoint:

$$f_i(\Lambda) \cap f_j(\Lambda) = \emptyset \qquad \text{for } i \neq j \ . \tag{3.2}$$

For a proof, let I denote the convex hull of Λ. The most natural cover of Λ is the cover by all n-cylinder intervals $\{I_{i_1\dots i_n}\}$ for some $n \in \mathbb{N}$, where $I_{i_1\dots i_n} := f_{i_1\dots i_n}(I)$. (Throughout the paper we use the notation $f_{i_1\dots i_n} := f_{i_1} \circ \dots \circ f_{i_n}$.) Let d_n be the exponent that corresponds to the cover $\{I_{i_1\dots i_n}\}$. That is, we define d_n as the solution of the equation:

$$\sum_{i_1\dots i_n} |I_{i_1\dots i_n}|^{d_n} = 1. \tag{3.3}$$

It is well known that

$$\dim_H \Lambda = \lim_{n\to\infty} d_n. \tag{3.4}$$

Obviously, in the special case when all the functions f_i are similarities, for every n we have $d_n = s$, the solution of (1.2). In the general case, unfortunately, the convergence in (3.4) is slow. Namely,

$$|\dim_H \Lambda - d_n| < \frac{const}{n}, \tag{3.5}$$

and in general there is no way to improve this. It is useful to express the Hausdorff dimension using topological pressure. Using that the Hausdorff dimension can be obtained as the root of some pressure equation, Pollicott and Jenkinson [22] found a way to approximate the Hausdorff dimension numerically with much higher speed than in (3.5).

3.1. Computation of the dimension by the pressure formula

Here we give yet again a very heuristic argument, explaining how the topological pressure appears in this field. In this argument we assume that the cover by the cylinder intervals is the cover at which the infimum is attained in (2.4). One can find detailed proofs in the books [9] and [21].

Assume that on the symbolic space $\Sigma := \{1,\dots,m\}^{\mathbb{N}}$ we are given a continuous function $\varphi : \Sigma \to \mathbb{R}$. Let $\sigma : \Sigma \to \Sigma$ denote the shift map and $[i_1, ..., i_n]$ the set of all sequences starting with $i_1, ..., i_n$.

Definition 3.1. The pressure of φ is defined by:

$$P(\varphi) := \lim_{n\to\infty} \frac{1}{n} \log \sum_{i_1\dots i_n} \exp\left\{ \sup_{\mathbf{j}\in[i_1\dots i_n]} \sum_{k=0}^{n-1} \varphi\left(\sigma^k \mathbf{j}\right) \right\}. \tag{3.6}$$

We use this for the function

$$\varphi_s(\mathbf{j}) := s \log \left| f'_{j_1}\left(\Pi(\sigma \mathbf{j})\right)\right|, \tag{3.7}$$

where $0 \leq s$ and $\Pi : \Sigma \to \Lambda$ is called the natural projection and is defined by:

$$\Pi(\mathbf{i}) := \lim_{n\to\infty} f_{i_1\dots i_n}(0). \tag{3.8}$$

By the chain rule we have

$$\sum_{k=0}^{n-1} \varphi_s\left(\sigma^k \mathbf{j}\right) = \log \left| f'_{i_1\dots i_n}\left(\Pi(\sigma^n \mathbf{j})\right)\right|, \tag{3.9}$$

holds for every $\mathbf{j} \in [i_1, \ldots, i_n]$. Using the so-called bounded distortion lemma and Lagrange Theorem we obtain that for all $\mathbf{j} \in [i_1, \ldots, i_n]$,

$$|I_{i_1 \ldots i_n}| = |f_{i_1 \ldots i_n}(I)| \approx \left| f'_{i_1 \ldots i_n}(\Pi(\sigma^n \mathbf{j})) \right|, \tag{3.10}$$

where $\approx$ means equality up to a multiplicative constant. So,

$$\exp\left[\sup_{\mathbf{j} \in (i_1, \ldots, i_n)} \sum_{k=0}^{n-1} \varphi_s\left(\sigma^k \mathbf{j}\right) \right] \approx |I_{i_1 \ldots i_n}|^s .$$

We see that the topological pressure is the exponential growth rate in n of the sum of the s-powers of the diameters of the elements of the cover $I_{i_1 \ldots i_n}$. Assuming again that this cover is the most economical cover in the sense of (2.4), we can see why the notion of topological pressure is so important in the computation of dimension. Namely, one can easily see that the function $s \mapsto P(\varphi_s)$ is a strictly decreasing smooth function, which is positive at 0 and negative at 1. So there is a unique t at which $P(\varphi_t) = 0$. When $s > t$ then $P(\varphi_s) < 0$. This means that the exponential growth rate of the sum $\sum_{i_1 \ldots i_n} |I_{i_1 \ldots i_n}|^s$ is negative, that is, from (2.4) we obtain that $\mathcal{H}^s(\Lambda) = 0$. Similarly, if $s < t$ then the exponential growth rate of $\sum_{i_1 \ldots i_n} |I_{i_1 \ldots i_n}|^s$ is positive. So, assuming that this is the most economical cover in the sense of (2.4), we get that the s-dimensional Hausdorff measure is infinity. Therefore we have

Theorem 3.2.

$$\dim_H(\Lambda) \quad \textit{equals the zero of the function: } s \mapsto P(\varphi_s), \tag{3.11}$$

where φ_s was defined in (3.7).

4. Uniformly hyperbolic invariant sets

The precise definition of an axiom A diffeomorphism can be found in the books [6, p.68] and [21, p.228]. An important consequence of the definition is that if Λ is an invariant set for a $\mathcal{C}^r$, $r \geq 2$ axiom A diffeomorphism $f : \mathcal{M} \to \mathcal{M}$, where $\mathcal{M}$ is a smooth Riemannian manifold, then for every $x \in \Lambda$ the tangent space $T_x M$ can be written as the direct product

$$T_x = E_x^u \bigoplus E_x^s,$$

such that the derivative of f^n contracts with an exponential speed in E_x^s direction and expands with exponential speed in E_x^u direction. That is, there exists $\lambda \in (0,1)$ such that for every $n \geq 0$:

$$\|Df^n(\mathbf{v})\| \leq \lambda^n \|\mathbf{v}\| \text{ for } \mathbf{v} \in E_x^s \text{ and } \|Df^{-n}(\mathbf{v})\| \leq \lambda^n \|\mathbf{v}\| \text{ for } \mathbf{v} \in E_x^u. \tag{4.1}$$

Furthermore, the subspaces E_x^u and E_x^s are invariant. That is:

$$Df(E_x^s) = E_{f(x)}^s \text{ and } Df(E_x^u) = E_{f(x)}^u.$$

The local stable and instable manifolds

$$W_\varepsilon^s := \{y \in \mathcal{M} : d(f^n(x), f^n(y)) \leq \varepsilon \ \forall n \geq 0\},$$

and

$$W_\varepsilon^u := \left\{y \in \mathcal{M} : d(f^{-n}(x), f^{-n}(y)) \leq \varepsilon \ \forall n \geq 0\right\},$$

are $\mathcal{C}^r$ disks tangent to E_x^s and E_x^u respectively. That is $T_x W_\varepsilon^x = E_x^u$ and $T_x W_\varepsilon^x = E_x^u$. Since locally Λ is the product of the sets $\Lambda \cap W_\varepsilon^s(x)$ and $\Lambda \cap W_\varepsilon^u(x)$ the Hausdorff dimension of Λ is

$$\dim_H(\Lambda \cap B(x,\varepsilon)) = \dim_H(\Lambda \cap W_\varepsilon^s(x)) + \dim_H(\Lambda \cap W_\varepsilon^u(x)), \tag{4.2}$$

where $B(x,\varepsilon)$ is the ball centered at x with radius ε. This is so, because $\dim_H(\Lambda \cap W_\varepsilon^u(x)) = \dim_B(\Lambda \cap W_\varepsilon^u(x))$ and $\dim_H(\Lambda \cap W_\varepsilon^s(x)) = \dim_B(\Lambda \cap W_\varepsilon^s(x))$ Then a well known theorem (see [8], p. 95) says that (4.2) holds.

5. Hausdorff dimension of Axiom A invariant sets on a surface

Assume that Λ is an invariant set of a $\mathcal{C}^2$ diffeomorphism on a surface. Then for all $x \in \Lambda$ the local stable and unstable manifolds $W_\varepsilon^s(x)$ and $W_\varepsilon^u(x)$ are smooth curves. Let $\Lambda^s(x) := \Lambda \cap W_\varepsilon^s(x)$ and $\Lambda^u(x) := \Lambda \cap W_\varepsilon^u(x)$ and write t_s and t_u for their Hausdorff dimensions respectively. Since the Cantor sets $\Lambda^s(x)$ and $\Lambda^u(x)$ are constructed in a very similar way to the Cantor set in Section 3, their Hausdorff dimension can be computed with essentially the same pressure formulas as in (3.11). Namely, the zeros of the functions

$$s \mapsto P(s \log \left|d_x f|_{E^s(x)}\right|) \quad \text{and} \quad s \mapsto P(-s \log \left|d_x f|_{E^u(x)}\right|) \tag{5.1}$$

are equal to t_s and t_u respectively. Then by (4.2) we obtain the Hausdorff dimension of Λ:

$$\dim_H(\Lambda) = t_s + t_u. \tag{5.2}$$

The same result was obtained by McCluskey and Manning [17] for C^1 axiom A diffeomorphisms. Their result is much more difficult than the argument which appears here, since we always assume that our maps are $\mathcal{C}^2$, so we have the bounded distortion lemma.

6. Hausdorff dimension of Axiom A invariant sets in space

We do not know in general how to compute the dimension of an axiom A invariant set on a three- (or higher) dimensional Riemannian manifold. Let f be a $\mathcal{C}^2$ diffeomorphism on a three-dimensional Riemannian manifold. We may assume that the local stable manifolds are two-dimensional and the local unstable manifolds are one-dimensional submanifolds. Using (4.2) we can compute the Hausdorff dimension as the sum of the dimensions of the parts of Λ falling into the local stable and unstable manifolds. Since by assumption the local unstable manifold is one-dimensional, for any x the dimension $\dim_H(\Lambda \cap W_\varepsilon^u(x))$ can be computed as the zero of the second function in (5.1). The problem is that, in general, we have a difficulty computing $\dim_H(\Lambda \cap W^s_\varepsilon(x))$. Since E^s_x is a plane, it can happen that there are two directions in E^s_x with different contraction coefficients. This is the non-conformal case, so here a difficulty appears similar to that in Section 2.

Since we cannot solve the problem in full generality, it is useful to try to give an answer in some special cases. The most natural axiom A attractor in space is the solenoid.

6.1. Solenoids with two different constant contracting coefficients

If in the definition of a solenoid (see below) the two contracting coefficients coincide, then the problem is easy (see e.g. [8]). Therefore the first non-trivial example is the solenoid with two different constant contracting coefficients. Y. Pesin in 1982 (in the same year when Manning and McCluskey solved the problem on a surface) asked the following question:

Question (Pesin). Let Λ be the solenoid with two different constant contraction coefficients. How can we compute the Hausdorff dimension of Λ?

The answer was given independently by Bothe [4] and the author [32]. First we give the definition of the solenoid and state the answer for Pesin's question.

Definition 6.1. Suppose $0 < \lambda_2 < \lambda_1 < \frac{1}{2}$. Let $T := S^1 \times D$, where D is the unit disk centered at the origin. We define the map (see Figure 1) $f : T \to T$ by

$$f(t,x,z) := \left(2t \pmod 1, \lambda_1 x + \frac{1}{2}\cos(2\pi t), \lambda_2 z + \frac{1}{2}\sin(2\pi t)\right). \tag{6.1}$$

We assumed that $\lambda_1 < \frac{1}{2}$ to make the map one-to-one on T. We call the attractor

$$\Lambda = \bigcap_{n=0}^{\infty} f^n(T)$$

the solenoid with contracting coefficients λ_1, λ_2.

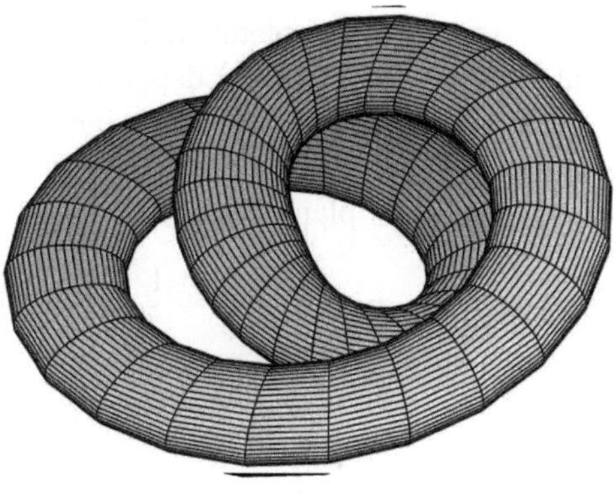

FIGURE 1. $f(T)$, the first image of solenoid map on the torus T

Obviously, on S^1 the map f is expanding with a constant expanding coefficient 2. On D, both in x and in z direction f is contractive with constant contracting coefficients λ_1 and λ_2 respectively.

Theorem 6.2 (Bothe [4], Simon [30, 32]). *We have* $\dim_H(\Lambda) = 1 + \frac{\log 2}{-\log \lambda_1}$.

Here we give a heuristic argument for the proof sketch. For an arbitrary $I_{i_n,\dots,i_1}$, a dyadic interval of S^1, we define

$$S_{i_1\dots i_n} := f^n(I_{i_n\dots i_1}).$$

Then $\Lambda = \bigcap_n \bigcup_{i_1\dots i_n} S_{i_1\dots i_n}$. In order to understand the structure of the solenoid, it is enough to understand the structure of its angular sections. Let us call the set $A_\theta := \{(t,x,z) \in T : t = \theta\}$ the θ-angular section of T. We want to study $A_\theta \cap \Lambda$ for an arbitrary θ. To do this, choose a $\theta \in S^1$. Let $t_1, t_2 \in S^1$ be the inverse images of θ for the function $t \mapsto 2t(\mod 1)$. Then $f(A_{t_i}) \subset A_\theta$ and $f(A_{t_i})$, $i = 1, 2$, are ellipses with semi-axes λ_1, λ_2. In the same way for an n and $i_1, \dots, i_n$ let $\{t_{i_1\dots i_n}\}_{i_1\dots i_n}$ be the inverse images of the function $t \mapsto 2^n t(\mod 1)$. Then $f(A_{t_{i_1\dots i_n}}) \subset A_\theta$ are 2^n congruent ellipses with semi-axes λ_1^n, λ_2^n. Let us call these 2^n ellipses the n-cylinders on A_θ. Obviously, $\Lambda \cap A_\theta \subset \bigcup_{i_1,\dots,i_n} f(A_{t_{i_1\dots i_n}})$. Using this, we obtain a natural cover of $\Lambda \cap A_\theta$ by the n-cylinders on A_θ. The most natural approach to the computation of the dimension of Λ is to compute first the dimension of $\Lambda \cap A_\theta$ by trying to prove that the cover above is the most economical one in the sense of minimizing the sum in the definition of Hausdorff measure in (2.4). Surprisingly, this most natural approach did not work. Namely, every n-cylinder ellipse contains two 'children' $(n+1)$-cylinder ellipses. Unfortunately, the relative positions of these $(n+1)$-cylinder ellipses are completely different in different n-cylinder ellipses. Therefore the structure of $\Lambda \cap A_\theta$ is different from a self-affine set and difficult to handle. Still we can see from this angular section approach that

$$\dim_H \Lambda \leq 1 + \frac{\log 2}{-\log \lambda_1}. \tag{6.2}$$

To get the lower estimate, we need a completely different approach. Observe that the first two component functions of f in (6.1) depend only on the first two variables t and x. (The roles of x and y are different since the contraction in the direction of y is stronger than in the direction of x.) We can consider the projection $\tilde{\Lambda}$ of the solenoid Λ to the (t,x) plane. Then $\tilde{\Lambda}$ is the attractor of

$$\tilde{f}(t,x) := \left(2t(\mod 1), \lambda_1 x + \frac{1}{2}\cos(2\pi t)\right). \tag{6.3}$$

That is

$$\tilde{\Lambda} = \bigcap_{n \geq 0} \tilde{f}^n(\tilde{T}), \tag{6.4}$$

where we denote the projection of T to the (t,x) plane by $\tilde{T}$ (see Figure 2).

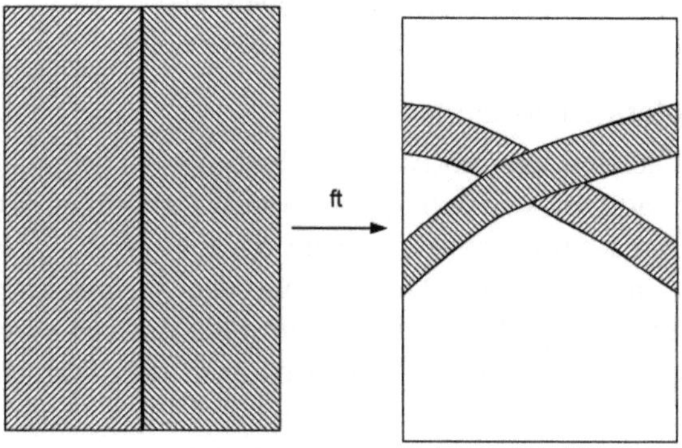

FIGURE 2. $\tilde{f}(\tilde{T})$ the first image of the plane solenoid map on the torus $\tilde{T}$

That is, $\tilde{T} = S^1 \times [-1,1]$. Clearly $\dim_H \tilde{\Lambda} \leq \dim_H \Lambda$. So it is enough to prove that $\dim_H \tilde{\Lambda} \geq 1 + \frac{\log 2}{-\log \lambda_1}$. Let $I_{i_n \dots i_1}$ be a dyadic interval on S^1. We write $\tilde{S}_{i_1 \dots i_n} := \tilde{f}^n(I_{i_n \dots i_1})$. Then $\left\{\tilde{S}_{i_1 \dots i_n}\right\}_{i_1 \dots i_n}$ are 2^n strips; each of them intersects every vertical line on the (t,x) plane with $0 \leq t \leq 1$, in an interval of length $2\lambda_1$. **IF** these strips were disjoint (they are not), then we could immediately see that the dimension of $\tilde{\Lambda}$ must be $1 + \frac{\log 2}{-\log \lambda_1}$. Namely, in that case, every vertical line would intersect $\cup_{i_1 \dots i_n} \tilde{S}_{i_1 \dots i_n}$ in a set which is a union of 2^n *disjoint* intervals of length λ_1^n. This would imply that the Hausdorff dimension of the intersection of $\tilde{\Lambda}$ with any vertical line would be $\frac{\log 2}{-\log \lambda_1}$. However, the strips $\tilde{S}_{i_1 \dots i_n}$ intersect each other. Therefore, the intersection of $\tilde{\Lambda}$ with a vertical line consists of 2^n *overlapping* intervals. This causes the difficulty of the problem.

The solution (see [4], [30]) comes from an idea that is now called the transversality condition. Some modifications of this idea for iterated function systems were used later in several other papers, e.g. [23, 37, 38, 35, 36], or see [20] for further applications. To apply this idea we have to check two assumptions:

(i) the contraction is uniformly stronger than the expansion,
(ii) for every $(i_1, i_2, \dots) \in \{1,2\}^{\mathbb{N}}$ we consider the curve $\Lambda_{i_1,i_2,\dots} := \bigcap_n S_{i_1 \dots i_n}$. If any two of these uncountably many curves intersect each other, then this intersection must be transversal.

It is proved in [32] that if these two conditions hold, then the overlapping between $S_{i_1 \dots i_n}$ does not affect the calculation of the dimension of the attractor. Condition (i) follows immediately from the assumption $\lambda_1 < \frac{1}{2}$. Condition (ii) is checked in [32]. In this way we can see that $\dim_H \tilde{\Lambda} \geq 1 + \frac{\log 2}{-\log \lambda_1}$, which completes the proof of Theorem 6.2.

Bothe [4] considered more general solenoids and proved a similar result assuming that the transversality condition (ii) above holds.

6.2. Baker's transformations in space

The result of this section was motivated by Falconer's Theorem [12] (see p. 80.). Observe that if we add one expanding direction to a self-affine set, then we obtain a very special axiom A invariant set. Namely, let $\mathcal{F} = \{f_1, \dots, f_m\}$ be a self-affine IFS on the plane, as in Section 2.1. Suppose that we are given a partition of $[0,1]$ into closed intervals $\{I_k\}_{k=1}^m$ with disjoint interiors and a strictly monotonic, expanding $\mathcal{C}^2$ map ψ with derivative separated from zero, such that $\psi(I_k) = [0,1]$ for every $k = 1, \dots, m$. We define Λ, the associated axiom A invariant set, as the attractor of the hyperbolic map which is defined for $z \in I_k$ as

$$F(x,y,z) := (f_k(x,y), \psi(z)). \tag{6.5}$$

Then the attractor Λ is the product of the attractor of the self-affine IFS $\mathcal{F} = \{f_1, \dots, f_m\}$ and the unit interval. Let us call $A_\theta := \{(x,y,z) \in \Lambda : z = \theta\}$ the θ section of Λ. In our paper with B. Solomyak [33] we considered more general baker's transformations, where the A_θ sections are different for different θ and none of the component functions are linear. In the rest of this section let $\{I_k\}_{k=1}^m$ and the map ψ be as above. We define $F : [0,1]^3 \to [0,1]^3$ by

$$F(x,y,z) := (\gamma(x,z), \tau(y,z), \psi(z)), \tag{6.6}$$

where the maps are chosen so that: $\gamma, \tau : [0,1] \times [0,1] \to [0,1]$ with

- $0 < \lambda_1 < |\gamma'_x|, |\tau'_y| < \lambda_2 < \frac{1}{2}$;
- $F|_{\Delta_k}$ a smooth 1-1 map, where $\Delta_k := [0,1]^2 \times I_k$.

We denote the non-constant contraction in x and y directions by

$$f_1 := \log|\gamma'_x|, \quad f_2 := \log|\tau'_y|.$$

To avoid the usage of singular value function we need four pressure formulas.

$$P_{F^{-1}}(s_1 f_1) = 0, \quad P_{F^{-1}}(s_2 f_2) = 0, \tag{6.7}$$

$$P_{F^{-1}}(f_1 + r_1 f_2) = 0, \quad P_{F^{-1}}(r_2 f_1 + f_2) = 0. \tag{6.8}$$

Let $s := \max\{s_1, s_2\}$ and $r := \max\{r_1, r_2\}$. Then

Theorem 6.3 (Simon, Solomyak). *For a typical (in the sense of Theorem* 2.1*) baker's map F of the form* (6.6), *the following holds:*

(i) *if* $s \le 1$ *then* $\dim_H(\Lambda) = \dim_B(\Lambda) = 1 + s$,
(ii) *if* $s > 1$ *then* $\dim_H(\Lambda) = \dim_B(\Lambda) = 2 + \min\{1, r\}$,
(iii) *if* $r > 1$ *then* $\mathcal{L}_3(\Lambda) > 0$.

The restriction in this theorem is that the directions of contraction and expansion are the same at every point. If we consider a more general situation then the problem becomes much more difficult, not only for technical reasons.

One way to try to get more general results is to study the dimension theory of non-conformal repellers. This was done by Falconer [12] and Barreira [1]. Let M

be an open subset of $\mathbb{R}^d$ and $f : M \to M$ be a C^2 map. (In Barreira's work $C^{1+\alpha}$ for $0 < \alpha \leq 1$ but here we use Falconer's set-up.) Let J be a mixing repeller for f. That is, J is a compact subset of M such that there exists a bounded neighborhood V of J such that

$$J = \{x \in V : f^n(x) \in V \text{ for all } n \geq 0\}.$$

Furthermore, f is topologically mixing and expanding on J. There is, however, one more very important assumption, which Barreira calls the 1-*bunched property*. Namely,

$$\left\| (D_x f)^{-1} \right\|^2 \cdot \|D_x f\| < 1 \text{ for } x \in J. \tag{6.9}$$

Then Falconer introduces the so-called subadditive pressure. Using that, he gives an upper bound for the Hausdorff dimension of J by an appropriate pressure formula. Here two problems have been left open: Although in some very special case Falconer proves that the upper bound is sharp, in general, this is still open. On the other hand, it is also a very interesting open problem to give an upper bound without assuming the 1-bunched property (if it is possible).

6.3. Geometric measure-theoretical properties of the SBR measure for solenoids

The SBR measure is the most natural invariant measure (sometimes it is called the natural measure). If the SBR measure μ exists, then the SBR measure of a set V is the portion of the orbit falling into V for a **LEBESGUE** typical initial point. That is,

$$\mu(V) = \lim_{n\to\infty} \frac{1}{n} \# \left\{0 \leq k < n : \; f^k(x) \in V\right\},$$

holds for Lebesgue typical x.

6.3.1. Solenoid with two different constant contractive coefficients. First we consider the solenoid map (6.1). We recall that here we assumed that $0 < \lambda_2 < \lambda_1 < \frac{1}{2}$. Above we saw that the Hausdorff dimension of the attractor Λ is $t = 1 + \frac{\log 2}{-\log \lambda_1}$. The SBR measure is supported by Λ. Then it is natural to ask if the SBR measure is equivalent to the appropriate (t-dimensional) Hausdorff measure? This question was answered negatively by M. Rams and the author.

Theorem 6.4 (M. Rams, K. Simon [26]). *Let Λ be the solenoid above. Then*

(i) $\mathcal{H}^t(\Lambda) = 0$,
(ii) $0 < \mathcal{P}^t(\Lambda) < \infty$,
(iii) *The SBR measure is equivalent to the t-dimensional packing measure $\mathcal{P}^t$.*

There have been quite a few examples in the literature of dynamically defined sets having zero Hausdorff but positive and finite packing measure. Our theorem above shows that such a phenomenon can occur even in the case of the most natural axiom A attractor.

6.3.2. Tsuji's result on SBR measures for 'fat' solenoids in the plane. M. Tsuji in [39] considers attractors for maps which are a little bit more general than the ones defined in (6.3). The very important difference is that Tsuji assumes that the solenoid is 'fat', that is, rather expanding than contracting, while in (6.3) the contraction was $\lambda_1 < \frac{1}{2}$ and the expansion was 2. Under this assumption, Tsuji proves that typically the SBR measure is absolutely continuous. More precisely, let $T : S^1 \times \mathbb{R} \to S^1 \times \mathbb{R}$,

$$T(x,y) := (\ell x, \lambda y + f(x)), \tag{6.10}$$

where $\ell \geq 2$; and $\frac{1}{\ell} < \lambda < 1$ and f is a C^2 function on S^1. Let

$$\mathcal{D} := \left\{(\lambda, f) : \text{ the SBR is absolutely continuous w.r.t. to } \mathcal{L}eb_2 \text{ with } L^2 \text{ density}\right\}.$$

Theorem 6.5 (Tsuji). *$\mathcal{D}$ contains an open and dense subset of $(\frac{1}{\ell}, 1) \times C^2(S^1, \mathbb{R})$.*

The method of the proof: Tsuji uses a clever, further developed version of the transversality condition. Originally the transversality condition would require that the contraction ratio be smaller than $\frac{1}{2}$. To avoid such an assumption, Tsuji considers iterates of some perturbation of the map T. Using the special form of the perturbed map, he proves that the iterates of the perturbed map have essentially the same form but have relatively small contraction ratio.

6.3.3. Rams' result on SBR measures for fat solenoids in space. M. Rams used the method of Tsuji and a method of Y. Peres to obtain the result below. For $k \geq 3$ let

$$\Phi : (t, x, z) \to (kt(\mod 1), \lambda_1 x + f(t), \lambda_2 z + g(t)),$$

where $0 < \lambda_2 < \lambda_1 < \frac{1}{2}$; $k\lambda_1 > 1$; $k\lambda_1\lambda_2 < 1$.

The map Φ is a self-map of the torus $T = S^1 \times D$, where D is the unit disk. The functions $f(t)$ and $g(t)$ are C^2 and chosen so that the mapping Φ is an injection. Let Λ be the attractor of Φ; we write μ for the SBR measure of ϕ and ν for the SBR measure of the map defined by the projection of Φ to the (t, x) plane. This solenoid is 'fat' because Φ is more expanding than contracting.

Put

$$t := 2 + \frac{\log k + \log \lambda_1}{-\log \lambda_2}.$$

Theorem 6.6 (M. Rams [27]). *We have $\dim_P(\Lambda) = t$. Furthermore, for generic (all but a set of first Baire category) f, g the following holds:*

- $\dim_H(\Lambda) = t$
- $\mathcal{P}^s(\Lambda) = \infty$,
- *if* $\frac{d\nu}{d\mathcal{L}eb_2}$ *is bounded then* $0 < \mathcal{H}^t(\Lambda) < \infty$ *and* $\mu \sim \mathcal{H}^t|_\Lambda$,
  if $\frac{d\nu}{d\mathcal{L}eb_2}$ *is unbounded then* $\mathcal{H}^t(\Lambda) = 0$.

The case when contracting coefficients are not constant looks more difficult.

References

[1] L. Barreira, *A non-additive thermodynamic formalism and applications to dimension theory of hyperbolic dynamical systems.* Ergodic Theory Dynam. Syst. **16** (1996), 871–927.

[2] L. Barreira, *Hyperbolicity and reccurence in dynamical systems: a survey of recent results.* Resenhas IME-USP **5** (2002), 171–230.

[3] T. Bedford and M. Urbański, *The box and Hausdorff dimension of self-affine sets.* Ergodic Theory Dynam. Syst. **10** (1990), 627–644.

[4] H.G.Bothe, *The Hausdorff-dimension of certain solenoids.* Ergodic Theory Dynam. Syst. **15** (1995) 449–474.

[5] R. Bowen, *Hausdorff dimension of quasi-circles* . Inst Hautes Études Sci. Publ. Math. **50** (1979), 259–273.

[6] R. Bowen, *Equilibrium states and the ergodic theory of Anosov diffeomorphisms.* Springer Lecture Notes in Mathematics, **470**, Springer 1979.

[7] K. Falconer, *The geometry of fractal sets*, Cambridge Univ. Press, 1985.

[8] K. Falconer, *Fractal geometry: mathematical foundations and applications.* John Wiley and Sons, 1990.

[9] K. Falconer, *Techniques in Fractal Geometry.* John Wiley and Sons, 1997.

[10] K. Falconer, *Hausdorff-dimension of some fractals.* J. Stat. Phys. **47** (1987), 123–132.

[11] K. Falconer. *The Hausdorff-dimension of self affine fractals.* Math. Proc. Camb. Phil. Soc. **103** (1988), 339–350.

[12] K. Falconer. *Bounded distorsion and dimension for non-conformal repellers.* Math. Proc. Cambridge Philos. Soc. **115** (1994), 315–334.

[13] J.E. Hutchinson, *Fractals and self-similarity.* Indiana University Mathematics Journal. **30** (1981), 713–747.

[14] P. Mattila, *Geometry of Sets and Measures in Euclidean Spaces.* Cambridge Univ. Press, 1995.

[15] P.A.P. Moran, *On plane sets of fractional dimensions.* Proc. London Math. Soc. **51** (1949). 415–423.

[16] C. McMullen, *The Hausdorff dimension of general Sierpiński gasket.* Nagoya Math.. J. **96** (1984), 1–9.

[17] H. McCluskey and M. Manning, *Hausdorff-dimension for horseshoes.* Ergodic Theory Dynam. Syst. **3** (1983), 251–260.

[18] J. Palis and F. Takens, *Hyperbolicity and sensitive chaotic dynamics at homoclinic bifurcations.* Cambridge Univ. Press, 1993.

[19] Y. Peres and W. Schlag, *Smoothness of projections, Bernoulli convolutions, and the dimension of exceptions.* Duke Math. J. **102** (2000), 193–251.

[20] Y. Peres, W. Schlag and B. Solomyak, *Sixty years of Bernoulli convolutions.* Fractal Geometry and Stochastics II, Birkhauser (1999), 39–65.

[21] Y. Pesin, *Dimension Theory in Dynamical Systems.* The University of Chicago Press, 1997.

[22] M. Polliott and O. Jenkinson, *Calculating Hausdorff dimension of Julia sets and Kleinian limit sets.* Amer. J. Math. **124** (2002), 495–545.

[23] M. Pollicott and K. Simon, *The Hausdorff-dimension of λ-expansions with deleted digits.* Trans. Amer. Math. Soc. **347** (1995), 967–983.

[24] M. Pollicott and H. Weiss, *The dimensions of some self affine limit sets in the plane.* J. Stat. Phys. **77** (1994), 841–866.

[25] F. Przyticki and M. Urbański, *On the Hausdorf dimension of some fractal sets.* Studia Math. **93** (1989) 155–186.

[26] M. Rams and K. Simon, *Hausdorff and packing measure for solenoids.* Ergodic Theory Dynam. Syst. **23** (2003) 273–291.

[27] M. Rams, *Absolute continuity for the SBR measure for non-linear fat baker maps.* Nonlinearity **16** (2003), 1649–1655.

[28] J. Schmeling and H. Weiss, *An overview of the dimension theory and dynamical systems.* AMS Symposia in Pure Math. # 69 (2001).

[29] M. Shub, *Global stability of dynamical systems.* Springer-Verlag, 1987.

[30] K. Simon, *Hausdorff-dimension for non-invertible maps.* Ergodic Theory Dynam. Syst. **13** (1993) 199–212.

[31] K. Simon, *Overlapping cylinders: the size of a dynamically defined set.* London Mathematical Society Lecture Notes Series **228** (1996), 259–272.

[32] K. Simon, *The Hausdorff-dimension of the Smale-Williams solenoid with different contraction coefficients.* Proc. Amer. Math. Soc. **125** (1997), 1221–1228.

[33] K. Simon and B. Solomyak, *Hausdorff-dimension for Horseshoes in R^3.* Ergodic Theory Dynam. Syst. **19** (1999), 1343–1363.

[34] K. Simon and B. Solomyak, *On the dimension of self-similar sets.* Fractals **10** (2002), 59–65.

[35] K. Simon, B. Solomyak and M. Urbanski, *Hausdorff-dimension for limit sets for parabolic IFS with overlaps. Pacific J. of Math.* **201** (2001) 441–477.

[36] K. Simon, B. Solomyak and M. Urbanski, *Invariant meausures for parabolic IFS with overlaps and random continued fractions.* Trans. Amer. Math. Soc. **353** (2001) 5145–5164.

[37] B. Solomyak, *On the random series $\Sigma \pm \lambda^n$ (an Erdős problem.* Annals of Math. **142** (1995), 611–625.

[38] B. Solomyak, *Measures and dimension for some fractal families.* Math. Proc. Cambridge Phil. Soc. **124** (1998), 531–546.

[39] M. Tsuji, *Fat solenoidal attractors.* Nonlinearity **14** (2001), 1011–1027.

Károly Simon
Department of Stochastics
Institute of Mathematics
Technical University of Budapest
P.O.Box 91, 1521 Budapest
Hungary
e-mail: simonk@math.bme.hu

Progress in Probability, Vol. 57, 93–107

The Exponent of Convergence of Kleinian Groups; on a Theorem of Bishop and Jones

Bernd O. Stratmann

Abstract. In this note we give a characterization of the Hausdorff dimensional significance of the exponent of convergence for any arbitrary Kleinian group. We show that this exponent is always equal to the Hausdorff dimension of the uniformly radial limit sets of the Kleinian group. We give a detailed and elementary proof of this important fact, clarifying and generalizing a result of Bishop and Jones.

Mathematics Subject Classification (2000). 20H10, 30F40, 37F99, 37F30, 28A80.

Keywords. Kleinian groups, exponent of convergence, fractal geometry.

1. The Exponent of Convergence versus Hausdorff Dimension

Already Hadamard observed that on a cusped hyperbolic surface the set of geodesic movements starting from an arbitrary fixed point and remaining in some bounded region of that surface represents a rather tricky and sophisticated set of directions, which he certainly would have called 'fractal' if in those days this term would have already been coined. Another important observation is due to Poincaré, who realised the significance of a certain series which is naturally associated with a hyperbolic surface. This series is now called the Poincaré series and its abzissa of convergence is referred to as the exponent of convergence or Poincaré exponent of the fundamental group of the surface. In this note we show in a far more general context how these two observations are related. Namely, we give a detailed proof of the fact that for *any* hyperbolic manifold the Poincaré exponent coincides with the Hausdorff dimension of the set of directions at an arbitrary point for which the associated geodesic movements remain in bounded regions of the manifold.

This fact was verified in many special cases using various different techniques, for instance for the modular group implicitly by Jarník [8] (see also Schmidt [11]), for general cofinite Fuchsian groups by Patterson [9], for cofinite Kleinian groups by Dani [4] and Stratmann [12], and for geometrically finite Kleinian groups by Stratmann [13] as well as Fernández and Melián [6]. More recently, Bishop and

Jones [2] gave an astonishingly elementary method which allows to derive this fact in its complete generality, that is for arbitrary non-elementary Kleinian groups. The main goal of this paper is to clarify the construction in [2]. We try to keep our presentation accessible to non-experts in the field.

Before we state the main theorem more explicitly, let us begin with explaining the setting. We consider non-elementary, discrete subgroups G of the group $Con(N)$ of isometries of the $(N+1)$-dimensional hyperbolic space. (Throughout, for the latter we shall use exclusively the Poincaré ball model $(\mathbb{D}^{N+1}, d)$.) Such a group G is called *Kleinian group*, and its associated *exponent of convergence* $\delta = \delta(G)$, often also referred to as the *Poincaré exponent*, is defined by

$$\delta(G) := \inf\{s : \sum_{g\in G} \exp(-sd(0, g(0))) \text{ converges}\}.$$

A Kleinian group G acts discontinuously on $\mathbb{D}^{N+1}$. Hence, the orbit $G(0)$ of the origin $0 \in \mathbb{D}^{N+1}$ under G forms a discrete point set which with respect to the Euclidean metric accumulates only at the boundary of hyperbolic space $\mathbb{S}^N = \partial\mathbb{D}^{N+1}$. The set $L(G)$ of accumulation points of $G(0)$ is called the *limit set of* G. Important subsets of $L(G)$ turn out to be the *radial limit set* $L_r(G)$ and the *uniformly radial limit set* $L_{ur}(G)$ which are given by the following definitions.

- A point $\xi \in L(G)$ is called *radial limit point* if there exists a positive constant $c = c(\xi)$ such that $s_\xi \cap b(g(0), c) \neq \emptyset$ for infinitely many different orbit points $g(0) \in G(0)$.
- A point $\xi \in L(G)$ is called *uniformly radial limit point* if for some positive $c = c(\xi)$ we have that $s_\xi \subset \bigcup_{g\in G} b(g(0), c)$.

Here s_ξ denotes the hyperbolic ray from 0 to ξ, and $b(x, r)$ the hyperbolic ball of radius r centred at x.

Note that we may project the ray s_ξ onto the associated hyperbolic manifold $M := \mathbb{D}^{N+1}/G$ where it becomes a geodesic ray starting from the point on M which corresponds to the origin. If $\xi \in L_r(G)$, then in general this ray performs a recurrent geodesic excursion on M, i.e. there exists a bounded region in M which gets visited infinitely often, If $\xi \in L_{ur}(G)$, then the ray describes a bounded excursion, i.e. in M each point on the ray is at most a bounded distance away from the starting point. Clearly, every uniformly radial point is radial (whereas the opposite is only true for cocompact and convex cocompact Kleinian groups).

In this note we shall give a detailed and elementary proof of the following result of Bishop and Jones. This result gives an ultimate clarification of the 'Hausdorff dimensional significance' of the exponent of convergence for an arbitrary Kleinian group. We restrict the theorem to non-elementary Kleinian groups, that is groups with limit sets of cardinality strictly greater than 2. The elementary cases are in fact trivial.

Theorem. *For a non-elementary Kleinian group G with exponent of convergence δ we have that*

$$\delta = \dim_H(L_r(G)) = \dim_H(L_{ur}(G)).$$

The paper is organized as follows. We begin with giving in section 2 a discussion of the geometries which are relevant in the proof of the theorem. This includes some well-known topics from hyperbolic geometry, conformal geometry and fractal geometry. In section 3 we introduce the concept of 'divergence points', and show that the set of these points is dense in the limit set. Using this, we then give a geometrisation of the Poincaré series for parameter values below the exponent of convergence. This then allows to construct Cantor subsets of the uniformly radial limit set and to give estimates for their Hausdorff dimensions in terms of δ.

We end this introduction by giving a brief description of the proof of the theorem in terms of the associated hyperbolic manifold M. Namely, the heart of the proof, if 'projected onto M', is to show that everywhere on any arbitrary M the following scenario is met.

The sketch on M. At each point $x_0 \in M$ there exist bouquets $\mathrm{B}_n^{(i)}$ (for $n \in \mathbb{N}$ and $i = 1, 2$) of closed loops starting and terminating at x_0 with the following properties. The length of each loop in $\mathrm{B}_n^{(i)}$ is roughly equal to $l_n^{(i)}$ (for some $l_n^{(i)} \nearrow \infty$, for $i = 1, 2$), and there exists a positive constant α_0 such that the angle between any of the initial directions in $\mathrm{B}_n^{(1)}$ and any of those in $\mathrm{B}_n^{(2)}$ is bounded from below by α_0. In combination with these bouquets there exists a positive constant σ and a sequence $\kappa_n \searrow 0$ such that if we consider all initial directions at x_0 giving rise to geodesic rays which stay always σ-close to elements of $\mathrm{B}_n^{(1)} \cup \mathrm{B}_n^{(2)}$, then the Hausdorff dimension of this set of directions is greater than $\delta - \kappa_n$.
Roughly speaking, the proof then follows by observing that trivially the Hausdorff dimension of all recurrent geodesic rays starting at x_0 does not exceed δ, and that the set of directions giving rise to bounded geodesic rays is contained in the set of those giving rise to recurrent geodesic rays.

Remark. The method of Bishop and Jones, as explained in this paper, has recently been employed in a straightforward manner to derive corresponding results also in slightly more general settings. For geometrically finite groups in rank 1 symmetric spaces a result similar to the above theorem has been obtained in [3]. Also, the Bishop-Jones argument has been adapted in [10] to hyperbolic groups in the sense of Gromov, and in [7] to the case of pinched negative curvature.

Acknowledgement. The author would like to thank Kurt H. Falk for enjoyable conversations on the subject matter of this paper and for being helpful with producing the pictures.

2. Background Geometry

2.1. Hyperbolic and Conformal Geometry

Recall that for the Poincaré ball model of hyperbolic space $(\mathbb{D}^{N+1}, d)$ the hyperbolic distance $d(v, w)$ between two arbitrary points $v, w \in \mathbb{D}^{N+1}$ is given by the expression

$$d(v,w) := \inf \int_\gamma \frac{2}{1-|z|^2}|dz|,$$

where the infimum is taken over all smooth curves connecting the points v and w. (For a comprehensive introduction into hyperbolic geometry the reader is referred to Beardon's book [1].)

For subsets $A \subset \mathbb{D}^{N+1}$, the *shadow map* $\Pi : \mathbb{D}^{N+1} \to \mathbb{S}^N$ is defined by

$$\Pi(A) := \{\xi \in \mathbb{S}^N : s_\xi \cap A \neq \emptyset\}.$$

Lemma 2.1. *For every $z \in \mathbb{D}^{N+1}$ and for each positive $\rho < d(0,z)$, the spherical diameter $|\Pi(b(z,\rho))|$ of the projection of the open hyperbolic ball $b(z,\rho)$ centred at z and of radius ρ has the property*

$$|\Pi(b(z,\rho))| \asymp_\rho e^{-d(0,z)}.$$

Here, $\asymp_\rho$ means that the quotient of the two quantities is bounded from below and above by some positive constants which depend only on ρ.

Proof. Using the definition of the hyperbolic metric above, a straight forward computation (see [1]) gives that the hyperbolic length of a circle C_t around the origin of hyperbolic radius t is equal to $2\pi \sinh t$, and hence it is asymptotic to πe^t. This implies that for t sufficiently large, we may cover nearly all of C_t by roughly $\pi e^t/(2\rho)$ pairwise disjoint balls of hyperbolic radius ρ which are centred at points of C_t. Since for large t and $z \in C_t$ we have that the Euclidean diameter of $b(z,\rho)$ is comparable to $|\Pi(b(z,\rho))|$, it follows that

$$|\Pi(b(z,\rho))| \asymp 2\rho e^{-t} \asymp_\rho e^{-t}.$$ □

Lemma 2.2 (Geometric Distortion Lemma). *Let $\sigma > 0$, and let $\gamma \in Con(N)$ be non-elliptic such that $d(0,\gamma(0)) > \sigma$. Then we have for all $z \in \mathbb{D}^N$ with $d(0,z) > \sigma$ and $\Pi(z) \in \gamma^{-1}(\Pi(b(\gamma(0),\sigma)))$ that*

$$d(0,\gamma(0)) + d(0,z) - 2\sigma < d(0,\gamma(z)) \le d(0,\gamma(0)) + d(0,z).$$

Proof. (see Figure 1) Raise the perpendicular from the origin onto the geodesic segment t between z and $\gamma^{-1}(0)$. Let p denote the point of intersection of this perpendicular with t. Note that by construction, the hyperbolic distance s of p to the origin is less than or equal to σ. Let t_1 and t_2 denote the hyperbolic lengths of the geodesic segments between p and $\gamma^{-1}(0)$, p and z respectively. Using the triangle inequality, we derive that

$$d((0,\gamma^{-1}(0)) \le \sigma + t_1 \text{ and } d(0,z) \le \sigma + t_2.$$

It hence follows that

$$d(0,\gamma(z)) = d(\gamma^{-1}(0),z) = t_1 + t_2 \ge d(0,\gamma^{-1}(0)) + d(0,z) - 2\sigma.$$

The second inequality in the lemma is an immediate consequence of the triangle inequality. □

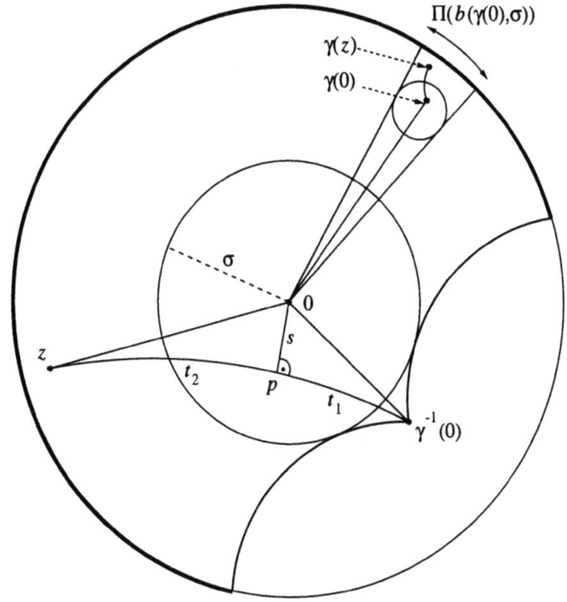

FIGURE 1.

Lemma 2.3 (Light Cone Lemma). *For all* $\sigma > \log 2$ *and for all non-elliptic* $\gamma \in Con(N)$ *for which* $d(0, \gamma(0)) > \sigma$, *we have that*

$$\mathbb{S}^N \setminus B\left(\Pi(\gamma^{-1}(0)), 2\pi e^{-\sigma}\right) \subset \gamma^{-1}\left(\Pi(b(\gamma(0), \sigma))\right) \subset \mathbb{S}^N \setminus B\left(\Pi(\gamma^{-1}(0)), e^{-\sigma}\right).$$

Here, $B(x, r)$ *denotes the spherical* N*-ball with centre* $x \in \mathbb{S}^N$ *and radius* r.

Proof. Let $\sigma > 0$ and let γ be chosen as stated in the lemma. For l denoting a geodesic in $\mathbb{D}^N$ which intersects $\partial b(0, \sigma)$ in exactly one point, say z_σ, let r_σ be the radius of the spherical disc $\Pi(l)$. Then it is geometrically evident that (see Figure 2)

$$\mathbb{S}^N \setminus B\left(\Pi(\gamma^{-1}(0)), 2r_\sigma\right) \subset \gamma^{-1}\left(\Pi(b(\gamma(0), \sigma))\right) \subset \mathbb{S}^N \setminus B\left(\Pi(\gamma^{-1}(0)), r_\sigma\right);$$

where we assumed that $2r_\sigma \leq \pi$. Hence it remains to find lower and upper bounds for r_σ. For this, note that by definition of the hyperbolic distance we have that $d(0, z_\sigma) = \log((1 + |z_\sigma|)/(1 - |z_\sigma|))$. An elementary calculation then gives

$$e^{-\sigma} \leq 1 - |z_\sigma| \leq r_\sigma \leq \frac{\pi}{2}(1 - |z_\sigma|) \leq \pi e^{-\sigma}.$$

Finally, since $2\pi e^{-\sigma} < \pi$ if and only if $\sigma > \log 2$, it follows that for σ in this range the condition $2r_\sigma \leq \pi$ is always satisfied. □

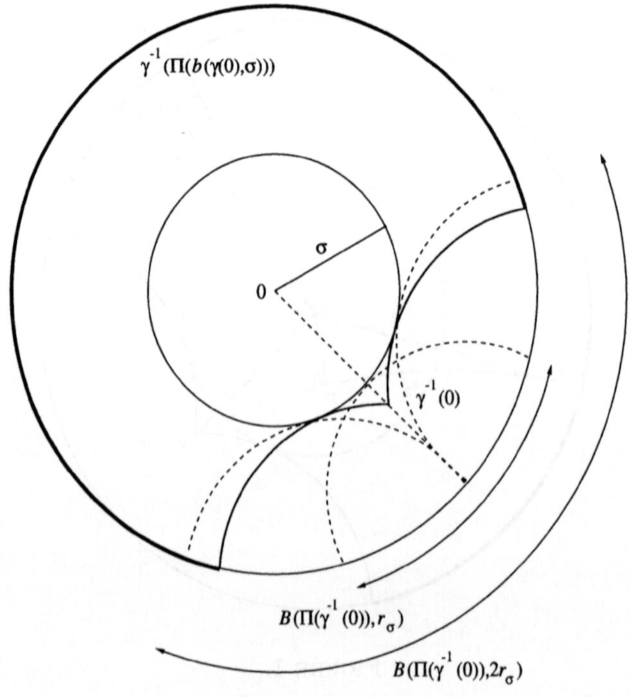

FIGURE 2.

2.2. Fractal Geometry

Recall that for a Borel set $E \subset \mathbb{S}^N$ the Hausdorff dimension $\dim_H(E)$ of E is defined by

$$\dim_H(E) := \inf\{s \geq 0 : \lim_{\rho\to 0} \inf_{U\in \mathrm{U}_\rho(E)} \sum_{u\in U} |u|^s < \infty\},$$

where $\mathrm{U}_\rho(E)$ denotes the set of coverings of E by balls with radii at most ρ. In computations of $\dim_H(E)$ to find lower bounds is usually the hardest part, and for this the following *mass distribution principle* very often turns out be useful. In here, $a \ll b$ refers to that a/b is uniformly bounded away from infinity.

- Let μ be a probability measure supported on E. If there exists $\tau \geq 0$ such that for each $x \in E$ we have $\limsup_{r\to 0} \mu(B(x,r))/r^\tau \ll 1$, then $\dim_H(E) \geq \tau$.

For a good introduction into fractal geometry the reader is referred to [5]. In this paper we shall require the following elementary facts from fractal geometry.

Lemma 2.4. *Let $\{B(x_i, r_i)\}_{i\in\mathbb{N}}$ be a family of spherical N-balls in $\mathbb{S}^N$ with centre x_i and radius r_i such that $\lim_{i\to\infty} r_i = 0$ and such that $\sum_{i\in\mathbb{N}} r_i^s$ has exponent of convergence $\tau \geq 0$. We then have that*

$$\dim_H\Big(\bigcup_{c>0} \limsup_{i\in\mathbb{N}}\{B(x_i, c\, r_i)\}\Big) \leq \tau.$$

Proof. For each $c, \rho > 0$ we have that $\{B(x_i, cr_i) : i \in \mathbb{N}, r_i \le \rho\}$ represents a ρ-covering of

$$\mathrm{B}_c := \limsup_i \{B(x_i, cr_i)\} = \{x \in \mathbb{S}^N : x \in B(x_j, cr_j) \text{ for infinitely many } j\}.$$

Hence, for arbitrary $\epsilon > 0$ we have that

$$\inf_{U \in \mathrm{U}_\rho(\mathrm{B}_c)} \sum_{u \in U} |u|^{\tau+\epsilon} \le \sum_{i \in \mathbb{N}} (cr_i)^{\tau+\epsilon} \le c^{\tau+\epsilon} \sum_{i \in \mathbb{N}} r_i^{\tau+\epsilon} < \infty.$$

By letting ρ tend to 0, this implies that $\dim_H(\mathrm{B}_c) \le \tau + \epsilon$. Since ϵ was chosen to be arbitrary, we deduce that $\dim_H(\mathrm{B}_c) \le \tau$. And finally, since $\dim_H(\mathrm{B}_{c_0}) \le \dim_H(\mathrm{B}_{c_1})$ for $c_0 < c_1$, the result follows by monotonicity of Hausdorff dimension. □

Lemma 2.5. *For a finite index set I containing at least two elements, let a descending sequence $A(i_1) \supset A(i_1, i_2) \supset \cdots$ of closed spherical N-balls in $\mathbb{S}^N$ be given for each $(i_1, i_2, \ldots) \in I^\infty$. Assume that there exist $0 < \beta < 1$ and $\tau \ge 0$ such that for each $n \in \mathbb{N}$ and for all distinct $(i_1, \ldots, i_n), (j_1, \ldots, j_n) \in I^n$ we have that*

(i) $|A(i_1, \ldots, i_n)| \asymp \beta^n$ *and* $A(i_1, \ldots, i_n) \cap A(j_1, \ldots, j_n) = \emptyset$;
(ii) $\sum_{j \in I} |A(i_1, \ldots, i_n, j)|^\tau \ge |A(i_1, \ldots, i_n)|^\tau$.

It then follows that

$$\dim_H\Big(\bigcap_{n \in \mathbb{N}} \bigcup_{(i_1,\ldots,i_n) \in I^n} A(i_1, \ldots, i_n)\Big) \ge \tau.$$

Proof. We sketch the proof which uses the mass distribution principle. Let μ be a probability measure with support $\mathrm{A} := \bigcap_{n \in \mathbb{N}} \bigcup_{(i_1,\ldots,i_n) \in I^n} A(i_1, \ldots, i_n)$ such that for each $n \in \mathbb{N}$ and $(i_1, \ldots, i_n) \in I^n$,

$$\mu(A(i_1, \ldots, i_n)) = \frac{|A(i_1, \ldots, i_n)|^\tau}{\sum_{(j_1,\ldots,j_n) \in I^n} |A(j_1, \ldots, j_n)|^\tau}.$$

Consider the spherical N-ball $B(x, r)$ centred at $x \in A$ of small radius $r > 0$. We then have $r \asymp \beta^n$ for some $n \in \mathbb{N}$, and condition (i) in the lemma guarantees that there exists a constant $c > 0$ such that for the number of balls of the n-th generation which intersect $B(x, r)$ we have

$$card(\{(i_1, \ldots, i_n) : B(x, r) \cap A(i_1, \ldots, i_n) \ne \emptyset\}) < c.$$

Using this observation and (ii) of the lemma, we obtain

$$\begin{aligned}
\mu(B(x,r)) &\le c \cdot \max\{\mu(A(i_1, \ldots, i_n)) : B(x,r) \cap A(i_1, \ldots, i_n) \ne \emptyset\} \\
&\ll \frac{\beta^{n\tau}}{\sum_{(j_1,\ldots,j_n) \in I^n} |A(j_1, \ldots, j_n)|^\tau} \ll \frac{r^\tau}{\sum_{i \in I} |A(i)|^\tau} \\
&\ll r^\tau.
\end{aligned}$$

Applying the mass distribution principle, the lemma follows. □

The following is an immediate consequence of the previous lemma. It in particular gives the model for the special type of Cantor set, which in the final section will be shown to exist inside the uniformly radial limit set of any arbitrary Kleinian group.

Corollary 2.6. *Let $I^{(1)}$ and $I^{(2)}$ be two finite index sets, each having at least two elements. Let* L *be an infinite subset of all finite words in the alphabet $I^{(1)} \cup I^{(2)}$, such that $j \in \mathrm{L}$ for all $j \in I^{(1)} \cup I^{(2)}$, and such that if $(i_1 \dots i_n) \in \mathrm{L}$ (for $n \geq 1$) then either $(i_1 \dots i_n j) \in \mathrm{L}$ for all $j \in I^{(1)}$ and $(i_1 \dots i_n k) \notin \mathrm{L}$ for all $k \in I^{(2)}$, or $(i_1 \dots i_n j) \in \mathrm{L}$ for all $j \in I^{(2)}$ and $(i_1 \dots i_n k) \notin \mathrm{L}$ for all $k \in I^{(1)}$. Assume that for each $n \in \mathbb{N}$ and for every $(i_1, i_2, \dots, i_n) \in \mathrm{L}$ we have a descending sequence $A(i_1) \supset A(i_1, i_2) \supset \dots \supset A(i_1, i_2, \dots, i_n)$ of closed spherical N-balls in $\mathbb{S}^N$ with the following properties. There exist $\tau \geq 0$ and $0 < \beta_i < 1$ for $i = 1, 2$, such that for each $n \in \mathbb{N}$ and for all distinct $(i_1, \dots, i_n), (j_1, \dots, j_n) \in \mathrm{L}$ we have that*

(i) $|A(i_1, \dots, i_n)| \asymp \beta_1^{\sharp\{k : i_k \in I^{(1)}\}} \beta_2^{\sharp\{k : i_k \in I^{(2)}\}}$; $A(i_1, \dots, i_n) \cap A(j_1, \dots, j_n) = \emptyset$;
(ii) $\sum_{j:(i_1,\dots,i_n,j) \in \mathrm{L}} |A(i_1, \dots, i_n, j)|^\tau \geq |A(i_1, \dots, i_n)|^\tau$.

It then follows that

$$\dim_H \Big(\bigcap_{n \in \mathbb{N}} \bigcup_{(i_1,\dots,i_n) \in \mathrm{L}} A(i_1, \dots, i_n) \Big) \geq \tau.$$

Proof. The proof is basically the same as the proof of Lemma 2.5. The only difference is that instead of using the 'lexicographical coverings' of the type $\{A(i_1, \dots, i_n) : (i_1, \dots, i_n) \in \mathrm{L}\}$, one uses the 'geometrical coverings' given by

$$\{A : A = A(i_1, \dots, i_k) \text{ for some } k \in \mathbb{N};\ (\beta_1 \beta_2)^{n+1} \leq |A| < (\beta_1 \beta_2)^n\}. \qquad \square$$

3. The Proof of the Theorem

3.1. A Geometrisation of the Poincaré Series

For $\xi \in \mathbb{S}^N$ and $0 < r < \pi$, let the *lense* $\lambda(\xi, r) \subset \mathbb{D}^{N+1}$ be defined as the interior of the $(N+1)$-dimensional hyperbolic half space for which $\Pi(\lambda(\xi, r))$ is the spherical N-ball of radius r centred at ξ (see Figure 3).

Definition 3.1. *For $\epsilon > 0$, an element $\xi \in L(G)$ is called ϵ-divergence point if and only if*

$$\sum_{g(0) \in \lambda(\xi, r)} e^{-(\delta - \epsilon) d(0, g(0))} = \infty \quad \textit{for all } r \in (0, 1).$$

Let $\mathrm{D}_\epsilon(G)$ refer to the set of all ϵ-divergence points.

Lemma 3.2. *For each $\epsilon > 0$, the set $\mathrm{D}_\epsilon(G)$ is dense in $L(G)$.*

Proof. Let ϵ be fixed, and suppose, by way of contradiction, that $\mathrm{D}_\epsilon(G) = \emptyset$. From this we see that for each $\eta \in L(G)$ there exists $r(\eta) > 0$ such that

$$\sum_{g(0) \in \lambda(\eta, r)} e^{-(\delta - \epsilon) d(0, g(0))} < \infty \quad \text{for all } r \in (0, r(\eta)).$$

The set $\{\Pi(\lambda(\eta, r(\eta))) : \eta \in L(G)\}$ provides a covering of $L(G)$, and since $L(G)$ is compact, there exist $\eta_1, \cdots, \eta_k \in L(G)$ such that

$$L(G) \subset \bigcup_{i=1}^{k} \Pi(\lambda(\eta_i, r(\eta_i))),$$

and hence

$$\sum_{i=1}^{k} \sum_{g(0)\in\lambda(\eta_i,r(\eta_i))} e^{-(\delta-\epsilon)d(0,g(0))} < \infty.$$

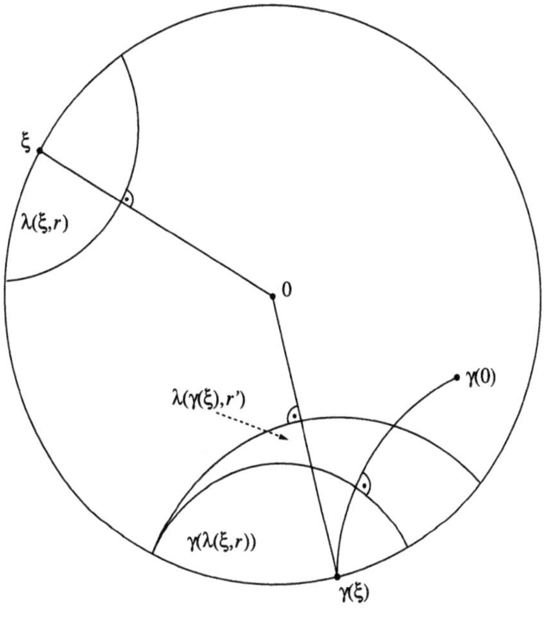

FIGURE 3.

Combining this with the fact that $G(0) \setminus \bigcup_{i=1}^{k} \lambda(\eta_i, r(\eta_i))$ is a finite set of orbit points, it follows that

$$\sum_{g\in G} e^{-(\delta-\epsilon)d(0,g(0))} < \infty,$$

which contradicts the fact that δ is the exponent of convergence of G. Hence, there exists at least one ϵ-divergence point $\xi \in L(G)$. Since $G(\xi)$ is a dense subset of $L(G)$, it is now sufficient to show that $\gamma(\xi) \in \mathrm{D}_\epsilon(G)$ for arbitrary $\gamma \in G$ not fixing ξ. In order to see this, note that $\gamma(s_\xi)$ is the ray from $\gamma(0)$ to $\gamma(\xi)$ and that for each sufficiently small positive r there exists a minimal r' such that $\gamma(\lambda(\xi, r)) \subset \lambda(\gamma(\xi), r')$ (see Figure 3). Clearly, if r decreases to 0, then the corresponding r'

also decreases to 0. Now, the claim follows from the following observation.

$$\begin{aligned}
\infty &= \sum_{g(0)\in\lambda(\xi,r)} e^{-(\delta-\epsilon)d(0,g(0))} \\
&= \sum_{g(0)\in\lambda(\xi,r)} e^{-(\delta-\epsilon)d(\gamma(0),\gamma g(0))} \\
&\asymp_\gamma \sum_{g(0)\in\lambda(\xi,r)} e^{-(\delta-\epsilon)d(0,\gamma g(0))} \\
&\leq \sum_{g(0)\in\lambda(\gamma(\xi),r')} e^{-(\delta-\epsilon)d(0,g(0))}.
\end{aligned}$$

□

For $n \in \mathbb{N}$ and $\tau > 0$, the hyperbolic n-annulus $A_n(\tau)$ of width τ is defined as follows

$$A_n(\tau) := \{g(0) \in G(0) : n\tau \leq d(0, g(0)) < (n+1)\tau\}.$$

Lemma 3.3. *Let $\epsilon, \tau > 0$ and $\xi \in \mathrm{D}_\epsilon(G)$ be given. Also, let $M > 0$ be some given large number. Then, for each $r > 0$ there exists an increasing sequence of positive integers $\{m_i\}_{i\in\mathbb{N}}$ such that for all $i \in \mathbb{N}$,*

$$\sum_{g(0)\in\lambda(\xi,r)\cap A_{m_i}(\tau)} e^{-(\delta-2\epsilon)d(0,g(0))} \geq M.$$

Proof. Assume that the statement of the lemma is false. Then there exists $r > 0$ such that for almost all $n \in \mathbb{N}$ (i.e. for all n apart from finitely many exceptions)

$$\sum_{g(0)\in\lambda(\xi,r)\cap A_n(\tau)} e^{-(\delta-2\epsilon)d(0,g(0))} < M.$$

In order to get rid of the 'finitely many exceptions', note that there exists some $0 < r^* \leq r$ such that for all $n \in \mathbb{N}$,

$$\sum_{g(0)\in\lambda(\xi,r^*)\cap A_n(\tau)} e^{-(\delta-2\epsilon)d(0,g(0))} < M.$$

Using this, it follows that

$$\begin{aligned}
\sum_{g(0)\in\lambda(\xi,r^*)} e^{-(\delta-\epsilon)d(0,g(0))} &\leq \sum_{n\in\mathbb{N}} \sum_{g(0)\in\lambda(\xi,r^*)\cap A_n(\tau)} e^{-\epsilon d(0,g(0))} e^{-(\delta-2\epsilon)d(0,g(0))} \\
&\leq M \sum_{n\in\mathbb{N}} e^{-\epsilon\tau n} \\
&< \infty,
\end{aligned}$$

which contradicts the fact that ξ is an ϵ-divergence point. □

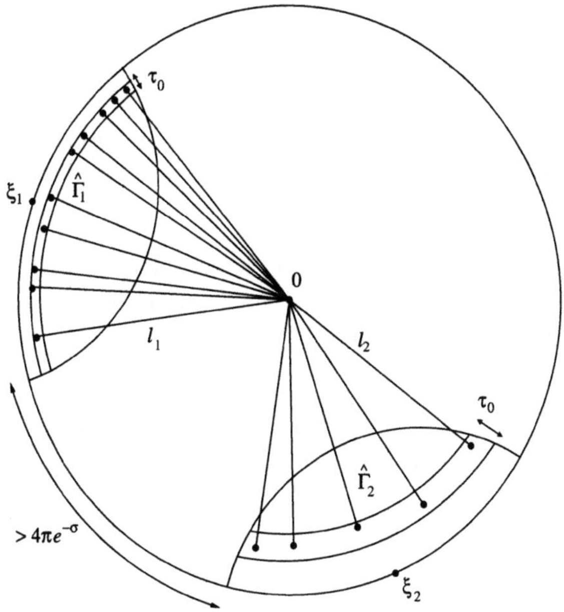

FIGURE 4.

3.2. The Building Stone for the Cantor Construction

Recall that the *injective radius* at the origin is defined by $\text{inj}(0) := \frac{1}{2}\inf\{d(0,\gamma(0)) : g \in G \setminus \{id.\}\}$. In the following put $\tau_0 := \frac{1}{2}\text{inj}(0)$, and let σ denote some fixed constant such that $e^\sigma > \max\{2; 4\pi/|L(G)|\}$.

Lemma 3.4. *For each $\epsilon > 0$ and for each arbitrarily large number $M > 0$ there exist $\xi_1, \xi_2 \in \mathrm{D}_\epsilon(G), r > 0$ and arbitrarily large $l_1, l_2 > 0$, as well as finite sets $\hat{\Gamma}_1 := \{h \in G : h(0) \in A_{l_1}(\tau_0) \cap \lambda(\xi_1, r)\}$ and $\hat{\Gamma}_2 := \{h \in G : h(0) \in A_{l_2}(\tau_0) \cap \lambda(\xi_2, r)\}$ such that the following properties are satisfied for $i = 1, 2$.*

(i) *The spherical distance between $B(\xi_1, r)$ and $B(\xi_2, r)$ is at least $4\pi e^{-\sigma}$.*
(ii) $\sum_{h\in\hat{\Gamma}_i} e^{-(\delta-2\epsilon)d(0,h(0))} \geq M$.

Proof. (see Figure 4) Fix $\epsilon > 0$ and $M > 0$. Next, choose $\xi_1, \xi_2 \in \mathrm{D}_\epsilon(G)$ and $r > 0$ sufficiently small such that the spherical distance between $B(\xi_1, r)$ and $B(\xi_2, r)$ is greater than $4\pi e^{-\sigma}$. Note that this is possible by choice of σ (recall that $|L(G)| > 4\pi e^{-\sigma}$) and since, by Lemma 3.2, the set $\mathrm{D}_\epsilon(G)$ is dense in $L(G)$. Now, using Lemma 3.3, we obtain that there exist arbitrarily large $l_1, l_2 \in \mathbb{N}$ such that

$$\sum_{h(0)\in\lambda(\xi_i,r)\cap A_{l_i}(\tau_0)} e^{-(\delta-2\epsilon)d(0,h(0))} \geq M \quad \text{for } i = 1, 2. \qquad \square$$

The following proposition gives the type of 'building stone' out of which in the final section we shall construct certain specific Cantor sets.

Proposition 3.5. *Let $\epsilon > 0$ be given. There exist $\Gamma_1 \subset \hat{\Gamma}_1$ and $\Gamma_2 \subset \hat{\Gamma}_2$, each containing at least two elements, with the following property. For each $g \in G$ with $d(0, g(0)) > \sigma$ there exists $\Gamma(g) \in \{\Gamma_i : i = 1, 2\}$ such that the following holds.*

(i) *The family $\mathrm{F}(g) := \{\Pi(b(g(h(0)), \sigma)) : h \in \Gamma(g)\}$ consists of pairwise disjoint balls which are of comparable size and which are contained in $\Pi(b(g(0), \sigma))$.*

(ii) $\sum_{b \in \mathrm{F}(g)} |b|^{\delta - 2\epsilon} \geq |\Pi(b(g(0), \sigma))|^{\delta - 2\epsilon}$.

Remark. Note that for a particular $g \in G$ it may happen that both sets Γ_1 and Γ_2 have the properties *(i)* and *(ii)* in the proposition. In this case one may think of $\Gamma(g)$ as being randomly selected.

Proof. Fix $\epsilon > 0$ and $g \in G$ such that $d(0, g(0)) > \sigma$. Also, let $M > 0$ be sufficiently large (which will be specified at the end of the proof). By Lemma 3.4, there exist $\xi_1, \xi_2 \in \mathrm{D}_\epsilon(G)$ and $r > 0$ such that $B(\xi_1, r)$ and $B(\xi_2, r)$ are at least the distance $4\pi e^{-\sigma}$ apart. Thus, by the Light Cone Lemma (Lemma 2.3) we get that at least one of the $B(\xi_i, r)$, say $B(\xi_1, r)$, is fully contained in $g^{-1}(\Pi(b(g(0), \sigma)))$. In particular, this means that $\Pi(h(0)) \in g^{-1}(\Pi(b(g(0), \sigma)))$, for each $h \in \hat{\Gamma}_1$. We can now apply the Geometric Distortion Lemma (Lemma 2.2), which gives that for each $h \in \hat{\Gamma}_1$ we have

$$d(0, g(0)) + d(0, h(0)) - 2\sigma < d(0, g(h(0))) \leq d(0, g(0)) + d(0, h(0)). \qquad (*)$$

Using this estimate and Lemma 3.4 *(ii)*, we obtain that

$$\begin{aligned}\sum_{h \in \hat{\Gamma}_1} e^{-(\delta - 2\epsilon) d(0, g(h(0)))} &> e^{-(\delta - 2\epsilon) d(0, g(0))} \sum_{h \in \hat{\Gamma}_1} e^{-(\delta - 2\epsilon) d(0, h(0))} \\ &\geq M\, e^{-(\delta - 2\epsilon) d(0, g(0))}.\end{aligned}$$

By combining this latter estimate and Lemma 2.1, we deduce that there exists a constant $C_1 > 0$ depending on σ, such that

$$\sum_{h \in \hat{\Gamma}_1} |\Pi(b(g(h(0)), \sigma))|^{\delta - 2\epsilon} \geq C_1\, M\, |\Pi(b(g(0), \sigma))|^{\delta - 2\epsilon}.$$

Now, recall that for $h \in \hat{\Gamma}_1$ we have that $l_1 \tau_0 \leq d(0, h(0)) < l_1 \tau_0 + \tau_0$. Combining this observation and estimate $(*)$, we obtain that all $g(h(0))$ with $h \in \hat{\Gamma}_1$ are contained in an annulus of constant hyperbolic width $2\sigma + \tau_0$, that is

$$\{g(h(0)) : h \in \hat{\Gamma}_1\} \subset \{z \in \mathbb{D}^{N+1} : l_1 \tau_0 - 2\sigma \leq d(0, z) - d(0, g(0)) < l_1 \tau_0 + \tau_0\}. \qquad (**)$$

Also, recall that, by choice of τ_0, the set $\{b(g(h(0)), \tau_0) : h \in \hat{\Gamma}_1\}$ comprises a family of pairwise disjoint hyperbolic balls. From these two latter observations we can now deduce that each $\xi \in \Pi(b(g(0), \sigma))$ is contained in at most a bounded number (independently of g) of balls $\Pi(b(g(h(0)), \sigma))$ with $h \in \hat{\Gamma}_1$, i.e. there exist a set $\Gamma_1 \subset \hat{\Gamma}_1$ and a constant $C_2 > 0$, which depends only on σ and τ_0, such that $\{\Pi(b(gh(0),\sigma)) : h \in \Gamma_1\}$ consists of pairwise disjoint balls which are contained in $\Pi(b(g(0),\sigma))$, and such that

$$\sum_{h\in\Gamma_1} |\Pi(b(g(h(0)),\sigma))|^{\delta-2\epsilon} \geq C_2\, C_1\, M\, |\Pi(b(g(0),\sigma))|^{\delta-2\epsilon}.$$

By choosing $M = (C_1 C_2)^{-1}$, the statement (ii) of the proposition follows. Finally, note that from Lemma 2.1 and the estimate $(**)$ we have for each $h \in \Gamma_1$ that

$$\operatorname{diam}(\Pi(b(g(h(0)),\sigma))) \asymp_{\sigma,\tau_0} e^{-(d(0,g(0))+l_1\tau_0)},$$

from which we deduce that

$$|\Pi(b(g(h_1(0)),\sigma))| \asymp_{\sigma,\tau_0} |\Pi(b(g(h_2(0)),\sigma))| \quad \text{for all } h_1, h_2 \in \Gamma_1.$$

Finally, note that by increasing the lengths l_1 and l_2, if necessary, it can always be guaranteed that Γ_1 and Γ_2 are both of cardinality at least two. □

3.3. End of the Proof

Upper bound estimate.

For the upper bound of $\dim_H(L_r(G))$ note that by Lemma 2.1, for each $\rho > 0$ there exists $c = c(\rho) > 0$ such that for every $s > \delta$ we have

$$\sum_{g\in G} |B(\Pi(g(0)), ce^{-d(0,g(0))})|^s \asymp \sum_{g\in G} |\Pi(b(g(0),\rho))|^s \asymp_\rho \sum_{g\in G} e^{-sd(0,g(0))}.$$

Hence, these three series have the same exponent of convergence equal to δ. Therefore, we can apply Lemma 2.4 to the family $\{B(\Pi(g(0)), ce^{-d(0,g(0))}) : g \in G\}$, which then gives that

$$\dim_H(L_r(G)) \leq \delta.$$

Lower bound estimate.

For the lower bound of $\dim_H(L_{ur}(G))$ we build up a Cantor subset of $L_{ur}(G)$ by induction as follows.

Let $\epsilon > 0$ be fixed, and let $\sigma > 0$ be chosen as in the previous section. Note that by Lemma 3.4 we can assume that the lengths l_1 and l_2 (in Proposition 3.5) are sufficiently large, that is we can assume without loss of generality that $d(0,\gamma_{i_1}(0)) > \sigma$ for all $\gamma_{i_1} \in \Gamma_1 \cup \Gamma_2$. Now, the 'first generation' in our Cantor set construction is given by

$$C(i_1) := \Pi(b(\gamma_{i_1}(0),\sigma)) \ \text{ for } \ \gamma_{i_1} \in \Gamma_1 \cup \Gamma_2.$$

Using Proposition 3.5, we obtain that each $C(i_1)$ contains pairwise disjoint siblings $C(i_1, i_2)$, that is

$$C(i_1, i_2) := \Pi(b(\gamma_{i_1}\gamma_{i_2}(0),\sigma)) \ \text{ for } \ \gamma_{i_2} \in \Gamma(\gamma_{i_1}).$$

This gives the 'second generation' in our Cantor set construction. Note that by Proposition 3.5, we have for each $\gamma_{i_1} \in \Gamma_1 \cup \Gamma_2$ that

$$\sum_{\gamma_{i_2}\in\Gamma(\gamma_{i_1})} |C(i_1,i_2)|^{\delta-2\epsilon} \geq |C(i_1)|^{\delta-2\epsilon}.$$

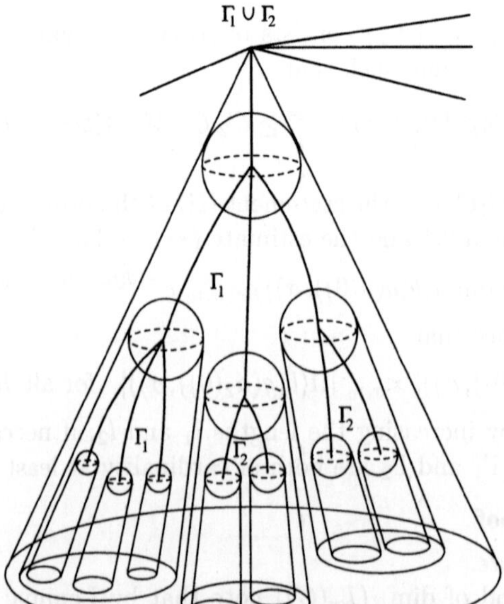

FIGURE 5.

Clearly, we can now proceed by induction, and in this way we derive for arbitrary $n \geq 2$ the 'n-th generation' which is then given by

$$C(i_1, \ldots, i_n) := \Pi(b(\gamma_{i_1} \cdots \gamma_{i_n}(0), \sigma)) \text{ for } \gamma_{i_n} \in \Gamma(\gamma_{i_1} \cdots \gamma_{i_{n-1}}).$$

By Proposition 3.5, the $C(i_1, \ldots, i_n)$ are pairwise disjoint and have the property that

$$\sum_{\gamma_{i_n} \in \Gamma(\gamma_{i_1} \cdots \gamma_{i_{n-1}})} |C(i_1, \ldots, i_n)|^{\delta - 2\epsilon} \geq |C(i_1, \ldots, i_{n-1})|^{\delta - 2\epsilon}.$$

Using Corollary 2.6, it now follows for $\mathrm{C}_\epsilon := \bigcap_{n \in \mathbb{N}} \bigcup_{(i_1, \ldots, i_n)} C(i_1, \ldots, i_n)$ that

$$\dim_H(\mathrm{C}_\epsilon) \geq \delta - 2\epsilon.$$

Finally, note that by construction we have for each $\xi \in \mathrm{C}_\epsilon$ that

$$s_\xi \subset \bigcup_{g \in G} b(g(0), c(\epsilon));$$

where we have set $c(\epsilon) := 3\sigma + \tau_0(1 + \max\{l_1, l_2\}/2)$ (see the inclusion (**) in proof of Proposition 3.5). Hence, we have that

$$\dim_H(L_{ur}(G)) \geq \dim_H(\mathrm{C}_\epsilon) \geq \delta - 2\epsilon.$$

Since ϵ was assumed to be arbitrary, it follows that

$$\dim_H(L_{ur}(G)) \geq \delta.$$

□

References

[1] A.F. Beardon, *The Geometry of Discrete Groups*, New York, 1983.

[2] C.J. Bishop and P.W. Jones, *Hausdorff dimension and Kleinian groups.* Acta Math. **179** (1995), 1–39.

[3] K. Corlette and A. Iozzi, *Limit sets of discrete groups of isometries of exotic hyperbolic spaces.* Trans. Amer. Math. Soc. **351** (1999), 1507–1530.

[4] S.G. Dani, *Bounded orbits of flows on homogeneous spaces.* Comment. Math. Helv. **61** (1986), 636–660.

[5] K. Falconer, *Fractal Geometry*, Wiley, New York, 1990.

[6] J.L. Fernández and M.V. Melián, *Bounded geodesics of Riemann surfaces and hyperbolic manifolds.* Trans. Amer. Math. Soc. **347** (1995), 3533–3549.

[7] S. Hersonsky and F. Paulin, *Hausdorff dimension of Diophantine geodesics in negatively curved manifolds.* J. Reine Angew. Math. **539** (2001), 29–43.

[8] V. Jarník, *Zur metrischen Theorie der Diophantischen Approximationen.* Prace Math.-Fiz. **36**, 2. Heft (1928).

[9] S.J. Patterson, *Diophantine approximation in Fuchsian groups.* Phil. Trans. Royal Soc. Lond. **282** (1976), 527–563.

[10] F. Paulin, *On the critical exponent of a discrete group of hyperbolic isometries.* Diff. Geom. Appl. **7** (1997), 231–236.

[11] W.M. Schmidt, *On badly approximable numbers and certain games.* Trans. Amer. Math. Soc. **123** (1966), 178–199.

[12] B.O. Stratmann, *Diophantine approximation in Kleinian groups.* Math. Proc. Camb. Phil. Soc. **116** (1994), 57–78.

[13] B.O. Stratmann, *The Hausdorff dimension of bounded geodesics on geometrically finite manifolds.* Ergod. Th. Dynam. Sys. **17** (1997), 227–246.

Bernd O. Stratmann
Mathematical Institute
University of St Andrews
St Andrews KY16 9SS
Scotland
e-mail: `bos@maths.st-and.ac.uk`

Progress in Probability, Vol. 57, 109–116

Lyapunov Exponents Are not Rigid with Respect to Arithmetic Subsequences

Amiran Ambroladze and Jörg Schmeling

Abstract. We construct a random walk on the integers tending with positive velocity to infinity while any walk along an arbitrary arithmetic subsequence is recurrent. This can be interpreted as an example of a product of random matrices with positive Lyapunov exponent while any subproduct along an arithmetic subsequence has zero exponent. This shows that one can have exponential stretching along an orbit of a dynamical system while any time series along an arithmetic subsequence indicates subexponential growth.

Mathematics Subject Classification (2000). Primary 37A30, 37A50.

Keywords. Oseledec's ergodic theorem, random matrix products, stationary stochastic processes.

1. Introduction

Let $T\colon X \to X$ be a dynamical system with invariant measure μ. Birkhoff's Ergodic Theorem tells that for a given L^1 function f for μ–a.e. $x \in X$ the limit

$$\lim_{N\to\infty} \frac{1}{N} \sum_{n=0}^{N-1} f(T^n x) = \hat{f}(x)$$

exists. The following obvious lemma shows that this limit is determined by its behavior on arithmetic subsequences.

Lemma 1.1. *Let (k_n) be a sequence of real (complex) numbers. If for any $a > 1$, $b \in \mathbb{N}$ the limit*

$$\lim_{N\to\infty} \frac{1}{N} \sum_{n=0}^{N-1} k_{an+b} = \hat{k}$$

then

$$\lim_{N\to\infty} \frac{1}{N} \sum_{n=0}^{N-1} k_n = \hat{k}.$$

We will show that this situation changes drastically when one considers matrix valued cocycles instead of real valued cocycles. Let $A\colon X \to Gl(n,\mathbb{R})$ be a matrix–valued function with

$$\log^+ \|A(x)\| \in L^1.$$

Then Oseledec' Theorem states that for μ–a.e. $x \in X$ the limit

$$\lim_{N\to\infty} \frac{1}{N} \log \left\| \prod_{n=0}^{N-1} A(T^n x) \right\| = \chi(x)$$

exists. This number is called the leading Lyapunov exponent. If it is positive it tells that the average stretching rate along an orbit is exponentially in some direction.

We will construct a random matrix product $A_1 A_2 \cdots A_N$ where the leading Lyapunov exponent is positive. On the other hand if we average the stretching rate along an arbitrary arithmetic subsequence we get subexponential growth:

$$\lim_{N\to\infty} \frac{1}{N} \log \left\| \prod_{n=0}^{N-1} A_{an+b} \right\| = 0.$$

This shows that the Lyapunov exponents cannot be determined by arithmetic subsequences.

This situation has also another interpretation. We construct a stochastic process (X_n), $n \in \mathbb{N}$ which leads to a random walk on $\mathbb{N}$ that tends with linear speed to $+\infty$. On the other hand any random walk generated by the subprocess (X_{an+b}) is recurrent. This illustrates that one cannot predict exponential stretching rates by averaging the local stretching rates only along subsequences.

2. The stochastic process and the generated random walk

In this section we are going to construct a matrix valued stochastic process and an associated random walk. Let

$$A = \begin{pmatrix} 2 & 0 \\ 0 & \frac{1}{2} \end{pmatrix} \quad B = A^{-1} = \begin{pmatrix} \frac{1}{2} & 0 \\ 0 & 2 \end{pmatrix} \quad C = \begin{pmatrix} 0 & 1 \\ -1 & 0 \end{pmatrix}.$$

We set $X_0 = A$ and for $n \in \mathbb{N}$

$$\mathbb{P}(X_n = C) = \mathbb{P}(X_n \neq C) = \frac{1}{2} \tag{1}$$

independent of all other times. This means the random variables

$$Y_n = \begin{cases} C & X_n = C \\ \text{not } C & X_n = A \text{ or } X_n = B \end{cases} \tag{2}$$

constitute a Bernoulli i.i.d. process. Then we set

$$\mathbb{P}(X_n = A \,|\, \text{the number of } 1 \le k < n \text{ with } X_k = C \text{ is even}) = \frac{1}{2} \tag{3}$$

and

$$\mathbb{P}(X_n = B \,|\, \text{the number of } 1 \leq k < n \text{ with } X_k = C \text{ is odd}) = \frac{1}{2}. \tag{4}$$

Therefore

$$\mathbb{P}(X_n = A \,|\, \#(1 \leq k < n \text{ with } X_k = C) \text{ is even and } X_n \neq C) = 1 \tag{5}$$

and

$$\mathbb{P}(X_n = B \,|\, \#(1 \leq k < n \text{ with } X_k = C) \text{ is odd and } X_n \neq C) = 1. \tag{6}$$

This means at the n-th moment we choose at random whether we put a C or not. If we have chosen not to put a C we chose A or B in a deterministic way depending on the parity of occurrences of C. This process is highly dependent!

Now we define a random walk generated by this process. Let

$$q_n = \frac{\log \|X_0 X_1 X_2 \cdots X_n\|}{\log 2}.$$

If we choose $\|(x,y)\| =: \max\{|x|,|y|\}$ and $\|A\| =: \max_{v \in \mathbb{S}^1} \frac{\|Av\|}{\|v\|}$ then 2^{q_n} is the length of the longest side of the rectangle

$$X_0 X_1 X_2 \cdots X_n([0,1] \times [0,1]).$$

Since each of the matrix products $X_0 X_1 X_2 \cdots X_n$ is area preserving and the side-length of the rectangles are of the form 2^l, $l \in \mathbb{Z}$ we have that $q_n \in \mathbb{N} \cup \{0\}$. Hence, (q_n) defines a random walk on $\mathbb{N} \cup \{0\}$.

Theorem 2.1.

$$\lim_{n \to \infty} \frac{1}{n} q_n = \frac{1}{2} \quad \mathbb{P}\text{-}a.s.$$

We immediately get the following corollary:

Corollary 2.2. *The leading Lyapunov exponent of the random matrix product is positive:*

$$\lim_{n \to \infty} \frac{1}{n} \log \|X_0 X_1 X_2 \cdots X_n\| = \chi_{\mathbb{P}} = \frac{1}{2} \log 2 > 0 \quad \mathbb{P}\text{-}a.e..$$

Theorem 2.3. *The process X_n is stationary.*

3. The process along arithmetic subsequences

Let $a > 1$ and b be two integers. We will consider the processes

$$(X_n^{a,b}) := (X_{an+b}).$$

The next theorem shows that for given a, b this process has much more independence.

Theorem 3.1. *Let $a > 1$, b be fixed. Let $\mathcal{F}_n^{a,b}$ be the σ-algebra generated by the random variables $\left(X_k^{a,b}\right)$, $0 \leq k < n$. Then for $n \geq 1$*

$$\mathbb{P}(X_n^{a,b} = C \mid \mathcal{F}_n^{a,b}) = \frac{1}{2},$$

$$\mathbb{P}(X_n^{a,b} = A \mid \mathcal{F}_n^{a,b}) = \frac{1}{4}$$

and

$$\mathbb{P}(X_n^{a,b} = B \mid \mathcal{F}_n^{a,b}) = \frac{1}{4}.$$

Now we consider the associated random walk

$$q_n^{a,b} = \frac{\log \|X_0^{a,b} X_1^{a,b} X_2^{a,b} \cdots X_n^{a,b}\|}{\log 2}.$$

For the same reasons as for (q_n) the random walk $q_n^{a,b}$ lives on $\mathbb{N} \cup \{0\}$.

Theorem 3.2. *Let $a > 1$, b be fixed. Then the random walk $q_n^{a,b}$ is recurrent. Moreover,*

$$\lim_{n\to\infty} \frac{1}{n} q_n^{a,b} = 0 \quad \mathbb{P}\text{-a.s.}$$

We immediately derive the following corollary:

Corollary 3.3. *The leading Lyapunov exponent of the random matrix product along the subsequence $(an + b)$ is zero:*

$$\lim_{n\to\infty} \frac{1}{n} \log \|X_0^{a,b} X_1^{a,b} X_2^{a,b} \cdots X_n^{a,b}\| = \chi_{\mathbb{P}}^{a,b} = 0 \quad \mathbb{P}\text{-a.e.}.$$

4. Proofs

Proof of Theorem 2.1. As explained above the random variable 2^{q_n} is the length of the longest side of the rectangle

$$X_0 X_1 X_2 \cdots X_n([0,1] \times [0,1]). \tag{7}$$

The matrix C is a rotation in the plane by angle $\frac{\pi}{2}$. In case the rectangle (7) is not the unit square the application of C interchanges the axes. By the construction laws (5) and (6) A is applied only if (7) has its longest side horizontally and B is applied only if (7) has its longest side vertically (we note that both cases may include that (7) is actually the unit square). Since $X_0([0,1] \times [0,1])$ is "aligned horizontally" each time A or B is applied the longest side of the rectangle (7) gets doubled. Therefore

$$q_n = \#\{1 \leq k \leq n : X_k \neq C\}.$$

By the definition (2) of the i.i.d. Bernoulli process (Y_n) we can rewrite

$$q_n = \#\{1 \leq k \leq n : Y_k \neq C\}.$$

We apply the strong law of large numbers to the bernoulli process (Y_n):

$$\lim_{n\to\infty} \frac{\#\{1 \le k \le n : Y_k \ne C\}}{n} = \frac{1}{2} \qquad \mathbb{P}-\text{a.s.}$$

Hence,

$$\lim_{n\to\infty} \frac{q_n}{n} = \frac{1}{2} \qquad \mathbb{P}-\text{a.s.} \qquad \square$$

Proof of Theorem 2.3. Let $m, n, k \in \mathbb{N}$ and $M_1, M_2, \cdots M_k \in \{C, A, B\}$. We assume that $M_1 = \cdots = M_{l-1} = C$ and $M_l = A$ for some $1 \le l \le k$. Then by (1)

$$\begin{aligned}&\mathbb{P}(X_n = M_1, \cdots, X_{n+k-1} = M_k)\\ &= \left(\frac{1}{2}\right)^{l-1} \mathbb{P}(X_{n+l-1} = M_l, \cdots, X_{n+k-1} = M_k)\end{aligned}$$

and

$$\begin{aligned}&\mathbb{P}(X_m = M_1, \cdots, X_{m+k-1} = M_k)\\ &= \left(\frac{1}{2}\right)^{l-1} \mathbb{P}(X_{m+l-1} = M_l, \cdots, X_{m+k-1} = M_k).\end{aligned}$$

Now

$$\begin{aligned}\mathbb{P}(X_{n+l-1} = M_l, \cdots, X_{n+k-1} = M_k) &= \left(\frac{1}{2}\right)^{\#\{l\le s\le k\,:\,M_s=C\}} \times\\ \times \mathbb{P}\Big(\bigcup_{l\le s\le k} [\#\{1 \le t < n+s-1 : X_t = C\} &\text{ is even if } M_s = A \text{ and}\\ \#\{1 \le t < n+s-1 : X_t = C\} &\text{ is odd if } M_s = B]\Big)\end{aligned}$$

and

$$\begin{aligned}\mathbb{P}(X_{m+l-1} = M_l, \cdots, X_{m+k-1} = M_k) &= \left(\frac{1}{2}\right)^{\#\{l\le s\le k\,:\,M_s=C\}} \times\\ \times \mathbb{P}\Big(\bigcup_{l\le s\le k} [\#\{1 \le t < m+s-1 : X_t = C\} &\text{ is even if } M_s = A \text{ and}\\ \#\{1 \le t < m+s-1 : X_t = C\} &\text{ is odd if } M_s = B]\Big).\end{aligned}$$

Since $M_l = A$ and by (5) we have that $\#\{1 \le t < l : X_t = C\}$ is even and hence,

$$\begin{gathered}\#\{1 \le t < n+s-1 : X_t = C\} \equiv 0 \mod 2\\ \Longleftrightarrow\\ \#\{l \le t < n+s-1 : X_t = C\} \equiv 0 \mod 2.\end{gathered}$$

This and the formulas above imply

$$\mathbb{P}(X_{n+l-1} = M_l, \cdots, X_{n+k-1} = M_k)$$
$$= \left(\frac{1}{2}\right)^{\#\{l\le s\le k\,:\,M_s=C\}} \times$$
$$\times\, \mathbb{P}\Big(\bigcup_{l\le s\le k} [\#\{n+l-1 \le t < n+s-1 \,:\, X_t = C\} \text{ is even if } M_s = A$$
$$\text{and } \#\{n+l-1 \le t < n+s-1 \,:\, X_t = C\} \text{ is odd if } M_s = B]\Big)$$
$$= \left(\frac{1}{2}\right)^{\#\{l\le s\le k\,:\,M_s=C\}} \times$$
$$\times\, \mathbb{P}\Big(\bigcup_{l\le s\le k} [\#\{l \le t < s \,:\, M_t = C\} \text{ is even if } M_s = A \text{ and}$$
$$\#\{l \le t < s \,:\, M_t = C\} \text{ is odd if } M_s = B]\Big)$$
$$= \mathbb{P}(X_{m+l-1} = M_l, \cdots, X_{m+k-1} = M_k).$$

Similar formulas can be derived for $M_l = B$. This implies that the process is stationary. □

Proof of Theorem 3.1. Let us fix $a > 1$ and b. Then we have that $a(n-1)+b < an+b-1$ and

$$\mathcal{F}_n^{a,b} \subset \mathcal{F}_{an+b-1} \tag{8}$$

where $\mathcal{F}_{an+b-1}$ is the σ–algebra generated by the random variables X_k, $1 \le k < an+b-1$.

Now the process (Y_n) is i.i.d. Bernoulli. Hence, the first statement of the theorem follows. Moreover,

$$\mathbb{P}\left(\#\{1\le k<m \,:\, Y_k = C\}\ \text{odd} \,|\, \#\{1\le k<l \,:\, Y_k = C\}\ \text{odd}\right)$$
$$= \mathbb{P}\left(\#\{1\le k<m \,:\, Y_k = C\}\ \text{odd} \,|\, \#\{1\le k<l \,:\, Y_k = C\}\ \text{even}\right)$$
$$= \mathbb{P}\left(\#\{1\le k<m \,:\, Y_k = C\}\ \text{even} \,|\, \#\{1\le k<l \,:\, Y_k = C\}\ \text{odd}\right)$$
$$= \mathbb{P}\left(\#\{1\le k<m \,:\, Y_k = C\}\ \text{even} \,|\, \#\{1\le k<l \,:\, Y_k = C\}\ \text{even}\right)$$
$$= \mathbb{P}\left(\#\{1\le k<l \,:\, Y_k = C\}\ \text{even}\right)$$
$$= \mathbb{P}\left(\#\{1\le k<l \,:\, Y_k = C\}\ \text{even}\right)$$
$$= \frac{1}{2}$$

if $l < m$. This and (8) imply that

$$\mathbb{P}\left(\#\{1\le k<an+b \,:\, Y_k = C\}\ \text{ is even} \,|\, \mathcal{F}_n^{a,b}\right) = \frac{1}{2} \tag{9}$$

and

$$\mathbb{P}\left(\#\{1\le k<an+b \,:\, Y_k = C\}\ \text{ is odd} \,|\, \mathcal{F}_n^{a,b}\right) = \frac{1}{2} \tag{10}$$

By equations (5), (6), (9) and (10) we have that

$$\begin{aligned}
\mathbb{P}(X_n^{a,b} = A) &= \mathbb{P}(\#\{1 \le k < an+b : X_k = C\} \text{is even and } X_n^{a,b} \ne C) \\
&= \mathbb{P}(\#\{1 \le k < an+b : Y_k = C\} \text{ is even})\, \mathbb{P}(Y_{an+b} = \text{not } C) \\
&= \mathbb{P}\left(\#\{1 \le k < an+b : Y_k = C\} \text{ is even} \,|\, \mathcal{F}_n^{a,b}\right) \times \\
&\qquad \times \mathbb{P}(Y_{an+b} = \text{not } C \,|\, \mathcal{F}_n^{a,b}) \\
&= \frac{1}{2} \cdot \frac{1}{2} = \frac{1}{4}
\end{aligned}$$

and

$$\begin{aligned}
\mathbb{P}(X_n^{a,b} = B) &= \mathbb{P}(\#\{1 \le k < an+b : X_k = C\} \text{ is odd and } X_n^{a,b} \ne C) \\
&= \mathbb{P}(\#\{1 \le k < an+b : Y_k = C\} \text{ is odd})\, \mathbb{P}(Y_{an+b} = \text{not } C) \\
&= \mathbb{P}\left(\#\{1 \le k < an+b : Y_k = C\} \text{ is odd} \,|\, \mathcal{F}_n^{a,b}\right) \times \\
&\qquad \times \mathbb{P}(Y_{an+b} = \text{not } C \,|\, \mathcal{F}_n^{a,b}) \\
&= \frac{1}{2} \cdot \frac{1}{2} = \frac{1}{4}.
\end{aligned}$$

□

Proof of Theorem 3.2. let us fix $a > 1$, b. We write

$$Z_n^{a,b} = \begin{cases} H & X_b X_{a+b} \cdots X_{an+b}([0,1] \times [0,1]) \text{ is horizontally aligned} \\ V & X_b X_{a+b} \cdots X_{an+b}([0,1] \times [0,1]) \text{ is vertically aligned} \end{cases}$$

According to Theorem 3.1 the increments of the random walk $q_n^{a,b}$ have the following distribution:

$$\mathbb{P}(q_n^{a,b} = q_{n-1}^{a,b}) = \mathbb{P}(X_n^{a,b} = C) = \frac{1}{2},$$

and for $q_{n-1}^{a,b} \ne 0$

$$\begin{aligned}
\mathbb{P}(q_n^{a,b} - q_{n-1}^{a,b} = 1) &= \mathbb{P}\left(Z_{n-1}^{a,b} = H \text{ and } X_n^{a,b} = A\right) + \\
&\qquad + \mathbb{P}\left(Z_{n-1}^{a,b} = V \text{ and } X_n^{a,b} = B\right) \\
&= \mathbb{P}(X_n^{a,b} = A \,|\, Z_{n-1}^{a,b} = H)\mathbb{P}(Z_n^{a,b} = H) + \\
&\qquad + \mathbb{P}(X_n^{a,b} = B \,|\, Z_{n-1}^{a,b} = V)\mathbb{P}(Z_{n-1}^{a,b} = V) \\
&= \mathbb{P}(X_n^{a,b} = A \,|\, \mathcal{F}_n^{a,b})\mathbb{P}(Z_{n-1}^{a,b} = H) + \\
&\qquad + \mathbb{P}(X_n^{a,b} = B \,|\, \mathcal{F}_n^{a,b})\mathbb{P}(Z_{n-1}^{a,b} = V) \\
&= \frac{1}{4}\left[\mathbb{P}(Z_{n-1}^{a,b} = H) + \mathbb{P}(Z_{n-1}^{a,b} = V)\right] = \frac{1}{4}
\end{aligned}$$

and

$$\begin{aligned}\mathbb{P}(q_n^{a,b} - q_{n-1}^{a,b} = -1) &= \mathbb{P}\left(Z_{n-1}^{a,b} = V \text{ and } X_n^{a,b} = A\right) + \\ &\qquad + \mathbb{P}\left(Z_{n-1}^{a,b} = H \text{ and } X_n^{a,b} = B\right) \\ &= \mathbb{P}(X_n^{a,b} = A \mid Z_{n-1}^{a,b} = V)\mathbb{P}(Z_{n-1}^{a,b} = V) + \\ &\qquad + \mathbb{P}(X_n^{a,b} = B \mid Z_{n-1}^{a,b} = H)\mathbb{P}(Z_{n-1}^{a,b} = H) \\ &= \mathbb{P}(X_n^{a,b} = A \mid \mathcal{F}_n^{a,b})\mathbb{P}(Z_{n-1}^{a,b} = V) + \\ &\qquad + \mathbb{P}(X_n^{a,b} = B \mid \mathcal{F}_n^{a,b})\mathbb{P}(Z_{n-1}^{a,b} = H) \\ &= \frac{1}{4}\left[\mathbb{P}(Z_{n-1}^{a,b} = H) + \mathbb{P}(Z_{n-1}^{a,b} = H)\right] = \frac{1}{4}\end{aligned}$$

Therefore $(q_n^{a,b})$ is a symmetric random walk. Moreover, if we denote by r_n the standard symmetric random walk we get

$$q_n^{a,b} \leq |r_n| \qquad \mathbb{P} - \text{a.s.}$$

This implies that $q_n^{a,b}$ is recurrent. Since the standard random walk fulfills the law of the iterated logarithm

$$\limsup_{n\to\infty} \frac{|r_n|}{\sqrt{n \log\log n}} = 1 \qquad \mathbb{P} - \text{a.s.}$$

we have

$$\lim_{n\to\infty} \frac{q_n^{a,b}}{n} = 0 \qquad \mathbb{P} - \text{a.s.} \qquad \square$$

Amiran Ambroladze and Jörg Schmeling
Centre for Mathematical Sciences
LTH, Lund University
Box 118
SE-221 00 Lund
Sweden
e-mail: amiran@maths.lth.se
joerg@maths.lth.se

Part 3

Stochastic Processes and Random Fractals

Progress in Probability, Vol. 57, 119–134

Some Topics in the Theory of Multiplicative Chaos

Ai Hua Fan

Abstract. We present the theory of multiplicative chaos and some related topics including Dvoretzky covering, percolation on trees, random cascades and Riesz products.

Mathematics Subject Classification (2000). Primary 60G57; Secondary 28A75.

Keywords. Multiplicative chaos, random covering, percolation, random cascades, Riesz product.

1. Theory of multiplicative chaos

Chaotic phenomena were observed a long time ago in physics, especially in the statistical thermodynamics and in the turbulence. A. Einstein's work (1905) on the brownian motion concerned the thermodynamics. The brownian motion is now a well established theory which interacts with many sciences (physics, mathematics and finance). Concerning the turbulences, there had been a constant interaction between observations, physical models and mathematical models since J. Boussinesq (1877) and O. Reynolds (1883). However, N. Wiener declared in 1938 "the demands of chaos theory go considerably beyond the best knowledge of the present day". Around 1940, A. Kolmogorov and A. Obukhoff, leading their Russian school, studied the velocity field of a turbulence as a random field. In 1961, Kolmogorov [28] returned back to the subject by studying the dissipation of the energy defined by

$$\epsilon = \frac{\nu}{2} \sum_{1 \leq i,j \leq 3} \left(\frac{\partial u_i}{\partial x_j} + \frac{\partial u_j}{\partial x_i} \right)^2$$

where ν is the viscosity and (u_1, u_2, u_3) is the velocity. He made the hypothesis that the logarithm of the spacial average ϵ_r over a small ball of radius r is a normal variable and its variance is of order $O(\log \frac{1}{r})$. In 1971, B. Mandelbrot [32] gave a refinement of the log-normal hypothesis by defining the random measure ϵ

as a martingale limit. Later in 1974, Mandelbrot [34, 35] provided a discontinuous model without log-normal hypothesis. This model, called random cascades, was well studied by J. P. Kahane and J. Peyrière [26]. In 1985, Kahane studied a limit log-normal model, called gaussian chaos [22]. A limit Lévy model was studied in [7, 12].

A general framework was set up in 1987 by Kahane [23], which is called the theory of multiplicative chaos. We will recall this general theory in the Section 1. The theory involves in the resolutions of many other mathematical problems, some of which have at first appearance nothing to do with the theory and some others are even deterministic (without randomness). We will present some topics together with recent developments. These topics include Dvoretzky covering (Section 2), percolation on trees (Section 3), random cascades (Section 4) and Riesz products (Section 5). In the last section, we address some unsolved problems. The reader may refer to the survey papers [25] and [40] for other interesting materials.

1.1. Construction of multiplicative chaos

Let (T, d) be a compact (or locally compact) metric space and $(\Omega, \mathcal{A}, P)$ be a probability space. We are given an increasing sequence $\{\mathcal{A}_n\}_{n\geq 1}$ of sub-σ-fields of $\mathcal{A}$ and a sequence of random functions $\{P_n(t,\omega)\}_{n\geq 1}$ $(t \in T, \omega \in \Omega)$ such that

(i) $P_n(t) = P_n(t,\omega)$ are non-negative and independent processes; $P_n(\cdot,\omega)$ is borelian for almost all ω; $P_n(t,\cdot)$ is $\mathcal{A}_n$-measurable for each t.

(ii) $\mathbb{E}P_n(t) = 1$ for all $t \in T$.

Such a sequence $\{P_n\}$ is called a sequence of *weights* adapted to $\{\mathcal{A}_n\}$. Let

$$Q_n(t) = Q_n(t,\omega) = \prod_{j=1}^{n} P_j(t,\omega).$$

For any $n \geq 1$ and any positive Radon measure σ on T (we write $\sigma \in \mathcal{M}^+(T)$), we consider the random measures $Q_n\sigma$ defined by

$$Q_n\sigma(A) = \int_A Q_n(t)d\sigma(t) \qquad (A \in \mathcal{B}(T))$$

where $\mathcal{B}(T)$ is the Borel field of T. It is clear that for any $A \in \mathcal{B}(T)$, $Q_n\sigma(A)$ is a positive martingale, so it converges almost surely (a.s. for short). So does $\int \phi(t)dQ_n\sigma(t)$ for any bounded borelian function ϕ. The following fundamental theorem is a consequence of the last fact.

Theorem 1.1. ([23]) *For any Radon measure $\sigma \in \mathcal{M}^+(T)$, almost surely the random measures $Q_n\sigma$ converge weakly to a random measure S.*

Let us write

$$Q\sigma = S = \lim_{n\to\infty} Q_n\sigma.$$

We may consider Q as an operator which maps measures into random measures. We call Q a *multiplicative chaos operator*. Sometimes, we call S a *multiplicative chaos measure* which, in some special cases, describes the limit energy state of a turbulence [34, 35, 22, 7]. See next sections for examples of such constructions.

There are two possible extreme cases. The first one is that $Q_n\sigma(T)$ converges a.s. to zero, i.e. $S = 0$ a.s. (the energy is totally dissipated). The second one is that $Q_n\sigma(T)$ converges in L^1 so that $\mathbb{E}S(T) = \sigma(T)$ (the energy is conserved). If the first case occurs, we say that Q *degenerates* on σ or σ is said to be *Q-singular.* If the second case occurs, we say that Q *fully acts* on σ or σ is said to be *Q-regular.*

We define a map $\mathbb{E}Q : \mathcal{M}^+(T) \to \mathcal{M}^+(T)$ by

$$(\mathbb{E}Q\sigma)(A) = \mathbb{E}(Q\sigma(A)) \qquad (A \in \mathcal{B}(T)).$$

That σ is Q-singular (resp. Q-regular) is equivalent to $\mathbb{E}Q\sigma = 0$ (resp. $\mathbb{E}Q\sigma = \sigma$).

Theorem 1.2. ([23]) *Any Radon measure $\sigma \in \mathcal{M}^+(T)$ can be uniquely decomposed into $\sigma = \sigma_r + \sigma_s$ where σ_r is a Q-regular measure and σ_s is a Q-singular measure.*

The operator $\mathbb{E}Q$ extended to the space $\mathcal{M}(T)$ is thus a projection whose image (resp. kernel) consists of Q-regular (resp. Q-singular) measures.

1.2. General results

We are concerned with properties of the random measure $S = Q\sigma$ or of the operator Q, some of which are related to the following questions:

Question 1. Does Q degenerate on σ?
Question 2. Does Q fully act on σ?
Question 3. Is the moment $\mathbb{E}(S(T)^p)$ finite for a positive number $p > 1$?
Question 4. What is the dimension of the measure $Q\sigma$?
Question 5. Is the measure $Q\sigma$ multifractal?
Question 6. What are the relations between two measures $Q'\sigma'$ and $Q''\sigma''$ for two different operators Q' and Q'' defined in the same way as Q?

In the following, we state some results in the general case. Either they are partial answers to one of these questions, or they provide some useful tools. Complete solutions may be obtained for special cases, as we shall show later.

Theorem 1.3 ([23]). *Suppose that $\mathcal{H}^\alpha(T) < \infty$ where $\mathcal{H}^\alpha$ denotes the α-dimensional Hausdorff measure and that there exist constants $0 < h < 1$ and $C > 0$ with the property that for any ball B with radius r there exists an integer $n = n(B)$ such that*

$$\mathbb{E}\left(\sup_{t\in B} Q_n(t)\right)^h \le Cr^{\alpha(1-h)}. \tag{1.1}$$

Then all Radon measures on T are Q-singular.

This provides a good tool to verify the Q-singularity of σ. In fact, the condition $\dim\sigma < \alpha$ together with (1.1) implies the Q-singularity of σ. On the other hand, the following gives a simple condition of Q-regularity.

Theorem 1.4 ([23]). *We have $\mathbb{E}(Q\sigma(T))^2 < \infty$ if and only if*

$$\int\int \prod_{n=1}^{\infty} \mathbb{E}P_n(t)P_n(s)d\sigma(t)d\sigma(s) < \infty. \tag{1.2}$$

In particular, (1.2) *implies the Q-regularity of σ.*

There is a similar condition for $\mathbb{E}(Q\sigma(T))^m < \infty$ with an integer order m.

Suppose that σ is a Q-regular probability measure. A useful tool for studying the measure $Q\sigma$ is the *Peyrière measure* $\mathcal{Q}$ on the product space $T \times \Omega$ defined by

$$\int_{T\times\Omega} \varphi(t,\omega)d\mathcal{Q}(t,\omega) = \mathbb{E}\int_T \varphi(t,\omega)dQ\sigma(t) \tag{1.3}$$

for non-negative measurable functions φ.

Theorem 1.5 ([26, 23]). *Suppose that σ is a Q-regular probability measure and that the probability law of the variable $P_n(t)$ for fixed t is independent of t. Then P_n's, considered as random variables with respect to $\mathcal{Q}$, are $\mathcal{Q}$-independent.*

A direct application of the Peyrère measure leads to that almost surely for $Q\sigma$-almost every $t \in T$ we have

$$\lim_{n\to\infty} \frac{1}{n}\sum_{k=1}^{n} \log P_k(t,\omega) = \lim_{n\to\infty} \frac{1}{n}\sum_{k=1}^{n} \mathbb{E}P_k \log P_k. \tag{1.4}$$

This result may be used to study the dimension and the multifractality of the measure $Q\sigma$.

Suppose that we are given two sequences of weights $\{P'_n(t)\}_{n\geq 1}$ and $\{P''_n(t)\}_{n\geq 1}$ adapted to $\{\mathcal{A}'_n\}$ and $\{\mathcal{A}''_n\}$, respectively defined on probability spaces $(\Omega', \mathcal{A}', P')$ and $(\Omega'', \mathcal{A}'', P'')$. It is obvious that $\{P_n\}$ defined by $P_n(t) = P''_n(t)P'_n(t)$ is a sequence of weights adapted to $\{\mathcal{A}'_n \otimes \mathcal{A}''_n\}$, defined on the product space $(\Omega, \mathcal{A}, P) = (\Omega' \times \Omega'', \mathcal{A}' \otimes \mathcal{A}'', P' \otimes P'')$. We denote by Q', Q'' and Q the three operators corresponding to the above three sequences of weights. The following decomposition principle establishes a relationship between Q', Q'' and Q.

Theorem 1.6 ([16]). *Under the above condition, we have*

(a) $Q\sigma = Q''(Q'\sigma)$ a.s. *for any measure $\sigma \in \mathcal{M}^+(T)$.*
(b) $\sigma \in \mathrm{Im}\mathbb{E}Q \Rightarrow Q'\sigma \in \mathrm{Im}\mathbb{E}Q''$ *for almost all $\omega' \in \Omega'$.*
(c) $\sigma \in \mathrm{Ker}\mathbb{E}Q \Rightarrow Q'\sigma \in \mathrm{Ker}\mathbb{E}Q''$ *for almost all $\omega' \in \Omega'$.*
(d) $\mathbb{E}Q = \mathbb{E}Q''\mathbb{E}Q'$.

Let Q' and Q'' be two operators associated respectively to $\{P'_n\}$ and $\{P''_n\}$. Now we don't suppose the independence of $\{P'_n\}$ and $\{P''_n\}$. But we suppose that the law of the vector $(P'_n(t), P''_n(t))$ is independent of t. We have a Kakutani type criterion for the dichotomy of the mutual absolute continuity of $Q'\sigma$ and $Q''\sigma$.

Theorem 1.7 ([10]). *Assume the above assumptions. Suppose furthermore that σ is a Q'-regular probability measure. We have*

(a) $\prod_{n=1}^{\infty} \mathbb{E}\sqrt{P'_nP''_n} > 0 \Rightarrow Q''\sigma \ll Q'\sigma$ *and σ is Q''-regular.*
(b) $\prod_{n=1}^{\infty} \mathbb{E}\sqrt{P'_nP''_n} = 0 \Rightarrow Q''\sigma \perp Q'\sigma$.

2. Dvoretzky covering

The Dvoretzky covering problem was raised in 1956 by A. Dvoretzky [5]. It attracted attentions of P. Lévy, J.P. Kahane, P. Erdös, P. Billard, B. Mandelbrot who made significant contributions (see [21]). In 1972, L. Shepp [41] obtained a complete solution to the problem. There were different generalizations and variants of the problem (see [21]). The most recent result is a solution of Carleson problem (Theorem 2.5).

2.1. Dvoretzky covering problem and solutions of Shepp and Kahane

Let $T = \mathbb{T} = \mathbb{R}/\mathbb{Z}$. Take a decreasing sequence $\{\ell_n\}_{n\geq 1}$ $(0 < \ell_n < 1)$ such that $\ell_n \downarrow 0$ and $\sum_{n=1}^{\infty} \ell_n = \infty$ and a sequence of i.i.d. random variables $\{\omega_n\}_{n\geq 1}$ of the uniform distribution on $\mathbb{T}$. Let $I_n = \omega_n + (0, \ell_n)$. It is easy to see that with probability one almost every point in $\mathbb{T}$ with respect to the Lebesgue measure is covered by an infinite number of intervals I_n. A. Dvoretzky [5] discovered that it is possible, in some cases, that $\mathbb{T}$ is not necessary covered. The Dvoretzky covering problem is to find conditions on the length sequence $\{\ell_n\}$ such that

$$\mathbb{P}\left(\mathbb{T} = \bigcap_{N=1}^{\infty} \bigcup_{n=N}^{\infty} I_n\right) = 1 \tag{2.1}$$

where $\mathbb{P}$ is the probability measure of the underlying probability space $(\Omega, \mathcal{A}, \mathbb{P})$.

Theorem 2.1 ([41]). *A necessary and sufficient condition for covering (i.e. for* (2.1) *to be realized) is*

$$\sum_{n=1}^{\infty} \frac{1}{n^2} \exp(\ell_1 + \cdots + \ell_n) = \infty. \tag{2.2}$$

The device of multiplicative chaos is a useful tool, which goes back to P. Billard [4]. The corresponding weights are

$$P_n(t) = \frac{1 - 1_{(0,\ell_n)}(t - \omega_n)}{1 - \ell_n}.$$

Notice that t is covered by I_n (i.e. $t \in I_n$) iff $P_n(t) = 0$. So, if the whole circle $\mathbb{T}$ is covered, then the martingale $Q_n\lambda(\mathbb{T})$ must tend a.s. to zero. Therefore a necessary condition for covering is $\lim_n \mathbb{E}(Q_n\lambda(\mathbb{T}))^2 = \infty$, which is equivalent to (2.2). The necessity of the following theorem may also be proved in the same way.

Theorem 2.2 ([24]). *Let K be a compact set in $\mathbb{T}$. Almost surely $K \subset \limsup I_n$ iff $\mathrm{Cap}_k K = 0$ where Cap_k denotes the capacity relative to the integral kernel*

$$k(t) = \exp \sum_{n=1}^{\infty} (\ell_n - |t|)_+ \tag{2.3}$$

(convention: x_+ means the positive part of a real number x).

We refer to [27] for the potential theory. See §3.2. for a similar theory on trees. J. P. Kahane realized that the Dvoretzky covering on the circle is equivalent to a similar Poisson covering on the real line, also called Mandelbrot random cutouts.

Combining this observation and an idea of stopping time due to S. Janson [19], he gave an elegant proof of the sufficiency of the above two theorems, which is sketched below.

2.2. Mandelbrot random cutouts and Janson stopping time

In 1972, B. Mandelbrot [33] introduced an analogous covering problem on $\mathbb{R}$ instead of $\mathbb{T}$, by replacing a random point on $\mathbb{T}$ by a Poisson point process with intensity λ (Lebesgue measure). Thus the random interval $(\omega_n, \omega_n + \ell_n)$ is replaced by the union of intervals $(x, x + \ell_n)$ with x belonging to the Poisson process.

The general problem is as follows. Let μ be a positive measure on the half line $\mathbb{R}^+$. We consider the random closed set

$$E^+ = \mathbb{R}^+ \setminus \bigcup (x, x+y)$$

where the union is taken over all points (x, y) of the Poisson process on $\mathbb{R}^+ \times \mathbb{R}^+$ with intensity $\lambda \otimes \mu$. We call E^+ a *Mandelbrot random cutout*, which is similar to Cantor sets. Under what condition on μ have we $E^+ = \emptyset$ a.s.? Otherwise, which properties does E^+ have?

The answer to the first question was given by L. Shepp [42] in 1972. Fitzsimmons, Fristedt and Shell [17] proved that E^+ coincides a.s. with the closure of the image of an increasing Lévy process (also called subordinator).

We just sketch a proof due to Kahane of the Shepp's result. For $\epsilon > 0$, consider the Poisson process with intensity $\lambda \otimes \mu_\epsilon$ where μ_ϵ is the restriction of μ on $[\epsilon, \infty)$. Let G_ϵ be the union of all intervals $(x, x + y)$ with (x, y) belonging to the truncated Poisson process. Following S. Janson [19], define the stopping time

$$\tau_\epsilon = \inf\{t \geq 0 : t \notin G_\epsilon\}.$$

That $E^+ = \emptyset$ a.s. is equivalent to $\lim_{\epsilon \to 0} \tau_\epsilon = \infty$. By considering the double integral $\mathbb{E} \int_0^\infty e^{-t} 1_{\{t \notin G_\epsilon\}} dt$ and calculating it in two different ways, we can deduce

$$\mathbb{E} e^{-\tau_\epsilon} = \left(\int_0^\infty e^{-s} k_\epsilon(s) ds \right)^{-1}$$

where

$$k_\epsilon(s) = \exp \int_{|s|}^\infty \mu_\epsilon(u, \infty) du \qquad (s \in \mathbb{R}).$$

So, letting $\epsilon \to 0$ we get $\lim_{\epsilon \to 0} \tau_\epsilon = \infty$ a.s. when the following Shepp condition is satisfied

$$\int_0^\infty \exp \left(\int_{|s|}^\infty \mu_\epsilon(u, \infty) du \right) e^{-s} ds = \infty. \tag{2.4}$$

2.3. Covering operator $\mathbb{E}Q$ and dimension of $Q\lambda$

Recall that k is the integral kernel defined by (2.3). A Borel measure μ on $\mathbb{T}$ is said to be *k-regular* if it is a countable sum of finite k-energy and it is said to be *k-singular* if it is supported by a null k-capacity Borel set. The following theorem is an easy consequence of Theorem 2.2.

Theorem 2.3. *A measure $\mu \in M^+(\mathbb{T})$ is Q-regular (resp. Q-singular) if and only if it is k-regular (resp. k-singular).*

Given another kernel $\tilde{k}$ defined in the same way as k. Using the above theorem and the decomposition principle (Theorem 1.6), we can determine the $\tilde{k}$-regularity of $Q\sigma$ for any $\sigma \in M^+(\mathbb{T})$:

(a) if σ is $k\tilde{k}$-regular, then $Q\sigma$ is a.s. $\tilde{k}$-regular;

(b) if σ is $k\tilde{k}$-singular, then $Q\sigma$ is a.s. $\tilde{k}$-singular.

The kernel corresponding to the sequence $\ell_n = \frac{\alpha}{n}$ is the Riesz kernel $k_\alpha(t) = |t|^{-\alpha}$. It is clear that

(c) $\dim \mu > \alpha$ if μ is k_α-regular; $\dim \mu < \alpha$ if μ is k_α-singular.

Theorem 2.4. *Suppose that the Shepp condition (2.2) is satisfied. Let λ be the Lebesgue measure. Then $\dim Q\lambda = 1 - \limsup_{n\to\infty} \frac{\ell_1+\cdots+\ell_n}{\log n}$ a.s.*

To prove this, it suffices to use the above assertions (a), (b) and (c) and to notice that λ is kk_α-regular or singular according to the convergence or divergence of the series $\sum_{n=1}^\infty n^{\alpha-2}e^{\ell_1+\cdots+\ell_n}$. So, the critical value α of convergence of the above series is equal to $\dim Q\lambda$, as announced.

2.4. Carleson problem and covering numbers

When Shepp's result is established, a natural problem, which was raised by L. Carleson, is to describe the infinity of the set of intervals covering a given point. First works in this direction appeared in [8, 15].

We describe the Carleson problem in the following manner. Define, for $n \geq 1$, the n-th *covering number* of $t \in \mathbb{T}$ by

$$N_n(t) = \mathrm{Card}\{1 \leq k \leq n : I_k \ni t\} = \sum_{k=1}^n 1_{(0,\ell_k)}(t - \omega_k).$$

For any $\beta \geq 0$, we define the (random) set

$$F_\beta = \left\{ t \in \mathbb{T} : \lim_{n\to\infty} \frac{N_n(t)}{L_n} = \beta \right\}$$

where $L_n = \ell_1 + \cdots + \ell_n$. Let

$$\overline{\alpha} = \limsup_{n\to\infty} \frac{\sum_{j=1}^n \ell_j}{-\log \ell_n}. \tag{2.5}$$

Introduce the function

$$d_\alpha(\beta) = 1 + \alpha(\beta - 1 - \beta\log\beta), \qquad (\alpha, \beta \geq 0). \tag{2.6}$$

Theorem 2.5 ([3]). *Under some regularity condition on $\{\ell_n\}$, we have*

(i) *Suppose $\overline{\alpha} = 0$. With probability one, for all $\beta \geq 0$ we have $\dim(F_\beta) = 1$.*

(ii) *Suppose $0 < \overline{\alpha} < \infty$. With probability one, for all $\beta \geq 0$ such that $d_{\overline{\alpha}}(\beta) \geq 0$, we have $\dim(F_\beta) = d_{\overline{\alpha}}(\beta)$, and $F_\beta = \emptyset$ for all $\beta \geq 0$ with $d_{\overline{\alpha}}(\beta) < 0$.*

(iii) *Suppose $\lim_{n\to\infty} n\ell_n = \infty$. With probability one, we have $F_1 = \mathbb{T}$.*

Let us exhibit three typical sequences corresponding to above three cases:

$$\ell_n = \frac{1}{n \log n}, \qquad \ell_n = \frac{\alpha}{n}, \qquad \ell_n = \frac{\log n}{n} \qquad (n \gg 1).$$

In order to prove the above theorem, we construct a multiplicative chaos measure with weights

$$P_n(t) = \frac{\beta^{1_{(0,\ell_n)}(t-\omega_n)}}{(1-\ell_n) + \beta \ell_n}.$$

In an earlier work, it was proved the following dichotomy.

Theorem 2.6 ([15]). *Suppose $\ell_n = \frac{\alpha}{n}$ with $\alpha > 1$. Let $\{a_n\}$ be a decreasing sequence of positive numbers such that $\sum a_n = \infty$. Then*

(i) *a.s. $\forall t \in \mathbb{T}$ $\sum_{n: I_n \ni t} a_n = \infty \Leftrightarrow \sum_{n=1}^{\infty} \frac{a_n}{n} = \infty$;*

(ii) *a.s. $\forall t \in \mathbb{T}$ $\sum_{n: I_n \ni t} a_n < \infty \Leftrightarrow \sum_{n=1}^{\infty} \frac{a_n}{n} < \infty$.*

3. Percolation on trees

Let $T = \partial \mathcal{T}$ be the boundary of an infinite and locally finite tree $\mathcal{T}$, which is a compact metric space with metric $d(\xi, \eta) = e^{-|\xi \wedge \eta|}$ where $\xi \wedge \eta$ is the number of vertices passed by both paths ξ and η belonging to $\partial \mathcal{T}$.

3.1. p-Bernoulli percolation on trees

Let $\mathbf{p} = (p_n)_{n \geq 1}$ be a sequence of positive numbers such that $0 < p_n \leq 1$. We remove or keep edges at random from the tree $\mathcal{T}$, keeping each edge in the n–th generation with probability p_n and removing it with probability $1-p_n$, and making decisions independently for all different edges of all generations. This procedure is called $\mathbf{p}$–*Bernoulli percolation*. If, with a positive probability, an infinite path emanating from the root remains in the tree, we say that the *percolation occurs*. When does the percolation occur?

Theorem 3.1 ([8, 9, 30, 31]). *A necessary and sufficient condition for the $\mathbf{p}$–Bernoulli percolation to occur is $\mathrm{Cap}_K \partial \mathcal{T} > 0$ where K is the kernel defined by*

$$K(t,s) := K_{\mathbf{p}}(t,s) := \prod_{n=1}^{|t \wedge s|} \frac{1}{p_n}. \tag{3.1}$$

A. H. Fan proved the last theorem by considering the multiplicative chaos defined by the weights

$$P_n(\xi) = \frac{1_{\{\text{the } n\text{–th edge of } \xi \text{ is kept}\}}}{p_n} \tag{3.2}$$

R. Lyons used the method of electrical network. See [38] for related topics on random walks on trees, where both methods are useful.

3.2. Potential analysis on trees

We now explain what $\mathrm{Cap}_K\partial\mathcal{T} > 0$ means. The potential theory on trees has its own interests. For an integer $m \geq 2$, consider the group $\mathbb{Z}_m = \mathbb{R}/m\mathbb{Z} = \{0, 1, \cdots, m-1\}$. Given a sequence of integers $\{m_n\}_{n\geq 1}$ ($m_n \geq 2$), we consider the infinite product group, called an *odometer group*,

$$G = \prod_{n=1}^{\infty} \mathbb{Z}_{m_n}.$$

The dual group $\widehat{\mathbb{Z}}_m$ of $\mathbb{Z}_m$ is represented by $\widehat{\mathbb{Z}}_m = \{1, \varphi, \cdots, \varphi^{m-1}\}$ where

$$\varphi(j) = \omega_m^j \quad (j \in \mathbb{Z}_m), \quad \text{with } \omega_m = e^{\frac{2\pi i}{m}}.$$

The character $\varphi : \mathbb{Z}_m \to \mathbb{R}/\mathbb{Z}$ is called the *Rademacher function* of the group $\mathbb{Z}_m$. Denote by $\{\varphi_n\}_{n\geq 1}$ the sequence of Rademacher functions of the groups $\mathbb{Z}_{m_n}$ ($n \geq 1$). Every function φ_n induces a function on the odometer group G if we define $\varphi_n(t) = \varphi_n(t_n)$ for $t = (t_k)_{k\geq 1} \in G$, called the n-th Rademacher function of G. It is well-known that the dual group $\widehat{G}$ of G is equal to

$$\widehat{G} = \{\varphi_1^{\epsilon_1}\varphi_2^{\epsilon_2}\cdots\varphi_n^{\epsilon_n} : \ n \geq 1; \ \ 0 \leq \epsilon_j < m_n \ \ (1 \leq j \leq n)\}.$$

It is easy to see that every character in $\widehat{G}$ corresponds uniquely to an integer $n \geq 0$, and vice versa. The correspondence is as follows. Every integer $n \geq 0$ has an unique representation

$$n = \epsilon_1 + \epsilon_2 m_1 + \epsilon_3 m_1 m_2 + \cdots + \epsilon_k m_1 m_2 \cdots m_{k-1}$$

($0 \leq \epsilon_j < m_j$ for $j = 1, 2, \cdots, k$). The character corresponding to n, called the n-th *Walsh function*, is

$$\psi_n = \varphi_1^{\epsilon_1}\varphi_2^{\epsilon_2}\cdots\varphi_k^{\epsilon_k}.$$

The Fourier transform of a Haar integral function f on G is defined by $\widehat{f}(n) = \int f(t)\overline{\psi}_n(t)dt$ for $n \geq 0$.

Theorem 3.2. *If $f \in L^\infty(G)$ and $\widehat{f}(n) \geq 0$ for all $n \geq 0$, then $\sum_{n=0}^{\infty}\widehat{f}(n) < \infty$.*

It suffices to show the boundedness of some subsequence of $S_n = \sum_{k=0}^{n}\widehat{f}(k)$. Consider the subsequence $\{n_k\}$ with

$$n_k = (m_1 - 1) + (m_2 - 1)m_1 + \cdots + (m_k - 1)m_1 \cdots m_{k-1}.$$

We have

$$S_{n_k} = \int_G f(t)\sum_{\ell=0}^{n_k-1}\psi_\ell(t)dt = \int_G f(t)\prod_{j=1}^{k}\left(1 + \varphi_j + \cdots + \varphi_j^{m_j-1}\right)dt$$

Notice that $1 + \varphi_j(t_j) + \cdots + \varphi_j(t_j)^{m_j-1} = m_j$ or 0 according to $t_j = 0$ or $t_j \neq 0$. It follows that $S_{n_k} = \int_{I_k(0)} f(t)m_1m_2\cdots m_k dt \leq \|f\|_\infty$.

The boundary $\partial\mathcal{T}$ of any infinite and locally finite tree $\mathcal{T}$ can be embedded into some group $G = \prod_{n=1}^{\infty}\mathbb{Z}_{m_n}$. The kernel K defined by (3.1) can be formally extended onto G. The extended kernel is still denoted by K which is translation invariant in the sense that $K(t+\tau, s+\tau) = K(t,s)$. So, $K(t,s) = K(t-s)$ where

$$K(t) = \prod_{n=1}^{|t\wedge 0|} \frac{1}{p_n}.$$

Theorem 3.3 ([8, 9]). *The function $K(\cdot)$ is integrable if and only if*

$$\sum_{n=1}^{\infty} \frac{1}{(p_1 \cdots p_n)(m_1 \cdots m_n)} < \infty.$$

Under this integrability, we have $\gamma_n = \widehat{K}(n) \geq 0$ for all $n \geq 0$.

Assume that K is integrable. Recall that the *potential* and the *energy* (relative to K) of a probability Borel measure μ on G are respectively defined by

$$U^\mu(t) = \int K(t-s)d\mu(s), \qquad I^\mu = \int U^\mu(t)d\mu(t).$$

The *capacity* (relative to K) of a compact set E in G est defined by $I(E)^{-1}$ with $I(E) = \inf I^\mu$ where the infimum is taken over all probability measures supported by E.

Theorem 3.4 ([8, 9]). *Suppose that K is integrable. Let E be a compact set in G.*

1) *If $I(E) < \infty$ (i.e. E is of positive capacity), then there exists a unique probability measure μ_e supported by E such that $I(E) = I^{\mu_e}$.*
2) *The set of points $t \in E$ such that $U^{\mu_e} < I(E)$ has null measure with respect to every measure of positive energy.*

3.3. Percolation operator $\mathbb{E}Q_{\mathbf{p}}$

Recall that K is the integral kernel defined by (3.1) corresponding to the **p**-Bernoulli percolation. For a Borel measure μ on ∂T, as in §2.3., we define its K-regularity and K-singularity. Let $Q_{\mathbf{p}}$ be the multiplicative chaos operator corresponding to weights define by (3.2).

Theorem 3.5. *A measure $\mu \in M^+(\partial T)$ is $Q_{\mathbf{p}}$-regular (resp. $Q_{\mathbf{p}}$-singular) if and only if it is K-regular (resp. K-singular).*

4. Random cascades

Random cascades were introduced by B. Mandelbrot as models of turbulence.

4.1. Mandelbrot cascades

Let $T_c = (\mathbb{Z}/c\mathbb{Z})^{\mathbb{N}}$ (c integer ≥ 2), a special odometer group. We represent T_c as the boundary of a tree and equip T_c with the ultrametric $d(t,s) = e^{-|t\wedge s|}$. We denote by $[t_1, t_2, \cdots, t_n]$ the ball of radius e^{-n} containing $t = (t_k)_{k\geq 1}$. Such a ball corresponds to the vertex $t_1 t_2 \cdots t_n$ of the tree. Let $\{W_{t_1,t_2,\cdots,t_n}\}$ be a family of i.i.d. non-negative random variables of expectation 1, indexed by the vertices of the tree. We denote the common law of the family by W. For $n \geq 1$, we define a weight

$$P_n(t) = W_{t_1,t_2,\cdots,t_n} \qquad (t = (t_k)_{k\geq 1} \in T_c).$$

The corresponding multiplicative chaos operator will be denoted by Q_W.

Let μ be an ergodic invariant Markov measure defined by its transition probability matrix $P = (p_{i,j})$. For any $h \in \mathbb{R}$, let $\rho(h)$ be the spectral radius of the matrix $P(h)$ defined by $(p_{i,j}^h)$ (by convention, $0^h = 0$ for any h), i.e.

$$P(h) = \begin{pmatrix} p_{0,0}^h & p_{0,1}^h & \cdots & p_{0,c-1}^h \\ p_{1,0}^h & p_{1,1}^h & \cdots & p_{1,c-1}^h \\ \vdots & \vdots & \ddots & \vdots \\ p_{c-1,0}^h & p_{c-1,1}^h & \cdots & p_{c-1,c-1}^h \end{pmatrix} \tag{4.1}$$

The function $\rho(h)$ is real analytic. For the Bernoulli product measure defined by $(p_0, p_1, \cdots, p_{c-1})$, the function $\rho(h)$ has the explicit expression $\sum_{j=0}^{c-1} p_j^h$.

Theorem 4.1. [26, 14] *Let μ be an ergodic invariant Markov measure with transition probability matrix P. Then*

(a) *μ is Q_W-regular if and only if $\mathbb{E}W \log W < -\rho'(1)$.*
(b) *μ is Q_W-singular if and only if $\mathbb{E}W \log W \geq -\rho'(1)$.*
(c) *$0 < \mathbb{E}(Q_W\mu(T_c))^h < \infty$ with $h > 1$ if and only if $\mathbb{E}W^h \rho(h) < 1$.*
(d) *If $\mathbb{E}W \log W < \rho'(1)$, then $\dim Q_W\mu = (-\rho'(1) - \mathbb{E}W \log W)/\log c$ a.s.*

For any measure μ defined on T_c and for $h > 0$, define

$$\rho(h) = \limsup_{n\to\infty} \frac{1}{n} \log \sum_{t_1,t_2,\cdots,t_n} \mu([t_1, t_2, \cdots, t_n])^h. \tag{4.2}$$

It is clear that ρ is convex such that $\rho(1) = 0$.

Theorem 4.2 ([14]). *Suppose $\mathbb{E}W^{1+\delta} < \infty$ for some $\delta > 0$. Then*

(a) *μ is Q_W-regular if $\mathbb{E}W \log W < -\rho'(1^+)$.*
(b) *μ is Q_W-singular if $\mathbb{E}W \log W > -\rho'(1^-)$.*
(c) *If $\mathbb{E}W \log W < -\rho'(1^+)$, then almost surely*

$$-\frac{\rho'(1^+) + \mathbb{E}W \log W}{\log c} \leq \dim Q_W\mu \leq -\frac{\rho'(1^-) + \mathbb{E}W \log W)}{\log c}. \tag{4.3}$$

The proof of Theorem 4.1 is subtle. However, it is easy to sketch a proof for Theorem 4.2.

(a) We write $Q_{n+1}\mu(\mathbb{T}) - Q_n\mu(\mathbb{T})$ as

$$\sum_{t_1,t_2,\cdots,t_{n+1}} \prod_{k=1}^{n} W_{t_1,t_2,\cdots,t_k} \left(W_{t_1,t_2,\cdots,t_{n+1}} - 1\right) \mu([t_1, t_2, \cdots, t_{n+1}]).$$

For $1 \leq h \leq \min(2, 1+\delta)$, applying the Von Bahr-Esseen inequality in [43] and then using the hypothesis, we get an $h > 1$ sufficiently close to 1 such that

$$\mathbb{E}|Q_{n+1}\mu(\mathbb{T}) - Q_n\mu(\mathbb{T})|^h \leq C\mathbb{E}|W-1|^h e^{-\frac{\epsilon n}{2}}.$$

Then $\sum_n \|Q_{n+1}\mu(\mathbb{T}) - Q_n\mu(\mathbb{T})\|_h < \infty$, which implies the L^h-convergence of $Q_n\mu(\mathbb{T})$, a fortiori, the L^1-convergence of $Q_n\mu(\mathbb{T})$.

(b) It is just a consequence of the Fatou lemma: for any $0 < h < 1$ we have

$$\mathbb{E}(Q_W\mu(\mathbb{T})^h) = \liminf_{n\to\infty}(\mathbb{E}W^h)^n \sum_{t_1,\cdots,t_n} \mu([t_1, t_2, \cdots, t_n])^h.$$

The last limit is zero by the hypothesis.

(c) Let W_β be a variable such that $P(W_\beta = c^\beta) = c^{-\beta}$ and $P(W_\beta = 0) = 1-c^{-\beta}$. Let Q_β be the operator associated to W_β (β-model). Suppose that Q_β is independent of Q_W. Then the product $Q_\beta Q_W$ is associated to $W_\beta W$ (Theorem 1.6). Notice first that

$$\mathbb{E}(W_\beta W)\log(W_\beta W) = \mathbb{E}W\log W + \beta\log c.$$

Suppose that $\mathbb{E}W\log W < -\rho'(1)$. For any β such that $\mathbb{E}W\log W + \beta\log c < -\rho'(1)$, μ is $Q_\beta Q_W$-regular (by (a)). Then $Q_W\mu$ is Q_β-regular a.s. (Theorem 1.6). So, $\dim Q\mu \geq \beta$ a.s. (Theorem 3.5). It follows that

$$\dim Q\mu \geq \frac{1}{\log c}(-\rho'(1) - \mathbb{E}W\log W) \qquad a.s.$$

If β is chosen so that $\beta\log c > -\rho'(1) - \mathbb{E}W\log W$. then μ is $Q_\beta Q_W$-singular. Similar arguments show the inverse inequality.

There are different variants of the Mandelbrot model. See [18, 29, 37].

4.2. Multifractal analysis

Let λ be the Lebesgue measure on T_c. The multifractality of $Q_W\lambda$ was well studied, under some restrictive conditions on W by R. Holley and E. C. Waymire [20], and by K. J. Falconer [6]. These conditions were weaken by G. M. Molchan [36] and J. Barral [1]. The key work is the study of negative moment $\mathbb{E}(Q_W\lambda(\mathbb{T}))^{-h}$ with $h > 0$.

5. Riesz products and its generalizations

We present a result concerning deterministic Riesz products by using Peyrière measure.

5.1. Riesz products and lacunary trigonometric series

Let $T = G$ be a compact Abelian group. Given a sequence of dissociated characters $\Lambda = (\gamma_n)_{n\geq 1}$ in the dual group Γ of G and a sequence of complex numbers $a = (a_n)_{n\geq 1}$ with $|a_n| \leq 1$, we define a probability measure, called Riesz product,

$$\mu_a = \prod_{n=1}^{\infty}(1 + \text{Re } a_n\gamma_n(t)) \tag{5.1}$$

as the weak *-limit of the partial products of the infinite product. The system $\{\gamma_n(t)-\frac{1}{2}\overline{a}_n)\}_{n\geq 1}$ is orthogonal in $L^2(\mu_a)$. This leads us to consider the orthogonal series

$$\sum_{n=1}^{\infty} \alpha_n(\gamma_n(t) - \frac{1}{2}\overline{a}_n).$$

A classical example is the Riesz products on $\mathbb{T}$ where the dissociated sequence is a sequence of integers $\{\lambda_n\}$ satisfying the strong Hadamard lacunarity $\lambda_{n+1} \geq 3\lambda_n$.

According to a famous Men'shov-Rademacher theorem, $\sum |\alpha_n|^2 \log^2 n < \infty$ is a sufficient condition for the series converges μ_a-almost everywhere. By introducing a random Riesz product and using the Peyrière measure, we prove the following theorem.

Theorem 5.1 ([39, 11]). *Let $S_N(t)$ be the N-th partial sum of the above series. For any $1 < p < \infty$, we have the maximal inequality*

$$\|\max_{N\geq 1} S_N(t)\|_{L^p(\mu_a)} \leq C\frac{p^{3/2}}{p-1}\sqrt{\sum_{n=1}^{\infty} |\alpha_n|^2}$$

Consequently, the series converges μ_a-almost everywhere iff $\sum |\alpha_n|^2 < +\infty$.

J. Peyrière [39] obtained the result for the group $\mathbb{T}$ with $p = 2$ by using a combinatorial method. The method of multiplicative chaos is powerful. We introduce the Riesz product

$$\mu_{a,\omega} = \prod_{n=1}^{\infty} (1 + \text{Re } a_n\gamma_n(t + \omega_n)). \tag{5.2}$$

It is a random Riesz product if we assume that the phases $\{\omega\}$ are i.i.d. variables with Haar distribution. Consider $P_n(t) = 1 + \text{Re } a_n\gamma_n(t + \omega_n)$ $(n \geq 1)$ as weights which produce a multiplicative chaos. The random Riesz product $\mu_{a,\omega}$ is nothing but the image of the Haar measure under the multiplicative chaos operator. One of our observations used in the proof is that $t + \omega_n$ $(n \geq 1)$ are independent with respect to the Peyrière measure.

5.2. A generalization

Let $\{f_n\}$ be a sequence of 1-periodic non-negative functions, $\{\lambda_n\} \subset \mathbb{N}$ be a lacunary sequence and $\{\omega_n\} \subset [0,1]^{\mathbb{N}}$. We may consider the infinite product

$$\prod_{n=1}^{\infty} f_n(\lambda_n(t + \omega_n))dt.$$

Under some regularity of f_n's, the product does define a measure [13]. See [2] for recent study of such products.

6. Unsolved problems

We end the paper by mentioning some problems.

Random translation. Let $\{f_n\}$ be a sequence of positive functions in $L^\infty(\mathbb{T})$. Find conditions on $\{f_n\}$ such that

$$a.s. \ \ \forall\, t \in \mathbb{T} \ \ \sum_{n=1}^{\infty} f_n(t - \omega_n) = \infty.$$

Theorem 2.6 deals with the case $f_n(t) = a_n 1_{(0,\ell_n)}(t)$.

Dynamical translation. Let $S : G \to G$ be a transformation on a compact abelian group G which preserves the Haar measure λ. Let $\{g_n\}$ be a sequence of positive functions in $L^\infty(G)$. Find conditions on $\{g_n\}$ such that

$$\text{for } \lambda\text{−almost all } x, \ \ \forall\, t \in G \ \sum_{n=1}^{\infty} g_n(t - S^n x) = \infty.$$

α-model. Suppose $P(W = a) = p$ and $P(W = b) = 1 - p$ with $0 < p < 1, a > 0, b > 0$ and $ap + b(1-p) = 1$. Determine $\mathrm{Im}\mathbb{E}Q_W$ and $\mathrm{Ker}\mathbb{E}Q_W$. When $b = 0$ we get the β-model which is well studied (See Theorem 3.5).

Multifractal analysis of $Q_W\mu$. Give a multifractal analysis for $Q_W\mu$ when μ is a Markov measure. The case of Lebesgue measure is well studied (see §4.2.).

Invariance of absolute continuity of Riesz product. Let μ_a and $\mu_{a,\omega}$ be two Riesz products defined respectively by (5.1) and (5.2). Let μ_b and $\mu_{b,\omega}$ be two Riesz products similarly defined. We conjecture that for a fixed ω, $\mu_a \ll \mu_b$ implies $\mu_{a,\omega} \ll \mu_{b,\omega}$. A positive answer will lead to a condition characterizing $\mu_a \ll \mu_b$, which may be stated as in Theorem 1.7.

References

[1] J. Barral, *Moments, continuité, et analyse multifractale des martingales de Mandelbrot*, Probab. Theory Relat. Fields **113** (1999), 535–569.

[2] J. Barral, M.O. Coppens and B. Mandelbrot, *Multiperiodic multifractal martingale measures*, preprint.

[3] J. Barral and A.H. Fan, *Covering numbers in the Dvoretzky covering*, Prépublication, University of Picardie (2003).

[4] P. Billard, *Séries de Fourier aléatoirement bornées, continues, uniformément convergentes* Ann. Scient. Ec. Norm. Sup. **82** (1965), 131–179.

[5] A. Dvoretzky, *On covering a circle by randomly placed arcs* Pro. Nat. Acad. Sci. USA **42** (1956), 199–203.

[6] K.J. Falconer, *The multifractal spectrum of statistically self-similar measures*, J. Theor. Prob. **7** (1994), 681–702.

[7] A.H. Fan, *Chaos additifs et chaos multiplicatifs de Lévy*, C. R. Acad. Sci., Paris, Sér.I. **308** (1989), 151–154.

[8] A.H. Fan, *Recouvrement aléatoire et décompositions de mesures*, Publication d'Orsay 1989.

[9] A.H. Fan, *Sur quelques processus de naissance et de mort*, C. R. Acad. Sci., Paris, Sér.I. **310** (1990), 441–444.

[10] A.H. Fan, *Equivalence et orthogonalité des mesures aléatoires engendrées par martingales positives homogènes*, Studia Math. **98 (3)** (1991), 249–266.

[11] A.H. Fan, *Quelques propriétés de produits de Riesz*, Bull. Sci. Math. **117 (3)** (1993), 421–439.

[12] A.H. Fan, *Sur le chaos de Lévy d'indice* $0 < \alpha < 1$, Ann. Sc. Math. Québec **21** No. 1 (1997), 53–66.

[13] A.H. Fan, *Multifractal analysis of infinite products*, J. Stat. Phys., **86** nos. 5/6 (1997), 1313–1336.

[14] A.H. Fan, *On Markov-Mandelbrot martingales*, J. Math. Pues Appli. **81** (2002), 967–982.

[15] A.H. Fan and J.P. Kahane, *Rareté des intervalles recouvrant un point dans un recouvrement aléatoire*, Ann. Inst. Henri Poincaré **29** N°. 3 (1993), 453–466.

[16] A.H. Fan and J.P. Kahane, *Decomposition principle in the theory of multiplicative chaos*, Bull. London Math. Soc., to appear.

[17] P.J. Fitzsimmons, B.E. Fristedt and L.A. Shepp, *The set of real numbers left uncovered by random covering intervals*, Z. Wahrsch. Verw. Gebiete **70** (1985), 175–189.

[18] S. Graf, R.D. Mauldin and S.C. Williams, *The exact Hausdorff dimension in random recursive constructions*, Mem. Amer. Math. Soc. **71** No. 381, 1988.

[19] S. Janson, *Random coverings of the circle with arcs of random lengths*, Probability and mathematical statistics, 62–73, Uppsala Univ., Uppsala, 1983.

[20] R. Holley and E.C. Waymire, *Multifractal dimensions and scaling exponents for strongly bounded random cascades*, Ann. Appl. Probab. **2** (1992), 819–845.

[21] J.P. Kahane, *Some series of functions*, Cambridge University Press, 1985.

[22] J.P. Kahane, *Sur le chaos multiplicatif*, Ann. Sciences Math. Québec **9** (1985), 105–150.

[23] J.P. Kahane, *Positive martingales and random measures*, Chin. Ann. Math. **8B** (1) (1987), 1–12.

[24] J.P. Kahane, *Intervalles aléatoires et décomposition des mesures*, C. R. Acad. Sci., Paris, Sér.I. **304** (1987), 551–554.

[25] J.P. Kahane, *Random covering and multiplicative processes*, in Fractal Geometry and Stochastics II, pp. 125–146, Progress in Probability vol. 46, Ed. Ch. Bandt, S. Graf and M. Zähle, Birkhäuser Verlag, 2000.

[26] J.P. Kahane and J. Peyrière, *Sur certaines martingales de Mandelbrot*, Advances in Math. **22** (1976), 131–145.

[27] J.P. Kahane and R. Salem, *Ensembles parfaits et séries trigonométriques*, Paris, Hermann 1963.

[28] A.N. Kolmogorov, *Précisions sur la structure locale de la turbulence dans un fluide visqueux aux nombres de Reynolds élevés*, Mécanique de la trubulence, Colloq. intern. CNRS Marseille 1961, Editions CNRS 1962, 447–451.

[29] Q. Liu, *Sur une équation fonctionnelle et ses applications: une extension du théorème de Kesten-Stigum concernant des processus de branchement*, Adv. Appl. Prob. **29** (1997), 353–373.

[30] R. Lyons, *Random walks and percolation on trees*, Ann. Probab. **18** (1990), 931–958.

[31] R. Lyons, *Random walks, capacity and percolation on trees*, Ann. Probab. **20** (1992), 2043–2088.

[32] B. Mandelbrot, *Possible refinement of the log-normal hypothesis concerning the distribution of energy dissipation in intermittent trubulence in statistical models and trubulence*, Symposium at U. C. San Diego 1971, Lecture Notes in Physics, Springer-Verlag 1972, 333–351.

[33] B. Mandelbrot, *Renewal sets and random cutouts*, Z. Wahrsch. v. Geb. **22** (1972), 145–157.

[34] B. Mandelbrot, *Multiplications aléatoires itérées et distributions invariantes par moyennes pondérée aléatoire*, C. R. Acad. Sci. Paris **278** (1974), 289–292.

[35] B. Mandelbrot, *Multiplications aléatoires itérées et distributions invariantes par moyennes pondérée aléatoire: quelques extensions*, C. R. Acad. Sci. Paris **278** (1974), 355–358.

[36] G.M. Molchan, *Scaling exponents and multifractal dimensions for independent random cascades*, Commun. Math. Phys. **179** (1996), 681–702.

[37] L. Olsen, *Random geometrically graph directed self-similar multifractals*, Pitman Res. Notes Math. Ser., Vol **307**, 1994.

[38] Y. Peres, *Probability on Trees: An Introductory Climb*, Lecture Notes in Math **1717** (1999), 193–280.

[39] J. Peyrière, *Almost everywhere convergence of lacunary trigonometric series with respect to Riesz products*, Australian J. Math., (Series A) **48** (1990), 376–383.

[40] J. Peyrière, *Recent results on Mandelbrot multiplicative cascades*, in Fractal Geometry and Stochastics II, pp. 147–159, Progress in Probability vol. 46, Ed. Ch. Bandt, S. Graf and M. Zähle, Birkhäuser Verlag, 2000.

[41] L. Shepp, *Covering the circle with random arc*, Israel J. Math. **11** (1972), 328–345.

[42] L. Shepp, *Covering the line with random arc*, Z. Wahrsch. v. Geb. **23** (1972), 163–170.

[43] B. von Bahr and C.-G. Esseen, *Inequalities for the rth absolute momen of sum of random variables,* $1 \le r \le 2$, Ann. Math. Statist. **36** (1965), 299–303.

Ai Hua Fan
CNRS UMR 6140
Faculté de Mathématiques et Informatique
Unversité de Picardie
80039 Amiens cedex
France
e-mail: `ai-hua.fan@u-picardie.fr`

Progress in Probability, Vol. 57, 135–150

Intersection Exponents and the Multifractal Spectrum for Measures on Brownian Paths

Peter Mörters

Abstract. The aim of this paper is to give a survey of recent developments in the multifractal analysis of measures arising in the study of Brownian motion. We first discuss the harmonic measure on a Brownian path, studied by Lawler, and the intersection local time on the intersection of two independent Brownian paths, studied by Klenke and Mörters. In both examples the *intersection exponents* of Brownian motion play a crucial role in the multifractal analysis. Whereas the first example is in line with the multifractal formalism, the second example is not and more subtle effects come into the picture. Then we discuss the occupation measure of Brownian motion, in which case a *refined* multifractal analysis was carried out by Dembo, Peres, Rosen and Zeitouni.

Mathematics Subject Classification (2000). Primary 60J65; Secondary 28A80.

Keywords. Brownian motion, occupation measure, harmonic measure, intersection local time, critical exponent, intersection exponent, multifractal spectrum, multifractal formalism, exceptional set, large deviation heuristic.

1. Introduction

An interesting fractal geometric characteristic for discrete and continuous models of statistical physics is the *multifractal spectrum*, which originated in the study of turbulence models, see e.g. [33]. Loosely speaking, the multifractal spectrum evaluates the degree of variation in the local intensity of a spatial distribution, or, in other words, it describes quantitatively how frequent a certain type of local irregularity appears in a fractal measure, set or function.

There are many different ways in which this concept can be made precise, see for example the contribution of Lévy-Vehel in this volume. In this paper I would like to focus on the multifractal spectrum as discussed e.g. in [43, 13, 40, 4], where irregularities are described in terms of local dimensions and the size of sets is measured by the Hausdorff dimension. There is a host of literature on this particular type of spectrum spanning almost two decades, the reader is referred to the survey [35] and the extensive list of references given there.

To give the definition of this type of multifractal spectrum, suppose that μ is a (fractal) measure on $\mathbb{R}^d$. One can define the *local dimension* of the measure μ at the point $x \in \mathbb{R}^d$ as

$$\lim_{r\downarrow 0} \frac{\log \mu(B(x,r))}{\log r} = a, \tag{1.1}$$

if the limit exists. In words, this is the rate of decay of the measure $\mu(B(x,r))$ as the concentric balls $B(x,r)$ of radius $r \downarrow 0$ zoom into the point x. The value $f(a)$ of the multifractal spectrum is the Hausdorff dimension of the set of points x with local dimension a, i.e.

$$\dim \Big\{ x \in \mathbb{R}^d \, : \, \lim_{r\downarrow 0} \frac{\log \mu(B(x,r))}{\log r} = a \Big\}. \tag{1.2}$$

In many cases of interest, the limit in (1.1) has to be replaced by liminf or limsup to obtain an interesting nontrivial spectrum. A measure is called *multifractal* if $f(a) > 0$ for more than one value of $a \in \mathbb{R}$, in any of these definitions.

Whereas much of the earlier literature on multifractal theory was focused on the analysis of (mostly deterministic) objects with strong features of self-similarity, and the verification of the multifractal formalism proposed, e.g. in [13], the interest has shifted more recently towards the analysis of stochastic processes by means of multifractal spectra. This trend can be seen in applied fields, like the analysis of time series, see for example [15, 42], as well as in mathematical physics, see [10] for a survey from a physics point of view. In the probability context new patterns have emerged and a lot of interest is going towards measures where the multifractal formalism fails, and towards refinements of the multifractal formalism.

In the present paper we shall review these developments by considering three natural measures on the path of a Brownian motion in dimension $d = 3$.[1] These examples are

- the *harmonic measure* on a Brownian path, studied by Lawler [24, 25, 26],
- the *intersection local times* on the intersection of two independent Brownian path, studied by Klenke and Mörters [18],
- the *occupation measure* on a Brownian path, studied by Dembo, Peres, Rosen and Zeitouni [6, 7], see [5] for an introduction.

The main emphasis will be on the first two examples, as they give us also the opportunity to highlight another important subject, which has emerged in recent years in fractal geometry, namely the relation of Hausdorff dimension and the critical exponents of statistical physics, in this case the *intersection exponents*.

We start the paper with an introduction of intersection exponents in the next section, followed by three sections discussing our three examples. A final section tries to wrap up the developments described in this paper, hinting also at some expectations for future developments.

---

[1] The focus is on $d = 3$ for better comparison, though some results extend to other dimensions.

2. Intersection exponents

The intersection exponents for random curves measure how likely these curves do not intersect, and maybe should therefore be called *non*-intersection exponents. They contain crucial information about the fractal geometry of the curves, and substantial recent research is devoted to exploiting this information.

To define the intersection exponents for Brownian motion in $d = 2, 3$, suppose $n, m \geq 1$ are integers and let $W_1, \ldots, W_{n+m}$ are independent Brownian motions in $\mathbb{R}^d$ started uniformly on $\partial B(0,1)$ and running up to the first exit times $T^i(r)$ from a large ball $B(0,r)$. We divide the motions into two packets and look at the union of the paths in each family,

$$\mathfrak{B}_1(r) = W_1[0, T^1(r)] \cup \ldots \cup W_n[0, T^n(r)] \text{ and}$$
$$\mathfrak{B}_2(r) = W_{n+1}[0, T^{n+1}(r)] \cup \ldots \cup W_{n+m}[0, T^{n+m}(r)].$$

The event that the two packets of Brownian paths fail to intersect has a decreasing probability as $r \uparrow \infty$. Indeed, it is easy, using subadditivity, to show that there exists a constant $0 < \xi_d(n,m) < \infty$ such that

$$\mathbb{P}\{\mathfrak{B}_1(r) \cap \mathfrak{B}_2(r) = \emptyset\} = r^{-\xi_d(n,m)+o(1)}, \text{ as } r \uparrow \infty. \tag{2.1}$$

The numbers $\xi_d(n,m)$ are called the *intersection exponents.* It is not hard to see that this definition is very robust under changes in the way the motions are started, for example if one starts all motions in a packet in the same fixed point (2.1) continues to hold if the starting points of the packets do not agree.

In fact, the approximation of non-intersection probabilities is stronger than (2.1) suggests. Lawler [23] has shown that there exists constants $0 < c < C < \infty$ such that

$$cr^{-\xi_d(n,m)} \leq \mathbb{P}\{\mathfrak{B}_1(r) \cap \mathfrak{B}_2(r) = \emptyset\} \leq Cr^{-\xi_d(n,m)} \text{ for all } r > 1.$$

The definition of intersection exponents can be extended naturally to one *real* argument. For this purpose define the random variable

$$Z_n(r) = \mathbb{P}\{\mathfrak{B}_1(r) \cap W_{n+1}[0, T^{n+1}(r)] = \emptyset \mid W_1, \ldots, W_n\}.$$

Again one can show that, for every $\lambda > 0$, there exists a constant $\xi_d(n, \lambda)$ such that

$$\mathbb{E}\{Z_n(r)^\lambda\} = r^{-\xi_d(n,\lambda)+o(1)}, \text{ as } r \uparrow \infty. \tag{2.2}$$

Clearly, this is consistent with the definition of $\xi_d(n, \lambda)$ in the case of an integer λ.

Finally, a natural way to extend the definition to the case $\lambda = 0$ is by letting

$$\mathbb{P}\{Z_n(r) > 0\} = r^{-\xi_d(n,0)+o(1)}, \text{ as } r \uparrow \infty.$$

Note that $Z_n(r) = 0$ is the event that the packet of n Brownian motion disconnects the origin from infinity, so that any further Brownian motion cannot reach $\partial B(0,r)$ without intersecting the packet. Therefore the exponents $\xi_d(n, 0)$ are also called *disconnection exponents.*

The first researchers interested in the values of the intersection exponents were physicists. They have made the striking conjecture that in the plane many of these exponents are rational numbers, see e.g. Duplantier and Kwon [11] and [10] for a survey. Very recently, Lawler, Schramm and Werner, in a seminal series of papers [28, 29, 30] have been able to verify this rigorously, see [27] for a survey over the series. They have shown that

$$\xi_2(n,\lambda) = \frac{\left(\sqrt{24n+1}+\sqrt{24\lambda+1}-2\right)^2-4}{48}, \quad \textit{for } n \in \mathbb{N}, \lambda \geq 0. \tag{2.3}$$

The proof of (2.3) is based on conformal invariance, and the *stochastic Löwner evolution (SLE).* I shall not go further into the exciting recent developments about SLE, as our examples are in $d = 3$, where no analogue of the stochastic Löwner evolution exists.

Indeed, in $d = 3$ we do not know the values of the intersection exponents explicitly, and it is very likely that they are typically irrational. I briefly review what is currently known about the values of the intersection exponents $\xi_d(n,\lambda)$. The two values explicitly known in dimension $d = 3$ are

- $\xi_3(1,2) = \xi_3(2,1) = 1$. This follows from properties of the harmonic measure, and an intuitive argument for this will be a spin-off of the next section. For a rigorous and accessible argument see [20, Chapter 3].
- $\xi_3(n,0) = 0$ as no disconnection is possible.

Estimates for other values can be derived from this using (2.2), monotonicity and concavity of $\lambda \mapsto \xi_3(n,\lambda)$. A deep result of Lawler [23] shows that this function is even *strictly concave.* As an example we get for $\xi_3(2,2)$ the estimates

$$\xi_3(2,2) > \xi_3(2,1) = 1 \text{ by monotonicity and strict concavity,}$$

and

$$\xi_3(2,2) < 2\xi_3(2,1) = 2 \text{ by strict concavity.}$$

The project of exploiting intersection exponents as a means to describe the fractal geometry of Brownian motion and random walks was pioneered by Lawler [21, 22]. To name two examples, the set of *cut points* of a Brownian path $B[0,1] \subset \mathbb{R}^d$ is the set of points $x = B(t) \in B[0,1]$ such that $B[0,t) \cap B(t,1] = \emptyset$. Lawler [21] showed that the Hausdorff dimension of this set is $2 - \xi_d(1,1)$. Intersection exponents are also crucial in the solution of the famous *Mandelbrot conjecture* about the Brownian frontier, the boundary of the unbounded component of the complement of a planar Brownian curve. Lawler showed in [22] that the dimension of this set is $2 - \xi_2(2,0)$ and together with the value of this exponent obtained 5 years later this shows that the dimension of the Brownian frontier is $4/3$, as predicted by Mandelbrot in 1983. Our first example below can be seen as a multifractal version of this result (however presented here in $d = 3$).

3. The multifractal spectrum of harmonic measure

Suppose $A \subset \mathbb{R}^3$ is a compact, nonpolar set. The harmonic measure on the set A is defined as follows. Suppose $B(y,R)$ is a ball such that A is contained inside this ball. Let $W : [0,\infty) \to \mathbb{R}^d$ be a Brownian motion started in a point chosen uniformly from the sphere $\partial B(y,R)$ and let

$$T = \inf\{t > 0 : W_t \in A\}$$

be the first hitting time of the set A by a Brownian motion. The law of $W(T)$ can be shown to be independent of the choice of $B(y,R)$. The *harmonic measure* h_A on A is defined by

$$h_A(B) = \mathbb{P}\{W_T \in B, T < \infty\} \text{ for all Borel sets } B \subset A.$$

Note that the support of the harmonic measure is contained in the boundary of the unbounded components of A^c, called the *frontier* of A.

Now we look at $A = B[0,1] \subset \mathbb{R}^d$, the path of a Brownian motion and study the harmonic measure $h = h_A$ on this set. The following result about the multifractal spectrum of the measure h_A is due to Lawler [25, 26].

Theorem 3.1 (Lawler 1999). *Define a convex function ξ^* as the Legendre transform of the the strictly concave function $\lambda \mapsto \xi_3(2,\lambda)$, i.e.*

$$\xi^*(a) = \sup_{\lambda>0} \{\xi_3(2,\lambda) - a\lambda\}.$$

Then, for every a such that $2 - \xi^*(a-1) > 0$, almost surely there exist points $x \in A$ with

$$\lim_{r\downarrow 0} \frac{\log h(B(x,r))}{\log r} = a.$$

Moreover, for these values of a, almost surely,

$$\dim\Big\{x \in A : \lim_{r\downarrow 0} \frac{\log h(B(x,r))}{\log r} = a\Big\} = 2 - \xi^*(a-1).$$

Remark. The fact that this spectrum is nontrivial is due to the *non-mean field* nature of the intersections, expressed in the *strict* concavity of $\xi_3(2,\cdot)$. ◇

Following [26] I now give a very rough heuristic argument for Theorem 3.1, which is based on the so-called *multifractal formalism* of [13]. Let

$$A(j,n) = B\Big[\frac{j-1}{e^{2n}}, \frac{j}{e^{2n}}\Big], \text{ for } 1 \le j \le e^{2n}.$$

Consider the quantity

$$Q(n,q) = \sum_{j=1}^{e^{2n}} h(A(j,n))^q.$$

This is a random variable, which is intimately related to the multifractal spectrum. Indeed, for large n, we can roughly assume that $A(j,n)$ is the intersection of the Brownian path A with a ball of radius e^{-n} and if the local dimension of the points in this ball is approximately a, then $h(A(j,n)) \approx e^{-an}$. Hence, if we denote by

$$\mathfrak{E}(a) = \Big\{1 \le j \le e^{2n} : h(A(j,n)) \approx e^{-an}\Big\},$$

we can write

$$Q(n,q) = \sum_a \#\mathfrak{E}(a) e^{-qan},$$

where the sum extends over an appropriate discretisation of the reals. The Laplace method suggests that, for

$$f(a) = \lim_{n\to\infty} \frac{1}{n} \log\left(\#\mathfrak{C}(a)\right)$$

we have

$$Q(n,q) \approx \exp\left(n \sup_a \{f(a) - qa\}\right).$$

Suppose now that we can show that, for some function τ,

$$Q(n,q) \approx \exp(n\tau(q)). \tag{3.1}$$

Then we get,

$$\sup_a \{f(a) - qa\} = \tau(q).$$

As the sets in $\mathfrak{C}(a)$ roughly cover the set of points of local dimension a, the quantity $f(a)$ is at least an upper bound for the dimension spectrum. If everything is sufficiently regular, this bound is sharp. To identify f note that

$$\tau(q) = \sup_a \{f(a) - qa\} = f^*(q).$$

If f is concave, we may invert this Legendre transform and get

$$f(a) = \tau^*(a) = \inf_{q>0} \{\tau(q) + qa\}. \tag{3.2}$$

Hence it boils down to finding the τ in (3.1).

Here the intersection exponents are coming into play. We sketch the argument for the upper bound for τ, which is based on the first moment method. First fix the set $A = B[0,1]$ and look at the harmonic measure of a small ball,

$$h_A(B(x,r)) = P\{W \textit{ hits } A \textit{ first in } B(x,r)\} \textit{ for } x \in A.$$

The probability that the path W hits $B(x,r)$ is $cr^{d-2} = cr$ for some constant $c > 0$. Given that W hits $B(x,r)$, we can look at the time-reversed path starting on the boundary $\partial B(x,r)$ and stopped upon hitting $\partial B(y,R)$. Though this path is strictly speaking no Brownian path, a sufficiently large part of it is equal to a Brownian path conditioned to exit the annulus $B(y,R) \setminus B(x,r)$ at $\partial B(y,R)$. This is good enough for our purpose, and hence the reversed path heuristically makes the *first packet* $\mathcal{B}_1'(r)$ consisting of one Brownian motion. Considering A as a random set, its distribution given $x \in A$ essentially coincides with a two-sided Brownian motion started in x, forming the *second packet* $\mathcal{B}_2'(r)$ of two independent Brownian motions. The Brownian motions in these packets are started on the sphere $\partial B(x,r)$ and stopped when they leave the fixed sphere $\partial B(y,R)$, and by Brownian scaling we easily map this onto the situation in the definition of the intersection exponents, where motions are started on the unit sphere and stopped upon leaving a large sphere of radius R/r. Hence we have

$$\mathbb{E}\{h_A(B(x,r))\} = cr\, P\{\mathcal{B}_1'(r) \cap \mathcal{B}_2'(r) = \emptyset\} = r^{1+\xi_3(2,1)+o(1)},$$

and more generally,

$$\mathbb{E}\big\{h_A(B(x,r))^q\big\} \approx r^{q+\xi_3(2,q)+o(1)}. \tag{3.3}$$

This gives

$$\mathbb{E}Q(n,q) \approx \exp\big(n(2-q-\xi_3(2,q)+o(1))\big).$$

By the Borel-Cantelli lemma we get from this that

$$Q(n,q) \le \exp\big(n(2-q-\xi_3(2,q)+o(1))\big)$$

for all sufficiently large n. Hence

$$\tau(q) \le 2-q-\xi_3(2,q).$$

To obtain the lower bound we have to add a second moment estimate to our first moment estimate (3.3). This is considerably harder, and after substantial work one arrives at

$$\tau(q) = \lim_{n\to\infty} \frac{1}{n} \log Q(n,q) = 2-q-\xi_3(2,q),$$

and hence, from (3.2),

$$f(a) = 2 - \sup_{q>0} \big\{\xi_3(2,q) - q(a-1)\big\}.$$

Remark. As the total harmonic measure is constant, we have $\tau(1)=0$ and hence our heuristics gives that $\xi_3(2,1) = 1$. For a rigorous argument using this idea, see [20]. ◇

4. The multifractal spectrum of intersection local time

Let W_1, W_2 be two independent Brownian motions in $\mathbb{R}^3$ with a joint starting point and running for one unit of time. Let $S = W_1[0,1] \cap W_2[0,1]$ be the intersection of the paths, which is a (random) set of Hausdorff dimension 1. This set can be equipped with a finite measure, the intersection local time ℓ, the *(projected) intersection local time*, which can be described symbolically by the formula

$$\ell(A) = \int_A dy \prod_{j=1}^{2} \int_0^1 ds\, \delta_y(W_j(s)), \ \textit{for } A \subset \mathbb{R}^d \textit{ Borel}. \tag{4.1}$$

Le Gall has shown in [31, 32] that ℓ is the φ-Hausdorff measure on the set S for the gauge function

$$\varphi(x) = cx \log\log(1/x)^2, \ \textit{for } x > 0,$$

where c is an unknown deterministic constant. Alternative rigorous constructions of the random measure ℓ are reviewed in [19]. It is well-known that the local dimension of ℓ-almost all points equals 1, which is also the Hausdorff dimension of the intersection set S.

Applying a multifractal formalism as in the previous example suggests that the measure ℓ is *not* multifractal, as the τ-function is linear. Indeed, points where the local dimension is strictly below the typical value 1 do not exist, even in the weaker sense of a liminf in (1.1). In other words there is no nontrivial multifractal spectrum to the left of 1.

In a recent paper, Klenke and Mörters [18] are investigating whether there are points where the local dimension exceeds 1. It turns out that this is indeed the case, if the local dimension is defined in the weaker sense of a limsup in (1.1). These exceptional points arise, loosely speaking, from random fluctuations in the intersection local time which are too subtle to be picked up by the multifractal formalism.

Theorem 4.1 (Klenke and Mörters 2003). *Suppose ℓ is the intersection local time of two Brownian motions in $\mathbb{R}^3$. For any $1 \le a \le \xi/(\xi-1)$, almost surely, there exist points $x \in S$ such that*

$$\limsup_{r \downarrow 0} \frac{\log \ell(B(x,r))}{\log r} = a,$$

for $\xi = \xi_3(2,2)$. Moreover, for these values of a, almost surely,

$$\dim\Big\{ x \in S \,:\, \limsup_{r\downarrow 0} \frac{\log \ell(B(x,r))}{\log r} = a \Big\} = \frac{\xi}{a} + 1 - \xi. \tag{4.2}$$

Remark. Here the multifractal spectrum is strictly *convex*, hence it cannot be found by means of the multifractal formalism of the last section, which by its nature always predicts a concave spectrum. ◇

In the following I give a heuristic argument for this result. The crucial ingredient in the proof is an investigation of the *strategy* used by Brownian motion in order to accumulate only a very small amount of mass in a small ball centred in an intersection point $x \in S$.

In order to find the probability of the event $\ell(B(x,r)) \approx r^a$ given $x \in S$ first note that it is intuitively clear that in a typical intersection point $x \in S$ a pair of two-sided Brownian motions are attached. As the motions start in the same point they typically have a lot of intersections right at the start. In order to avoid intersections to the extent that $\ell(B(x,r)) \approx r^a$ it is optimal for the Brownian motions to use the following strategy:

(1) Run normally until you hit $\partial B(x,r^a)$. The typical intersection local time until then is $\approx r^a$, and now one has to act.

(2) Do not intersect before leaving $B(x,r^{1-\varepsilon})$. The probability that this happens is, by scaling and the definition of the intersection exponents,
$$r^{-\xi(1-\varepsilon-a)+o(1)}.$$

(3) Once you reach $\partial B(x,r^{1-\varepsilon})$ never return to $B(x,r)$. The probability that this happens is $\approx 1 - r^{\varepsilon}$.

The total probability (or "cost") of this strategy is of order $r^{-\xi(1-a)+o(1)}$, essentially all of which is incurred in the second step. Note that it leads to an empty annulus $B(x,r) \setminus B(x,r^a)$. It can be shown that this strategy is optimal, i.e. no other strategy to get $\ell(B(x,r)) \approx r^a$ has a significantly higher probability.

We now sketch how the upper bound in our spectrum can be deduced from this. As $\dim S = 1$ we can cover the set S by approximately r^{-a} balls of radius r^a. Using the modulus of continuity result for Brownian motion one can see that, in a ball of radius r^{-a}, the probability that the ball contains a point $x \in S$ with $\ell(B(x,r)) \approx r^a$ is approximately the same as the probability that the centre of the ball has this property. Hence, for each ball in our collection, the probability of containing a point $x \in S$ with $\ell(B(x,r)) \approx r^a$ is approximately $r^{-\xi(1-a)}$. Choosing a covering of the set

$$\Big\{x \in S \,:\, \limsup_{r\downarrow 0} \frac{\log \ell(B(x,r))}{\log r} \geq a\Big\}$$

consisting of all these balls for the radii $r = 2^{-n}, n \geq m$ and m fixed, leads to a covering

$$B(x_1, r_1^a), B(x_2, r_2^a), B(x_3, r_3^a), \ldots$$

with expected s-value

$$\mathbb{E}\Big\{\sum_{i=1}^{\infty} (r_i^a)^s\Big\} = \sum_{n=m}^{\infty} 2^{-nas} 2^{na} 2^{n\xi(1-a)},$$

which is finite iff $s > \xi/a + 1 - \xi$. Note that this argument for the upper bound is particularly simple as it uses only a *first moment method.*

It is more difficult to establish a lower bound, as second moments (or independence in some form) have to be used. A particularly powerful technique to establish lower bounds is the *stochastic co-dimension* method. Based on earlier ideas of Taylor these techniques were developed further, first in [1, 38] and later more systematically by Peres and coauthors, see e.g. [17, 8] or [19]. Nowadays the method represents an elegant and powerful tool in fractal geometry, further evidence of this can be found in the contribution of Xiao in this volume.

The stochastic co-dimension approach suggests that the dimension of a set $A \subset \mathbb{R}^d$ is at least γ if it intersects a certain compact random set $\Gamma[\gamma]$ of dimension $d - \gamma$ with positive probability, and conversely if $\dim A > \gamma$ then A intersects the test set $\Gamma[\gamma]$ with positive probability. Suitable *test sets* are, for example, the *percolation limit sets* with retention parameter $2^{-\gamma}$, see [12] and also [37].

In our proof we assume that the test sets are independent of our Brownian motions and construct, for any $\gamma \in (0,1)$, a compact random set $S^* \subset S \cap \Gamma[\gamma]$, which is nonempty with positive probability, such that

$$\dim U \cap S^* \geq 1 - \gamma \textit{ whenever } U \textit{ open and } U \cap S^* \neq \emptyset.$$

Fix $\delta > 0$ and define open sets

$$U(m,\delta) = \Big\{x \in S \,:\, \ell\big(B(x,r)\big) < r^{a-\delta} \textit{ for some } 0 < r < \tfrac{1}{m}\Big\}.$$

The regularity of S^* is then used to show that for $\gamma = \xi/a + 1 - \xi$ the sets $U(m,\delta) \cap S^*$ are dense in S^*. To get an impression of this, fix $r = 2^{-n}$ and note that given $S^* \neq \emptyset$ by the regularity property, for every open set U which intersects S^* there is a packing of $S^* \cap U$ consisting of approximately $r^{\gamma-1}$ balls of radius r. We now *localise* the problem and look at each ball in the packing separately. For both motions let S be the last entry of the path into the ball before hitting the centre and T the first exit from the ball afterwards. We denote by ℓ^* the intersection local time of the paths in the time intervals $[S,T]$. Then the probability that inside our ball there is some $x \in S^*$ such that

$$\ell^*\big(B(x,r)\big) < r^{a-\delta} \tag{4.3}$$

is of order

$$r^{-\xi(1-a+\delta)} r^{(\gamma+2)(a-\delta-1)} r^{-3(a-\delta-1)}.$$

Here the first factor comes from the probability that a point x, and by continuity every point in a ball around x of radius $\approx r^{a-\delta}$, satisfies (4.3), the second factor comes from the requirement that this ball intersects S^*, and the third factor accounts for the number of such small balls in the fixed ball of radius r. Moreover, thanks to the localisation these events are essentially independent for different balls. Hence the probability that we do *not* find a point $x \in S^*$ satisfying (4.3) for $r = 2^{-n}$ is

$$\big(1 - 2^{n(\xi+\gamma-1)(1-a+\delta)}\big)^{2^{n(1-\gamma)}} \longrightarrow 0, \ \ as\ n \to \infty.$$

To complete the argument one has to show that for most points the localised intersection local time ℓ^* is not different from ℓ. In $d \geq 3$ this can be shown using the transience of Brownian motion

Now, as $U(m,\delta) \cap S^*$, for $m \in \mathbb{N}$, $\delta > 0$, are a nested family of open dense subsets of the complete metric space S^*, by Baire's theorem, the intersection is also dense, and hence nonempty if S^* is nonempty. Hence the intersection set

$$\mathcal{T}(a) := \bigcap_{\substack{m\in\mathbb{N}\\ \delta>0}} U(m,\delta) = \Big\{ x \in S \,:\, \limsup_{r\downarrow 0} \frac{\log \ell(B(x,r))}{\log r} \geq a \Big\}$$

intersects $\Gamma[\gamma]$ with positive probability. On the other hand almost surely $\mathcal{T}(a+\varepsilon) \cap \Gamma[\gamma] = \emptyset$ for all $\varepsilon > 0$ by the upper bound. This shows that (4.2) holds with positive probability, and the full result follows from a zero-one law argument.

5. The refined multifractal spectrum of occupation measure

The most natural measure on the Brownian path $\{W(t) : t \in [0,1]\}$ is of course the occupation measure μ, given by

$$\mu(A) = \int_0^1 1_A(W_s)\, ds.$$

We have neglected this measure so far in this paper, as it is not multifractal. Indeed, one can show that, almost surely,

$$\lim_{r\downarrow 0} \frac{\log \mu(B(x,r))}{\log r} = 2, \ \ for\ all\ x \in W[0,1].$$

This has been known for some time, for example it is implicit in the work of Kaufman [16]. As there is no variation in the local dimension of μ it is natural to look at a finer scale and ask for an almost sure upper and lower hull of $\mu(B(x,r))$ when the point x is varied along the support of μ. More precisely, looking at a the lower hull first, we are asking for a gauge function $\phi : (0,\varepsilon) \to (0,\infty)$ such that

$$0 < \inf_{x\in W[0,1]} \liminf_{r\downarrow 0} \frac{\mu(B(x,r))}{\phi(r)} < \infty.$$

A point $x \in W[0,1]$ is called a *thin point* with thinness $a > 0$ if

$$\liminf_{r\downarrow 0} \frac{\mu(B(x,r))}{\phi(r)} = a.$$

Once ϕ is identified, the *dimension spectrum of thin points* is given by

$$f(a) = \dim\Big\{x \in W[0,1] \,:\, \liminf_{r\downarrow 0} \frac{\mu(B(x,r))}{\phi(r)} = a\Big\}.$$

In [7] the correct gauge function is found, and the spectrum of thin points is identified for the occupation measure of a Brownian path in $d = 3$.

Theorem 5.1 (Dembo, Peres, Rosen and Zeitouni 2000). *Almost surely, we have*

$$1 = \inf_{x\in W[0,1]} \liminf_{r\downarrow 0} \frac{\mu(B(x,r))}{r^2[\log(1/r)]^{-1}},$$

and, for every $a \geq 1$, almost surely,

$$\dim\Big\{x \in W[0,1] \,:\, \liminf_{r\downarrow 0} \frac{\mu(B(x,r))}{r^2[\log(1/r)]^{-1}} = a\Big\} = 2 - \frac{2}{a}.$$

Passing to the upper hull, we are looking for a gauge function ψ such that

$$0 < \sup_{x\in W[0,1]} \limsup_{r\downarrow 0} \frac{\mu(B(x,r))}{\psi(r)} < \infty. \tag{5.1}$$

A point $x \in W[0,1]$ is called a *thick point* with thickness $a > 0$ if

$$\limsup_{r\downarrow 0} \frac{\mu(B(x,r))}{\psi(r)} = a,$$

and a *dimension spectrum of thick points* is defined accordingly. In [6] the correct gauge function is found, and the spectrum of thick points is identified for the occupation measure of a Brownian path in $d = 3$.

Theorem 5.2 (Dembo, Peres, Rosen and Zeitouni 2000). *Almost surely,*

$$\sup_{x\in W[0,1]} \limsup_{r\downarrow 0} \frac{\mu(B(x,r))}{r^2[\log(1/r)]} = \frac{16}{\pi^2}, \tag{5.2}$$

and, for all $0 \leq a \leq \frac{16}{\pi^2}$, almost surely,

$$\dim\Big\{x \in \mathbb{R}^d \,:\, \limsup_{r\downarrow 0} \frac{\mu(B(x,r))}{r^2[\log(1/r)]} = a\Big\} = 2 - \frac{\pi^2}{8}\,a. \tag{5.3}$$

For the proof of these two theorems the authors developed substantial refinements of the tools available for getting lower bounds for the Hausdorff dimension, some of which we have already benefited from in the previous section.

We only indicate briefly the key point in the proof of Theorem 5.2, which, as in Theorem 4.1, is the *strategy* used by the Brownian motion to accumulate exceptionally large mass in a small ball. For any $T > 0$ we look at the occupation measures μ_T for Brownian motion running for T time units,

$$\mu_T(A) = \int_0^T 1_A(W_s)\, ds, \text{ for } A \subset \mathbb{R}^3 \text{ Borel.}$$

Classical results of Ciesielski and Taylor give the upper tail behaviour of the random variable $\mu_\infty(B(0,1))$: For a positive constant $c > 0$,

$$\mathbb{P}\Big\{\mu_\infty(B(0,1)) \geq x\Big\} \sim c\, e^{-x\pi^2/8}, \quad \text{as } x \uparrow \infty, \tag{5.4}$$

hence, by scaling, we obtain for small balls $B(0,r)$ and fixed $a > 0$,

$$\mathbb{P}\Big\{\mu_\infty(B(0,r)) \geq ar^2 \log(1/r)\Big\} \sim c\, r^{a\pi^2/8}, \quad \text{as } r \downarrow 0.$$

The principal question is, which strategy is used by the Brownian path to accumulate an occupation time of order $r^2 \log(1/r)$ in $B(0,r)$. The answer found in [6] is that this mass is accumulated in a very short time interval after the start of the motion and eventual later returns to this ball do not contribute essentially. Indeed, for $h(r) = r^2 \log(1/r)^6$ they obtain

$$\mathbb{P}\Big\{\mu_{h(r)}(B(0,r)) \geq ar^2 \log(1/r)\Big\} \sim c\, r^{a\pi^2/8}, \quad \text{as } r \downarrow 0. \tag{5.5}$$

This *localisation effect* is the crucial effect which leads to Theorem 5.2, we note here that the effect breaks down in the recurrent case of $d = 2$, and in this dimension it is a complete different effect that leads to the dimension spectrum, which in this case was found in [8].

It is also worthwhile to look at a *refined* multifractal spectrum, when the measure under investigation has only a one-sided multifractal spectrum. This is the case, for example, in the situation of the intersection local time, studied in the previous section. Recall that in this case there exist points with local dimension exceeding the typical value 1 (these points are exceptionally *thin*!) but no points where the local dimension is smaller than 1. It therefore makes sense to ask for a dimension spectrum of thick points as defined above. This question is answered by König and Mörters in [19].

Theorem 5.3 (König and Mörters 2002). *Almost surely,*

$$\sup_{x\in S} \limsup_{r\downarrow 0} \frac{\ell(B(x,r))}{r[\log(1/r)]^2} = \frac{1}{\theta^2}, \tag{5.6}$$

for a constant θ given by

$$\theta = \inf\Big\{ \|\nabla\psi\|_2^2 \, : \, \psi \in C_0^2(\mathbb{R}^d), \int_{B(0,1)} \psi^4 = 1 \Big\}.$$

Moreover, for all $0 \le a \le \varrho^2/4$, almost surely,

$$\dim\Big\{ x \in S \, : \, \limsup_{r \downarrow 0} \frac{\ell(B(x,r))}{r[\log(1/r)]^2} = a \Big\} = 1 - \sqrt{a}\theta\,.$$

Finally, looking again at the occupation measure μ, recall that our definition of thickness requires large mass only along a sequence of radii $r_n \downarrow 0$. This sequence might be arbitrarily sparse. What happens if we replace the limsup in (5.1) by a liminf? In other words, we are looking for *consistently thick points.* A first task is to identify a gauge function $\varphi : (0,\varepsilon) \to (0,\varepsilon)$ such that

$$0 < \sup_{x \in W[0,1]} \liminf_{r \downarrow 0} \frac{\mu(B(x,r))}{\varphi(r)} < \infty.$$

Dembo, Peres, Rosen and Zeitouni have shown [6, Theorem 1.6] that the right gauge function here is $\varphi(r) = r^2$, which is different from the gauge function $\psi(r) = r^2 \log(1/r)$ in (5.2), but the problem of the dimension spectrum of consistently thick points is still *open.*

6. Summary and Outlook

By looking at three recently studied examples we have illuminated the following three important directions in modern fractal geometry.

- **Multifractal geometry beyond the multifractal formalism.** [9, 14, 39, 18] Interesting stochastic examples show that the multifractal formalism fails if the mass in typical small balls is a *heavy-tailed* random variable. In some instances we see a strictly convex spectrum, a behaviour which cannot be picked up by the formalism.
- **Refined multifractal spectrum.** [41, 6, 7, 8, 34, 19, 5] In some cases of measures which are not multifractal, or have only a one-sided multifractal spectrum, one can study a refined spectrum and get very detailed information about exceptional sets.
- **Critical exponents and fractal geometry.** [21, 25, 26, 27, 19] Critical exponents like the intersection exponents of Brownian motion contain crucial information about the fractal geometry of the models involved. In some cases this information can be extracted in terms of Hausdorff dimension spectra involving the exponents.

Trying to look into the future of the area, I believe that the following topics deserve attention before the next "Fractal Geometry and Stochastics" conference will reunite many researchers in several years time.

- **Fractal geometry of discrete models.**
  Almost all problems discussed here have a discrete analogue, and in some cases this has already been made explicit, see e.g. [25] for a discrete version of the multifractal spectrum of harmonic measure, and [8] for a discrete version of the refined multifractal spectrum of the occupation measure of a Brownian path, both in $d = 2$. In the latter case this discretisation solved a long-standing conjecture of Erdös and Taylor about the simple random walk in the plane. I believe that in the future fractal geometers will increasingly look into "discrete fractals" and apply their techniques to the fractal phenomena arising, for example, in interacting particle systems and statistical physics.
- **Increasingly complex dimension spectra.**
  Looking at recent results and techniques, for example, of the papers [7, 8] or [17], one cannot help noticing the enormous improvement in the technical standard and the sophistication of the tools presently available. However, I do not believe that an end to this development is reached yet, as important challenges remain open. Among these is the problem of consistently thick points for Brownian motion in $d \geq 3$, mentioned in Section 5, and the problem of the dimension spectrum of local times for Brownian motion in the plane raised in [2].
- **Fractal geometry of nonintersecting curves.**
  The stochastic Löwner evolution introduces a new class of universal non-intersecting random curves in the plane, which arise naturally as scaling limits of important physical models. These objects have a fractal nature which needs to be investigated. This programme has already started and recently Beffara [3] has calculated the Hausdorff dimension of these curves. Still, the fractal geometry of these models will capture some researcher's imaginations for a good while in the future.

References

[1] M.T. Barlow and E. Perkins *Levels at which every Brownian excursion is exceptional.* Seminar on probability XVIII, Springer LNM **1059** (1984) 1–28.

[2] R. Bass, K. Burdzy and D. Khoshnevisan. *Intersection local time for points of infinite multiplicity.* Ann. Probab. **22** (1994), 566–625.

[3] V. Beffara. *The dimension of the SLE curves.* Preprint arXiv: Math.PR/0211322 (2003).

[4] R. Cawley and R. D. Mauldin. *Multifractal decompositions of Moran fractals.* Adv. Math. **92** (1992), 196–236.

[5] A. Dembo. *Favourite points, cover times and fractals.* Lecture notes, Ecole d'été de St. Flour (2003), Springer, to appear.

[6] A. Dembo, Y. Peres, J. Rosen and O. Zeitouni. *Thick points for spatial Brownian motion: multifractal analysis of occupation measure.* Ann. Probab. **28** (2000), 1–35.

[7] A. Dembo, Y. Peres, J. Rosen and O. Zeitouni. *Thin points for Brownian motion.* Ann. Inst. Henri Poincaré, Probab. Stat. **36** (2000), 749–774.

[8] A. Dembo, Y. Peres, J. Rosen and O. Zeitouni. *Thick points for planar Brownian motion and the Erdös-Taylor conjecture on random walk.* Acta Math. **186** (2001), 239–270.

[9] D. Dolgopyat and V. Sidorov. *Multifractal properties of the sets of zeroes of Brownian paths.* Fund. Math. **147** (1995), 157–171.

[10] B. Duplantier. *Conformal fractal geometry and boundary quantum gravity.* arXiv:math-ph/0303034, 118pp. (2003).

[11] B. Duplantier and K. H. Kwon. *Conformal invariance and intersections of random walks.* Phys. Rev. Lett. **61** (1988), 2514–1517.

[12] J. Hawkes. *Trees generated by a simple branching process.* J. London Math. Soc. **24** (1981) 373–384.

[13] T. C. Halsey, M. N. Jensen, L. P. Kadanoff, I. Procaccia and B. I. Shraiman. *Fractal measures and their singularities.* Phys. Rev. **A33** (1986), 1141–1151.

[14] X. Hu and S. J. Taylor. *The multifractal structure of stable occupation measure.* Stoch. Processes Appl. **66** (1997) 283–299.

[15] P. Ch. Ivanov, L. A. N. Amaral, A. L. Goldberger, S. Havlin, M. G. Rosenblum, Z. Struzik, and H. E. Stanley. *Multifractality in human heartbeat dynamics.* Nature **399** (1999), 461–465.

[16] R. Kaufman. *Une propriété métrique du mouvement Brownien.* C. R. Acad. Sci. Paris **268** (1969), 727–728.

[17] D. Khoshnevisan, Y. Peres, and Y. Xiao. *Limsup random fractals.* Electron. J. Probab. **5**, Paper No. 4, (2000).

[18] A. Klenke and P. Mörters. *The multifractal spectrum of Brownian intersection local times.* Preprint, University of Bath (2004).

[19] W. König and P. Mörters. *Brownian intersection local times: upper tails and thick points.* Ann. Probab. **30** (2002), 1605–1656.

[20] G. F. Lawler. *Intersections of random walks.* Birkhäuser, Basel (1991).

[21] G. F. Lawler. *Hausdorff dimension of cut points for Brownian motion.* Elect. Journal Probab. **1**, Paper No. 2, (1996).

[22] G. F. Lawler. *The dimension of the frontier of planar Brownian motion.* Elect. Comm. in Probab. **1** (1996), 29–47.

[23] G. F. Lawler. *Strict concavity of the intersection exponent for Brownian motion in two and three dimensions.* Math. Phys. Electron. Journal **4**, Paper No.5, (1998).

[24] G. F. Lawler. *Geometric and fractal properties of Brownian motion and random walk paths in two and three dimensions.* Bolyai Math. Soc. Studies, **9** (1999), 219–258.

[25] G. F. Lawler. *Multifractal nature of two dimensional simple random walk paths.* Sympos. Math. XXXIX, 231–264, Cambridge Univ. Press, Cambridge, (1999).

[26] G. F. Lawler. *The frontier of a Brownian path is multifractal.* Unpublished (1999).

[27] G. F. Lawler, O. Schramm and W. Werner. *The dimension of the Brownian frontier is 4/3.* Math. Res. Lett. **8** (2001), 401–411.

[28] G. F. Lawler, O. Schramm and W. Werner. *Values of Brownian intersection exponents I: Half-plane exponents.* Acta Math. **187** (2001), 237–273.

[29] G. F. Lawler, O. Schramm and W. Werner. *Values of Brownian intersection exponents, II: Plane exponents.* Acta Math. **187** (2001), 275–308.

[30] G. F. Lawler, O. Schramm and W. Werner. *Values of Brownian intersection exponents. III: Two-sided exponents.* Ann. Inst. Henri Poincaré **38** (2002), 109–123.

[31] J.-F. Le Gall. *The exact Hausdorff measure of Brownian multiple points I. In: Seminar on Stochastic Processes 1986*, 107–137, Birkhäuser, Boston (1987).

[32] J.-F. Le Gall. *The exact Hausdorff measure of Brownian multiple points II. In: Seminar on Stochastic Processes 1988*, 193–197, Birkhäuser, Boston (1989).

[33] B. B. Mandelbrot. *Intermittent turbulence in self-similar cascades: Divergence of high moments and dimension of the carrier.* J. Fluid Mech. **62** (1974), 331–358.

[34] P. Mörters and N. R. Shieh. *Thin and thick points for branching measure on a Galton-Watson tree.* Stat. and Probab. Letters **58** (2002) 13–22.

[35] L. Olsen. *Multifractal geometry.* In: Fractal Geometry and Stochastics II. C. Bandt, S. Graf, M. Zähle (Eds), Birkhäuser, Basel (2000), pp. 3–37.

[36] S. Orey and S. J. Taylor. *How often on a Brownian path does the law of the iterated logarithm fail?* Proc. London Math. Soc. **28** (1974), 174–192.

[37] Y. Peres. *Remarks on intersection-equivalence and capacity-equivalence.* Ann. Inst. Henri Poincaré (Physique Théorique) **64** (1996) , 339–347.

[38] E.A. Perkins and S.J. Taylor. *Measuring close approaches on a Brownian path.* Ann. Probab. **16** (1988), 1458–1480.

[39] E.A. Perkins and S.J. Taylor. *The multifractal structure of super-Brownian motion.* Ann. Inst. Henri Poincaré. **34** (1998), 97–138.

[40] D. A. Rand. *The singularity spectrum $f(\alpha)$ for cookie-cutters.* Ergod. Th. & Dynam. Sys. **9** (1989), 527–541.

[41] N. R. Shieh and S. J. Taylor. *Logarithmic multifractal spectrum of stable occupation measure.* Stoch. Processes Appl. **75** (1998), 249–261.

[42] A. Turiel and C. Pérez Vicente. *Multifractal geometry in stock market time series.* Physica A **322** (2003), 629–649.

[43] E. B. Vul, Ya. G. Sinai and K. M. Khanin, *Feigenbaum universality and the thermodynamic formalism.* Russian Math. Surveys **39** (1984), 1–40.

Acknowledgement

I would like to thank the organisers of the conference, Christoph Bandt, Umberto Mosco, and Martina Zähle for their hospitality.

Peter Mörters
Department of Mathematical Sciences, University of Bath
Claverton Down, Bath BA2 6PA, United Kingdom
e-mail: `maspm@bath.ac.uk`

Progress in Probability, Vol. 57, 151–170

Additive Lévy Processes: Capacity and Hausdorff Dimension

Davar Khoshnevisan and Yimin Xiao

Abstract. This is a survey on recently-developed potential theory of additive Lévy processes and its applications to fractal geometry of Lévy processes.

Additive Lévy processes arise naturally in the studies of the Brownian sheet, intersections of Lévy processes and so on. We first summarize some recent results on the novel connections between an additive Lévy process X in $\mathbb{R}^d$, and a natural class of energy forms and their corresponding capacities. We then apply these results to study the Hausdorff dimension of the range and self-intersections of an ordinary Lévy process, solving several long-standing problems in the folklore of the theory of Lévy processes. We also list several open problems in this area.

Mathematics Subject Classification (2000). Primary 60J25, 28A80; Secondary 60G51, 60G17.

Keywords. Additive Lévy processes, Lévy sheets, Brownian sheet, capacity, Hausdorff dimension, multiple points.

1. Introduction

Classical potential theory has been very useful in studying the sample path properties of Markov processes, especially those of Brownian motion and Lévy processes. An important connection between classical potential theory and fractals is the fundamental result of Frostman (1935) [cf. Kahane (1985a) or Khoshnevisan (2002)] which states that the capacity dimension and the Hausdorff dimension [which will be denoted by $\dim_{\mathrm{H}}$] are the same. Hence, in order to prove $\dim_{\mathrm{H}} E \geq \beta$, it is sufficient to show that E has positive capacity with respect to the Bessel-Riesz kernel $f(x) = \|x\|^{-\beta}$. Taylor (1953, 1955) applied this idea to determine the Hausdorff dimension of the range and graph of a Brownian motion in $\mathbb{R}^d$. Since then, this method has become one of the standard tools in obtaining lower bounds for

Research supported in part by a grant from the NSF.

the Hausdorff dimension of random sets. We refer to the survey papers of Fristedt (1974), Taylor (1986) and Xiao (2003) for further information about the results and techniques for Lévy processes, and extensive lists of references.

There have been much effort in extending the potential theory for Markov processes to multiparameter processes of the forms $X(t) = X_1(t_1) + \cdots + X_N(t_N)$ or $Y(t) = (X_1(t_1), \ldots, X_N(t_N))$ $(t \in \mathbb{R}_+^N)$, where $X_1, \ldots, X_N$ are independent Markov processes; see e.g., Evans (1987a, b), Fitzsimmons and Salisbury (1989), Hirsch (1995), Hirsch and Song (1995a, b), Khoshnevisan (2002), Khoshnevisan and Xiao (2002, 2003a, b), Khoshnevisan, Xiao and Zhong (2003a). Besides of importance and interest in its own right, the potential theory of such multiparameter processes has found various applications in studying the fractal properties of ordinary Lévy processes, as well as their multiparameter analogue – the Lévy sheets.

The objective of this paper is to give a survey on the recently-developed potential theory of additive Lévy processes and its applications to fractal geometry of ordinary Lévy processes. In Section 2, we first recall the definition of Lévy processes and then introduce the definition of additive Lévy processes. We give two examples to show the close connections of additive Lévy processes to the Lévy sheets and ordinary Lévy processes. In Section 3, we summarize systematically the recent results on potential theory of additive Lévy processes. In Section 4, we show how the results in Section 3 can be applied to solve several long-standing problems about the ordinary Lévy processes. At the end of Sections 3 and 4, we list some open problems in the areas.

In the rest of this section, we give some general notation that will be used throughout. The underlying parameter space is $\mathbb{R}^N$, or $\mathbb{R}_+^N = [0, \infty)^N$. A typical parameter, $t \in \mathbb{R}^N$ is written as $t = (t_1, \ldots, t_N)$, sometimes also written as $\langle t_j \rangle$, or $\langle c \rangle$, if $t_1 = \cdots = t_N = c \in \mathbb{R}$.

There is a natural partial order, "$\preccurlyeq$", on $\mathbb{R}^N$. Namely, $s \preccurlyeq t$ if and only if $s_j \le t_j$ for all $j = 1, \ldots, N$. When it is the case that $s \preccurlyeq t$, we define the *interval*, $[s, t] = \prod_{j=1}^N [s_j, t_j]$.

We will let $\mathcal{A}$ denote the class of all N-dimensional intervals $I \subset \mathbb{R}^N$ of the form $I = [s, t]$. We always write λ_m for Lebesgue's measure on $\mathbb{R}^m$, no matter the value of the integer m. The state space, $\mathbb{R}^d$, is endowed with the ℓ^2 Euclidean norm $\|\cdot\|$, and the corresponding dot product $\langle x, y \rangle = \sum_{j=1}^d x_j y_j \ \ (x, y \in \mathbb{R}^d)$.

We write $\mathcal{P}(F)$ for the collection of all Borel-regular probability measures on a given Borel space F. Given a Borel measurable function $f : \mathbb{R}^d \to [0, \infty]$, we define the '$f$-energy' (of some $\mu \in \mathcal{P}(\mathbb{R}^d)$) and '$f$-capacity' (of some measurable $G \subset \mathbb{R}^d$) as follows:

$$\mathcal{E}_f(\mu) := \iint f(x-y)\, \mu(dx)\, \mu(dy), \qquad \mathcal{C}_f(G) := \left[\inf_{\mu \in \mathcal{P}(G)} \mathcal{E}_f(\mu) \right]^{-1}. \tag{1.1}$$

We refer to such a function f as a *gauge* function.

Given a number $\beta > 0$, we reserve $\mathcal{C}_\beta$ and $\mathcal{E}_\beta$ for $\mathcal{C}_f$ and $\mathcal{E}_f$ respectively, where the gauge function f is $f(t) := \|t\|^{-\beta}$. $\mathcal{C}_\beta$ and $\mathcal{E}_\beta$ are respectively the (β-dimensional) Bessel–Riesz *capacity* and *energy* to which some references were made earlier. More information about the Bessel–Riesz capacity and its connection to fractals can be found in Falconer (1990), Mattila (1995), Kahane (1985a), and Khoshnevisan (2002).

2. Additive Lévy processes

We first recall briefly the definition of a Lévy process and refer to Bertoin (1996) and Sato (1999) for the general theory. A stochastic process $Z = \{Z(t),\ t \geq 0\}$, with values in $\mathbb{R}^d$, is called a *Lévy process*, if it has stationary and independent increments, and such that $t \mapsto Z(t)$ is continuous in probability. It is well known that for $t \geq s \geq 0$, the characteristic function of $Z(t) - Z(s)$ is given by

$$\mathbb{E}\left[e^{i\langle \xi, Z(t)-Z(s)\rangle}\right] = e^{-(t-s)\Psi(\xi)},$$

where, by the Lévy-Khintchine formula,

$$\Psi(\xi) = i\langle \mathsf{a}, \xi\rangle + \frac{1}{2}\langle \xi, \Sigma\xi'\rangle + \int_{\mathbb{R}^d}\left[1 - e^{i\langle x,\xi\rangle} + \frac{i\langle x, \xi\rangle}{1+\|x\|^2}\right] \mathsf{L}(dx), \qquad \forall \xi \in \mathbb{R}^d,$$

and $\mathsf{a} \in \mathbb{R}^d$ is fixed, Σ is a non-negative definite, symmetric, $(d \times d)$ matrix, and L is a Borel measure on $\mathbb{R}^d \setminus \{0\}$ that satisfies

$$\int_{\mathbb{R}^d} \frac{\|x\|^2}{1+\|x\|^2}\, \mathsf{L}(dx) < \infty.$$

The function Ψ is the *Lévy exponent* of Z, and L is the corresponding *Lévy measure*. There are several different characterizations for the exponent Ψ. Note that $\Psi(0) = 0$ and that by Bochner's theorem the function $\xi \mapsto e^{-t\Psi(\xi)}$ is continuous and *positive definite* for each $t \geq 0$ since it is the Fourier transform of a probability measure. Hence, the Lévy exponent Ψ is a continuous *negative definite function*. See Berg and Frost (1975) for a systematic account on negative definite functions.

A Lévy process Z is *symmetric* if $-Z$ and Z have the same finite dimensional distributions. It is clear that Z is symmetric if and only if $\Psi(\xi) \geq 0$, for all $\xi \in \mathbb{R}^d$. In particular, if the Lévy exponent Ψ is of the form

$$\Psi(\xi) = -\sigma^\alpha \|\xi\|^\alpha \qquad \forall\, \xi \in \mathbb{R}^d, \tag{2.1}$$

where $\alpha \in (0, 2]$ and $\sigma > 0$ are constants, then Z is called an isotropic stable Lévy process of index α.

An N-parameter, $\mathbb{R}^d$-valued, *additive Lévy process* $X = \{X(t),\ t \in \mathbb{R}_+^N\}$ is a multiparameter stochastic process defined by

$$X(t) = \sum_{j=1}^N X_j(t_j), \qquad t \in \mathbb{R}_+^N, \tag{2.2}$$

where $X_1, \ldots, X_N$ denote independent Lévy processes that take their values in $\mathbb{R}^d$ and $X_j(0) = 0$ for $j = 1, \ldots, N$. Using tensor notation, we will often write $X = X_1 \oplus \cdots \oplus X_N$ for brevity. If X_j has Lévy exponent Ψ_j, then $\Psi = (\Psi_1, \ldots, \Psi_N)$ is called the Lévy exponent of X. When $X_1, \ldots, X_N$ are isotropic stable processes in $\mathbb{R}^d$ of index $\alpha \in (0, 2]$, then X is called an *additive stable process*. The case $\alpha = 2$ yields the additive Brownian motion. Similarly, one can define an *multiplicative Lévy process* $X = X_1 \otimes \cdots \otimes X_N$ by $X(t) = (X_1(t_1), \ldots, X_N(t_N))$.

In the following we give two examples to show that additive Lévy processes arise naturally in the analysis of multiparameter processes such as Lévy's sheets and in the analysis of ordinary Lévy processes. As such, the potential theory of additive Lévy processes can also be used as a powerful tool to study the Hausdorff dimension and capacity of random fractals determined by Lévy processes.

Example 2.1 [The Lévy sheets and Brownian sheet] A Lévy sheet is a multiparameter extension of an ordinary Lévy process. For any given negative definite function Ψ in $\mathbb{R}^d$, there is a random field $Z = \{Z(t),\ t \in \mathbb{R}_+^N\}$ taking values in $\mathbb{R}^d$ such that

(i) for any choice of disjoint intervals $Q_j \in \mathcal{A}$ $(n = 1, \ldots, n)$, the increments $Z(Q_1), \ldots, Z(Q_n)$ are independent; and
(ii) for each $Q \in \mathcal{A}$, the characteristic function of $Z(Q)$ is given by

$$\mathbb{E}\Big[\exp\big(i\langle \xi, Z(Q)\rangle\big)\Big] = \exp\big(- \lambda_N(Q)\Psi(\xi)\big), \tag{2.3}$$

where $Z(Q)$ is the increment of Z on $Q = [s, t]$ which is defined by

$$Z(Q) = \sum_{r \in \{0,1\}^N} (-1)^{N - \sum_\ell r_\ell} Z\big(\langle s_j + r_j(t_j - s_j)\rangle\big).$$

A random field $Z = \{Z(t),\ t \in \mathbb{R}_+^N\}$ taking values in $\mathbb{R}^d$ that is continuous in probability and satisfies the above conditions (i) and (ii) is called an (N, d)-Lévy sheet with exponent Ψ. Some general properties of Lévy sheets can be found in Adler et al. (1983), Dalang and Walsh (1992), as well as in Ehm (1981) and Vares (1983). When the Lévy exponent Ψ is given by (2.1), Z is called an isotropic stable sheet. In particular, when $\alpha = 2$ and $\sigma = 1/\sqrt{2}$, then Z is the N-parameter Brownian sheet in $\mathbb{R}^d$ and is usually denoted by $W = \{W(t),\ t \in \mathbb{R}_+^N\}$. We should mention that the Brownian sheet arises naturally in the stochastic partial differential equations and in statistical analysis [cf. Walsh (1986), Dalang (2003)] and is one of the most fundamental Gaussian random fields. By (2.1), we see that the components $W_1, \ldots, W_d$ of W are independent, centered Gaussian random fields in $\mathbb{R}$ with the covariance function $\mathbb{E}\big[W_1(s)W_1(t)\big] = \prod_{j=1}^N s_j \wedge t_j$, where $a \wedge b = \min\{a, b\}$.

It follows from the conditions (i) and (ii) that a Lévy sheet Z has independent and stationary increments. Along lines which are parallel to the axes, Z is an ordinary Lévy process in $\mathbb{R}^d$ with a constant speed. More precisely, for any fixed $a = \langle a_j \rangle \in \mathbb{R}_+^N$ and every $k = 1, \ldots, N$, the process $X_k = \{X_k(t_k),\ t_k \in \mathbb{R}_+\}$

defined by

$$X_k(t_k) = Z(a_1, \ldots, a_{k-1}, a_k + t_k, a_{k+1}, \ldots, a_N) - Z(a)$$

is a Lévy process in $\mathbb{R}^d$ with exponent $\Psi_k(\xi) = \big(\prod_{j \neq k} a_j\big)\Psi(\xi)$. This leads to the following very useful decomposition of Z:

$$Z(a+t) = Z(a) + \sum_{k=1}^{N} X_k(t_k) + \tilde{Z}(t), \quad t \in \mathbb{R}_+^N, \tag{2.4}$$

where $\tilde{Z}$ is a Lévy sheet in $\mathbb{R}^d$ with exponent Ψ, and all the processes on the right hand side of (2.4) are independent.

For many purposes such as in the studies of fractal properties of Z, the effects of $\tilde{Z}$ can sometimes be neglected. Hence the Lévy sheet Z can be approximated locally by the additive Lévy process $X = X_1 \oplus \cdots \oplus X_N$ which is easier to analyze. This approach has been exploited by Ehm (1981), Vares (1983), Rosen (1984), Zhong and Xiao (1995), Khoshnevisan, Xiao and Zhong (2003b) to study the local times and self-intersection local times of Lévy sheets; by Dalang and Mountford (1996, 1997, 2001, 2002, 2003), Dalang and Walsh (1993a, b) and Khoshnevisan (1995) to study the level sets, excursions and points of increase of the Brownian sheets [see also Dalang (2003)]; and by Khoshnevisan and Shi (1999), Khoshnevisan (1999) to establish the connection between hitting probabilities of the Brownian sheet and the Bessel-Riesz capacity.

Example 2.2 [Intersections of independent Lévy processes] Let $X_1, \ldots, X_k$ be independent Lévy processes in $\mathbb{R}^d$. It has been of interest to study whether their sample paths can intersect and, when they do, the fractal properties of the set of intersections; see Taylor (1986a), Xiao (2003) for further information. We say that the sample paths of $X_1, \ldots, X_k$ intersect if

$$\exists \text{ distinct points } t_1, \ldots, t_k \in \mathbb{R}_+ \text{ such that } X_1(t_1) = \cdots = X_k(t_k). \tag{2.5}$$

Define the multiparameter process $Z = \{Z(t),\ t \in \mathbb{R}_+^k\}$ by

$$Z(t) = \big(X_2(t_2) - X_1(t_1), \ldots, X_k(t_k) - X_{k-1}(t_{k-1})\big). \tag{2.6}$$

Then (2.5) is equivalent to $Z^{-1}(0) \neq \emptyset$, where $Z^{-1}(0) = \{t \in \mathbb{R}_+^k : Z(t) = 0\}$ is the zero set of Z. Rearranging the components of Z, it is easy to see that Z is an additive Lévy process with values in $\mathbb{R}^{(k-1)d}$. Hence the potential theory of additive Lévy processes can be applied; see Theorem 3.6 below.

3. Potential theory for additive Lévy processes

Potential theory of multiparameter Markov processes have been studied by Evans (1987b), Fitzsimmons and Salisbury (1989) for multiplicative Markov processes and by Hirsch (1995), Hirsch and Song (1995a, b) [see also the recent book of Khoshnevisan (2002) for a systematic account on the earlier results], Khoshnevisan and Xiao (2002, 2003a, b), Khoshnevisan, Xiao and Zhong (2003a) for additive Lévy processes in $\mathbb{R}^d$.

The following are some natural questions that the potential theory of X seeks to answer.

(a) Given a Borel set $F \subseteq \mathbb{R}^d$, when can $\mathbb{P}\{X(\mathbb{R}_+^N) \cap F \neq \emptyset\} > 0$?
(b) For any fixed $a \in \mathbb{R}^d$ and a Borel set $E \subset \mathbb{R}_+^N$, when can $\mathbb{P}\{X^{-1}(a) \cap E \neq \emptyset\} > 0$?
(c) When can the range of X, $X(\mathbb{R}_+^N)$ [or more generally $X(E)$, where $E \subset \mathbb{R}_+^N$], have positive Lebesgue measure?

For additive Lévy processes, the following lemma from Khoshnevisan and Xiao (2003b) shows that the above questions are closely related. For $N = 1$, it is due to Kahane (1972).

Lemma 3.1. *Let X be an N-parameter additive Lévy process in $\mathbb{R}^d$. We assume that, for every $t \in (0,\infty)^N$, the distribution of $X(t)$ is mutually absolutely continuous with respect to λ_d. Then for any Borel sets $E \subset (0,\infty)^N$ and $F \subset \mathbb{R}^d$, the following are equivalent:*

1. *With positive probability, $E \cap X^{-1}(F) \neq \emptyset$;*
2. *With positive probability, $F \cap X(E) \neq \emptyset$;*
3. *With positive probability, $\lambda_d(F \ominus X(E)) > 0$,*

where $A \ominus B := \{x - y : x \in A, y \in B\}$.

Under some mild conditions, Questions (a), (b) and (c) are answered in Khoshnevisan, Xiao and Zhong (2003a), and Khoshnevisan and Xiao (2002, 2003a, b). In this section, we summarize some of their results.
Let X be an additive Lévy process in $\mathbb{R}^d$ with Lévy exponent $\Psi = (\Psi_1, \ldots, \Psi_N)$. It induces an *energy form* $\mathcal{E}_\Psi$ that can be described as follows: For all finite measures μ on $\mathbb{R}^d$, and/or all integrable functions $\mu : \mathbb{R}^d \to \mathbb{R}$,

$$\mathcal{E}_\Psi(\mu) = (2\pi)^{-d} \int_{\mathbb{R}^d} |\widehat{\mu}(\xi)|^2 \prod_{j=1}^N \operatorname{Re}\left(\frac{1}{1+\Psi_j(\xi)}\right) d\xi, \tag{3.1}$$

where $\widehat{\ }$ denotes the Fourier transform normalized as $\widehat{f}(\xi) = \int_{\mathbb{R}^d} e^{i\xi\cdot x} f(x)\,dx$ [$f \in L^1(\mathbb{R}^d)$].

Under the assumption that there exists a positive constant $\vartheta > 0$ such that

$$\operatorname{Re}\left(\prod_{j=1}^N \frac{1}{1+\Psi_j(\xi)}\right) \geq \vartheta \prod_{j=1}^N \operatorname{Re}\left(\frac{1}{1+\Psi_j(\xi)}\right), \tag{3.2}$$

Khoshnevisan, Xiao and Zhong (2003a) have proved the following theorem.

Theorem 3.2. *Consider any d-dimensional additive Lévy process X, whose Lévy exponent Ψ satisfies* (3.2). *Then, given any nonrandom compact set $F \subset \mathbb{R}^d$, $\mathbb{E}\{\lambda_d(X(\mathbb{R}_+^N) \oplus F)\} > 0$ if and only if F carries a finite measure of finite energy.*

The proof of Theorem 3.2 consists of two parts. The sufficiency is established by using the Paley-Zygmund inequality [cf. Kahane (1985a,), p.8] and second moment estimates of the occupation measure of X. However the proof of the necessity is much harder and relies on developing a kind of multiparameter martingale theory in measure-spaces where the underlying measure has infinite mass.

By letting $F = \{0\}$, Theorem 3.2 gives the following answer to Question (c) for $X(\mathbb{R}_+^N)$, which extends the remarkable results of Kesten (1969) and Bretagnolle (1971) for $N = 1$.

Corollary 3.3. *Let X be an additive Lévy process in $\mathbb{R}^d$ with Lévy exponent $(\Psi_1, \ldots, \Psi_N)$, and suppose that Condition (3.2) holds. Then,*

$$\mathbb{E}\{\lambda_d(X(\mathbb{R}_+^N))\} > 0 \iff \int_{\mathbb{R}^d} \prod_{j=1}^N \operatorname{Re}\left(\frac{1}{1+\Psi_j(\xi)}\right) d\xi < +\infty.$$

It is known that the positiveness of $\mathbb{E}\{\lambda_d(X(\mathbb{R}_+^N))\}$ is closely related to the existence of local times of X. The precise connection for additive Lévy processes has been established by Khoshnevisan, Xiao and Zhong (2003b): under the condition (3.2),

$$X \text{ has local times if and only if } \mathbb{E}\{\lambda_d(X(\mathbb{R}_+^N))\} > 0.$$

They have also proved results on the joint continuity, local and uniform Hölder conditions for the local times of additive Lévy processes, extending the results of Ehm (1981) and Vares (1983) for Lévy sheets.

More generally, for every compact $E \subset \mathbb{R}_+$, a necessary and sufficient condition in terms of the Lévy exponent $(\Psi_1, \ldots, \Psi_N)$ for $\mathbb{E}\{\lambda_d(X(E))\} > 0$ has been given by Theorem 2.1 in Khoshnevisan, Xiao and Zhong (2003a). A different condition in terms of the gauge function is given by Theorem 3.6 below.

Applying Lemma 3.1 and Theorem 3.2 to additive stable Lévy processes, we have the following result which improves the earlier results of Hirsch (1995), Hirsch and Song (1995a, b), Khoshnevisan (2002).

Corollary 3.4. *Suppose $X = X_1 \oplus \cdots \oplus X_N$ is an additive stable process in $\mathbb{R}^d$ of index $\alpha \in (0, 2]$. Then for any Borel set $F \subset \mathbb{R}^d$, the following are equivalent:*

(i) $\mathcal{C}_{d-N\alpha}(F) > 0$;
(ii) $\mathbb{P}\{\lambda_d\{F \oplus X(\mathbb{R}_+^N)\} > 0\} > 0$;
(iii) *F is not polar for X in the sense that* $\mathbb{P}\{F \cap X(\mathbb{R}_+^N) \neq \emptyset\} > 0$.

Remark 3.5. Upon varying $d, N \in \mathbb{N}$, and $\alpha \in (0, 2]$, we see that this theorem associates an additive Lévy process to any Bessel–Riesz capacity, including those with dimension > 2. This connection is very useful in calculating the Hausdorff dimension of the range $X(\mathbb{R}_+^N)$ and various random fractals related to Lévy processes, via the co-dimension argument; see Section 4.

Note that, under the assumptions of Lemma 3.1, the conditions (3.2) and $\prod_{j=1}^N \operatorname{Re}\left(1+\Psi_j(\xi)\right)^{-1} \in L^1(\mathbb{R}^d)$ imply that X hits points, i.e. $\mathbb{P}\{X^{-1}(a) \neq \emptyset\} > 0$

for all $a \in \mathbb{R}^d$. In this case, it is natural to consider Question (b) and study the hitting probability of the level set $X^{-1}(a)$ for a fixed $a \in \mathbb{R}^d$. Moreover, when X satisfies the conditions of Lemma 3.1, Question (b) is equivalent to the question (c) for $X(E)$. This latter problem for Brownian motion was considered by Hawkes (1977) and Kahane (1983, 1985a, b), and for additive Lévy processes by Khoshnevisan, Xiao and Zhong (2003a, Theorem 2.1). For symmetric additive Lévy processes, Question (b) is answered by the following theorem of Khoshnevisan and Xiao (2002, 2003a). Recall that the additive Lévy process X is *absolutely continuous* if for each $t \in (0,\infty)^N$, the function $\xi \mapsto \exp\{-\sum_{j=1}^N t_j \Psi_j(\xi)\} \in L^1(\mathbb{R}^d)$. In this case, for every $t \in (0,\infty)^N$, $X(t)$ has a density function $p(t;\bullet)$ that is given by the formula

$$p(t;x) = (2\pi)^{-d} \int_{\mathbb{R}^d} e^{-i\langle \xi, x\rangle} \exp\Big(-\sum_{j=1}^N t_j \Psi_j(\xi)\Big)\, d\xi, \qquad x \in \mathbb{R}^d.$$

The function Φ defined by

$$\Phi(s) = p(\bar{s}, 0) \qquad s \in \mathbb{R}^N, \tag{3.3}$$

where $\bar{s} = \langle |s_j| \rangle \in \mathbb{R}^N_+$, is called the *gauge function* for X. Khoshnevisan and Xiao (2002, Theorem 2.9) have shown that $\Phi \in L^1_{loc}(\mathbb{R}^N)$ is equivalent to $X^{-1}(a) \neq \emptyset$.

Theorem 3.6. *Let $X_1, \ldots, X_N$ be N independent symmetric Lévy processes on $\mathbb{R}^d$ and let $X = X_1 \oplus \cdots \oplus X_N$. Suppose X is absolutely continuous with the gauge function Φ. Then for all $a \in \mathbb{R}^d$, $c > 0$ and for all compact sets $E \subset [c, \infty[^N$,*

$$A_1 \mathcal{C}_\Phi(E) \leq \mathbb{P}\Big\{X^{-1}(a) \cap E \neq \emptyset\Big\} \leq A_2 \mathcal{C}_\Phi(E) \quad , \text{ where}$$

$A_1 = \kappa^{-2} 2^{-d} \{\Phi(\langle c \rangle)\}^{-1} I_E^2(a)$, $A_2 = \kappa^3 2^{5d+3N} \Phi(\langle c \rangle)$ and $I_E(a) = \inf_{s \in E} p(s;a)$.

Theorem 2.10 in Khoshnevisan and Xiao (2002) also gives a formula for $\dim_{\rm H} X^{-1}(0)$. In particular, when X is an additive stable process, we have

Corollary 3.7. *Suppose $X_1, \ldots, X_N$ are independent isotropic stable Lévy processes in $\mathbb{R}^d$ with index $\alpha \in]0,2]$ and $X = X_1 \oplus \cdots \oplus X_N$. Then,*

(i) $\mathbb{P}\{X^{-1}(0) \neq \varnothing\} > 0$ *if and only if $N\alpha > d$; and*
(ii) *if $N\alpha > d$, then $\mathbb{P}\{\dim_{\rm H} X^{-1}(0) = N - \frac{d}{\alpha}\} = 1$.*

Furthermore, for each $M > 1$, there exists a constant $A > 1$, such that simultaneously for all compact sets $E \subset [M^{-1}, M]^N$, and for all $a \in [-M, M]^d$,

$$\frac{1}{A}\, \mathcal{C}_{d/\alpha}(E) \leq \mathbb{P}\Big\{X^{-1}(a) \cap E\Big\} \leq A\, \mathcal{C}_{d/\alpha}(E).$$

Remark 3.8. In Theorem 1.1 of Khoshnevisan and Xiao (2002), it is proved that (ii) in Corollary 3.7 holds with positive probability. It can be strengthened to a probability 1 result by using the conditional Borel-Cantelli lemma.

There are several interesting open questions in this area. In the following, we list some of them.

Problem 3.9. Can the condition (3.2) be dropped from Theorem 3.2?

A positive resolution to this question would lead to the following:

(i) A complete resolution of a conjecture regarding intersection of independent regenerative sets of J. Bertoin (1999, p. 49).
(ii) The computation of the Hausdorff dimension of the range of an arbitrary additive Lévy process; see Theorem 4.1 below for a partial result.

A resolution of item (ii) would go a long way toward solving the corresponding open problem for an arbitrary Lévy sheet (Straf 1972; Ehm 1981; Vares 1983).

Problem 3.10. Can the symmetry and absolute continuity assumptions be dropped from Theorem 3.6?

A positive solution to Problem 3.10 will have interesting consequences on the intersections of Lévy processes and on the Hausdorff dimension of the multiple points.

Another natural question is to characterize the polar sets for the space-time process $\{(t, X(t)); t \in \mathbb{R}_+^N\}$:

Problem 3.11. Given Borel sets $E \subseteq \mathbb{R}_+^N$ and $F \subseteq \mathbb{R}^d \backslash \{0\}$, find a necessary and sufficient condition for $\mathbb{P}\{X(E) \cap F \neq \emptyset\} > 0$?

In the case that $N = 1$ and X is a d-dimensional Brownian motion, the above problem is equivalent to the polarity of the set $E \times F$ relative to the Dirichlet problem for the full heat operator $\partial_t - \frac{1}{2}\Delta$; see Doob (1984, p. 637) and Watson (1976, 1978). In particular, Watson has proven that a set $G \subset \mathbb{R}_+ \times \mathbb{R}^d$ is polar for the heat equation if and only if it has zero *thermal capacity*. See also Kaufman and Wu (1982) and Taylor and Watson (1985) for further information.

A resolution of Problem 3.11 will contribute to developing a complete parabolic potential theory for various multiparameter Lévy processes. This is not only interesting of itself, but will also solve a number of old problems in the theory of one-parameter Markov processes (including one-dimensional Brownian motion). For instance, problems about the Hausdorff dimension of the set $B(E) \cap F$, where E and F are, respectively, compact subsets of $\mathbb{R}_+$ and $\mathbb{R}^d \setminus \{0\}$. This later problem was first considered by Kaufman (1972), who obtained upper and lower bounds. Hawkes (1978a) has found an extension of Kaufman's theorem to stable processes. However, in general, the Hausdorff dimension of $B(E) \cap F$ is unknown even for Brownian motion.

We remark that Problem 3.11 is intimately related to the existing potential theory of additive Lévy processes. Indeed, in Khoshnevisan and Xiao (2002, 2003a), we find a necessary and sufficient condition for $\mathbb{P}\{X(E) \cap \{x_0\} \neq \emptyset\}$ to be positive for an arbitrary symmetric additive Lévy process X that has a density for all $t > 0$. On the other hand, Khoshnevisan, Xiao and Zhong (2003a) have derived a necessary and sufficient condition for $\mathbb{P}\{X(\mathbb{R}_+^N) \cap F \neq \emptyset\}$ to be positive; see also Khoshnevisan and Xiao (2003b). These results can all be considered as partial solutions to Problem 3.11.

We believe that once Problem 3.11 is solved, the methods of Khoshnevisan and Xiao (2002, 2003a, 2003b) can be used to calculate the Hausdorff dimensions of the random sets $X(E) \cap F$ and $X^{-1}(F) \cap E$ not only for Brownian motion, but also for a large class of Lévy processes.

There are many other difficult open problems of similar parabolic nature in the literature. For example, let $F(\lambda)$ denote the set of λ-fast points of Brownian motion,

$$F(\lambda) = \left\{ t \in [0,1] : \limsup_{h \to 0+} \frac{|B(t+h) - B(t)|}{\sqrt{2h|\log h|}} \geq \lambda \right\}.$$

Then, given a Borel set $E \subset \mathbb{R}_+$, what is the Hausdorff dimension of the set $F(\lambda) \cap E$? See Orey and Taylor (1974), Kaufman (1975), Khoshnevisan, Peres and Xiao (2000), and Khoshnevisan and Shi (2000) for some information. The same question can be asked when $F(\lambda)$ denotes the set of thick points of the occupation measure of a Brownian motion in $\mathbb{R}^d$ (Dembo, Peres, Rosen and Zeitouni, 2000). We believe that a sufficiently powerful connection to parabolic potential theory may unravel some of these problems.

Finally, Khoshnevisan and Shi (1999) have proven that the (N,d)-Brownian sheet W and the (N,d)-additive Brownian motion $B = B_1 \oplus \cdots \oplus B_N$ are intersection equivalent, i.e. for any compact set $F \subset \mathbb{R}^d$,

$$\mathbb{P}\{W(\mathbb{R}^N_+) \cap F \neq \emptyset\} > 0 \iff \mathbb{P}\{B(\mathbb{R}^N_+) \cap F \neq \emptyset\} > 0,$$

see Corollary 6.2 of Khoshnevisan and Shi (1999). It would be interesting to solve the analogous problem for the (N,d)-stable sheets [cf. Ehm (1981)]. In particular, we believe that

Conjecture 3.12. Corollary 3.4 holds for an (N,d)-stable sheet of type A.

4. Applications to fractal properties of Lévy processes

By Corollary 3.4 and the Frostman's theorem [cf. Kahane (1985, p.133) or Khoshnevisan (2002, p.521)], we can use the range $X(\mathbb{R}^N_+)$ as a tool to determine the Hausdorff dimension of any Borel set F in $\mathbb{R}^d$. The original idea goes back to Taylor (1966), who showed that for any Borel set $F \subset \mathbb{R}^d$ with $\dim_{\mathrm{H}} F \geq d-2$,

$$\dim_{\mathrm{H}} F = d - \inf\{\alpha > 0 : \ F \text{ is not polar for } X_\alpha\},$$

where X_α is an isotropic stable Lévy process in $\mathbb{R}^d$ of index $\alpha \in (0,2]$. With the help of additive stable processes, the restriction on F can be removed. Corollary 3.4 implies that for any Borel set $F \subset \mathbb{R}^d$,

$$\dim_{\mathrm{H}} F = d - \inf\{N\alpha > 0 : \ F \text{ is not polar for } X_{\alpha,N}\},$$

where $X_{\alpha,N}$ denotes an N-parameter additive stable process in $\mathbb{R}^d$ of index $\alpha \in (0,2]$. Following Khoshnevisan and Shi (2000), this method for calculating $\dim_{\mathrm{H}} F$ is called the co-dimension argument, which is specially effective for determining the Hausdorff dimension of a random set. See Hawkes (1981), Lyons (1990),

Peres (1996, 1999), Khoshnevisan, Peres and Xiao (2000), Khoshnevisan (2002) for related results.

4.1. Hausdorff dimension of the range

As an application of the co-dimension argument, we derive the Hausdorff dimension of the range of an additive Lévy process, which is proved in Khoshnevisan, Xiao and Zhong (2003a).

Theorem 4.1. *Given an additive Lévy process X in $\mathbb{R}^d$ with Lévy exponent $(\Psi_1, \ldots, \Psi_N)$ that satisfies* (3.2),

$$\dim_{\mathrm{H}} \left(X(\mathbb{R}_+^N)\right) = d - \eta, \qquad \mathbb{P}\text{-}a.s., \tag{4.1}$$

where

$$\eta = \sup\left\{\gamma > 0 : \int_{\xi \in \mathbb{R}^d:\ \|\xi\|>1} \prod_{j=1}^{N} \mathrm{Re}\left(\frac{1}{1+\Psi_j(\xi)}\right) \frac{d\xi}{\|\xi\|^{\gamma}} = +\infty\right\}.$$

Here $\sup \emptyset = 0$. *In particular, if X is an additive stable process in $\mathbb{R}^d$ of index $\alpha \in (0, 2]$, then* $\dim_{\mathrm{H}} X(\mathbb{R}_+^N) = \min\{d, \alpha N\}$ *a.s.*

Sketch of Proof. We introduce an M-parameter additive stable process Y in $\mathbb{R}^d$ with index $\alpha \in (0, 2]$. The process Y is totally independent of X, and the constants M and α will be chosen appropriately. Note that $X \oplus Y$ is an $(N + M)$-parameter additive Lévy process in $\mathbb{R}^d$ whose Lévy exponent $\Phi = (\Phi_1, \ldots, \Phi_{N+M})$ is given by

$$\Phi_j(\xi) = \begin{cases} \Psi_j(\xi), & \text{if } j = 1, \ldots, N, \\ \frac{1}{2}\|\xi\|^{\alpha}, & \text{if } j = N+1, \ldots, N+M. \end{cases}$$

Then by Theorem 3.2, Corollary 3.4 and the zero-one law that $\mathbb{P}\{\mathcal{C}_{\beta}(X(\mathbb{R}_+^N)) > 0\} \in \{0, 1\}$ for any $\beta > 0$ [whose proof will be given elsewhere], we have

$$\mathcal{C}_{d-M\alpha}\left(X(\mathbb{R}_+^N)\right) > 0 \; a.s. \iff \int_{\xi \in \mathbb{R}^d:\ \|\xi\|>1} \prod_{j=1}^{N} \mathrm{Re}\left(\frac{1}{1+\Psi_j(\xi)}\right) \|\xi\|^{-M\alpha}\, d\xi < \infty. \tag{4.2}$$

From this follows (4.1). □

When $N = 1$, i.e., when X is an ordinary Lévy process in $\mathbb{R}^d$, Pruitt (1969) has shown that the Hausdorff dimension of the range $X(\mathbb{R}_+)$ is

$$\gamma = \sup\left\{\alpha \geq 0 : \limsup_{r \to 0}\, r^{-\alpha} \int_0^1 \mathbb{P}\{|X(t)| \leq r\}\, dt < +\infty\right\}.$$

In general, this formula is not satisfying, since the above lim sup is not easy to evaluate. Pruitt (1969, Theorem 5) addresses this issue by showing that if $\mathrm{Re}\,\Psi(\xi) \geq 2 \log \|\xi\|$ (for all $\|\xi\|$ large), then,

$$\gamma = \sup\left\{\alpha < d : \int_{\mathbb{R}^d} \mathrm{Re}\left(\frac{1 - e^{-\Psi(\xi)}}{\Psi(\xi)}\right) \frac{d\xi}{\|\xi\|^{d-\alpha}} < +\infty\right\}.$$

See Fristedt (1974, 377–378) for further discussions on Pruitt's work in this area.

Theorem 4.1 implies the following representation for the index γ in terms of the Lévy exponent Ψ of X that holds under no restrictions, thus solving the above mentioned problem of Pruitt (1969).

Corollary 4.2. *If X denotes a Lévy process in $\mathbb{R}^d$ with Lévy exponent Ψ, then*

$$\dim_{\mathrm{H}} X([0,1]) = \sup\left\{\alpha < d : \int_{\xi\in\mathbb{R}^d:\ \|\xi\|>1} \mathrm{Re}\left(\frac{1}{1+\Psi(\xi)}\right) \frac{d\xi}{\|\xi\|^{d-\alpha}} < +\infty\right\}. \tag{4.3}$$

Formula (4.3) gives an analytic way to compute the Hausdorff dimension of $X(\mathbb{R}_+)$ for Lévy processes. Its usefulness is shown by the following result from Khoshnevisan and Xiao (2003b).

Corollary 4.3. *Let $X = \{X(t), t \in \mathbb{R}_+\}$ be a Lévy process in $\mathbb{R}^d$ with Lévy exponent Ψ. If Ψ satisfies the following condition: for any $\epsilon > 0$, there exist positive constants K_1 and K_2 such that*

$$\frac{K_1}{\sum_{j=1}^d |\xi_j|^{\alpha_j+\epsilon}} \le \mathrm{Re}\left(\frac{1}{1+\Psi(\xi)}\right) \le \frac{K_2}{\sum_{j=1}^d |\xi_j|^{\alpha_j-\epsilon}}, \quad \forall \xi \in \mathbb{R}^d \text{ with } \min_{1\le j\le d} |\xi_j| \ge 1, \tag{4.4}$$

then almost surely,

$$\dim_{\mathrm{H}} X(\mathbb{R}_+) = \begin{cases} \alpha_1 & \text{if } \alpha_1 \le d_1, \\ 1 + \alpha_2(1 - 1/\alpha_1) & \text{otherwise.} \end{cases} \tag{4.5}$$

It is easily seen that the Lévy processes with independent stable components [cf. Hendricks (1973, 1974)] satisfy (4.4). Meerchaert and Xiao (2003) have recently shown that every operator stable Lévy process X in $\mathbb{R}^d$ with exponent B satisfies (4.4) with $\alpha_1, \ldots, \alpha_d$ being the real parts of the eigenvalues of B. Hence the Hausdorff dimension of the range $X(\mathbb{R}_+)$ is given by (4.5). This solves a problem of Becker–Kern, Meerschaert and Scheffler (2003).

When X is a Brownian motion or a stable Lévy process in $\mathbb{R}^d$, McKean (1955) and Blumenthal and Getoor (1960) have demonstrated that for all Borel sets $E \subset \mathbb{R}_+$,

$$\dim_{\mathrm{H}} X(E) = d \wedge \alpha \dim_{\mathrm{H}} E, \quad \text{a.s.}, \tag{4.6}$$

Blumenthal and Getoor (1961) extended (4.6) to a broad class of Lévy processes. For this purpose, they introduced the upper index β and lower indices β', β'' of a general Lévy process X and, in addition, the lower index σ of a subordinator. Blumenthal and Getoor (1961, Theorems 8.1 and 8.5) established the following upper and lower bounds for $\dim_{\mathrm{H}} X(E)$ in terms of the upper index β and lower indices β' and β'' of X: For every $E \subset \mathbb{R}_+$, almost surely

$$\dim_{\mathrm{H}} X(E) \le \beta \dim_{\mathrm{H}} E, \quad \text{if } \beta < 1, \tag{4.7}$$

and

$$\dim_{\mathrm{H}} X(E) \ge \begin{cases} \beta' \dim_{\mathrm{H}} E, & \text{if } \beta' \le d, \\ 1 \wedge \beta'' \dim_{\mathrm{H}} E, & \text{if } \beta' > d = 1. \end{cases}$$

They showed, in addition, that when X is a subordinator, then

$$\sigma \dim_{\rm H} E \le \dim_{\rm H} X(E) \le \beta \dim_{\rm H} E, \quad \text{a.s.} \tag{4.8}$$

The restriction $\beta < 1$ of (4.7) was removed subsequently by Millar (1971, Theorem 5.1). Blumenthal and Getoor (1961, p.512) conjectured that, given a Borel set $E \subset [0,1]$, there exists a constant $\lambda(X,E)$ such that

$$\dim_{\rm H} X(E) = \lambda(X,E), \quad \text{a.s.} \tag{4.9}$$

Except in the cases that X is a Lévy process with stable components [Hendricks (1973)], or a subordinator [Hawkes (1978b)], the question whether the above conjecture holds or not had remained unanswered. Recently, by using potential of additive Lévy processes, Khoshnevisan and Xiao (2003b) have verified the above conjecture by proving the following theorem.

Theorem 4.4. *Suppose $X = \{X(t); t \in \mathbb{R}_+\}$ is a Lévy process in $\mathbb{R}^d$, and denote its Lévy exponent by Ψ. Then for any Borel set $E \subset \mathbb{R}_+$, a.s.*

$$\begin{aligned}&\dim_{\rm H} X(E)\\ &= \sup\left\{\beta \in (0,d) : \inf_{\mu \in \mathcal{P}(E)} \int_{\mathbb{R}^d} \iint e^{-|s-t|\Psi\left(\operatorname{sgn}(s-t)\xi\right)} \|\xi\|^{\beta-d} \mu(ds)\mu(dt)d\xi < +\infty\right\}.\end{aligned} \tag{4.10}$$

If X is symmetric, then (4.10) *is equivalent to the following:*

$$\dim_{\rm H} X(E) = \sup\left\{\beta \in (0,d) : \mathcal{C}_{f_{d-\beta}}(E) > 0\right\}, \quad a.s., \tag{4.11}$$

where f_γ is defined as

$$f_\gamma(x) := \int_{\mathbb{R}^d} e^{-|x|\Psi(\xi)} \|\xi\|^{-\gamma}\, d\xi, \qquad \forall x \in \mathbb{R}, \gamma \in (0,d).$$

Formula (4.10) shows that $\dim_{\rm H} X(E)$ may depend on other characteristics of the set E than its Hausdorff dimension and it is not easy to use. Only in a few cases, we can express $\dim_{\rm H} X(E)$ in terms of $\dim_{\rm H} E$; see Blumenthal and Getoor (1960), Hendricks (1973), Meerchaert and Xiao (2003).

After the Hausdorff dimension of $X(\mathbb{R}_+)$ or $X(E)$ is known, it is natural to study its exact Hausdorff measure or exact capacity. The former problem has been studied extensively; see Taylor (1986a) or Xiao (2003) for a survey. The latter problem has been investigated by Kahane (1985b) and Hawkes (1998) for symmetric stable Lévy processes and subordinators, respectively. Applying potential theory of additive Lévy processes, Khoshnevisan and Xiao (2003b) have recently extended their results to general Lévy processes.

On the other hand, Pemantle, Peres and Shapiro (1996) have shown that the range of Brownian motion B in $\mathbb{R}^d$ ($d \ge 3$) is *capacity equivalent* to $[0,1]^2$. Rosen (2000) extends their results to a class of Lévy processes and introduces the notion of capacitary modulus. A function $h(x) : \mathbb{R}^d \to \mathbb{R}_+$ is called a *capacitary modulus* for $\Lambda \subset \mathbb{R}^d$ if there exist constants $0 < K_3 \le K_4 < \infty$ such that

$$\left[K_4 \int_{\mathbb{R}^d} f(|x|)h(x)dx\right]^{-1} \le \mathcal{C}_f(\Lambda) \le \left[K_3 \int_{\mathbb{R}^d} f(|x|)h(x)dx\right]^{-1}$$

for all non-increasing $f : \mathbb{R}_+ \to [0, \infty]$. Thus, the results of Pemantle, Peres and Shapiro (1996) imply that the function

$$h(x) = \begin{cases} |x|^{-(d-2)} & \text{if } d \geq 3 \\ |\log x| & \text{if } d = 2 \end{cases} \tag{4.12}$$

is a capacity modulus for $B([0,1])$. It would be interesting to study capacitary modulus for the range of an additive Lévy processes.

4.2. Existence and Hausdorff dimension of multiple points

Let $X = \{X(t), t \in \mathbb{R}_+\}$ be a d-dimensional Lévy process. The existence of k-multiple points of X has been solved by LeGall et al. (1989), Evans (1987a), Fitzsimmons and Salisbury (1989). In this subsection, we show that the potential theory of additive Lévy processes can be applied to study the existence of self-intersections of $X(t)$ when t is restricted to disjoint compact subsets $E_1, E_2, \ldots, E_k$ of $\mathbb{R}_+$ ($k \geq 2$). This problem was originally considered by Kahane (1983) who showed that if X is a symmetric stable Lévy process in $\mathbb{R}^d$ of index α and $k = 2$, then

$$\mathcal{C}_{d/\alpha}(E_1 \times E_2) > 0 \Rightarrow \mathbb{P}\{X(E_1) \cap X(E_2) \neq \emptyset\} > 0 \Rightarrow \lambda_{d/\alpha}(E_1 \times E_2) > 0. \tag{4.13}$$

Kahane (1983, p. 90) conjectured that $\mathcal{C}_{d/\alpha}(E_1 \times E_2) > 0$ is necessary and sufficient for $\mathbb{P}\{X(E_1) \cap X(E_2) \neq \emptyset\} > 0$. This conjecture was verified by Khoshnevisan (1999) for Brownian motion. The following theorem of Khoshnevisan and Xiao (2003b) establishes a necessary and sufficient condition for $\mathbb{P}\{X(E_1) \cap \cdots \cap X(E_k) \neq \emptyset\} > 0$ and implies that Kahane's conjecture is true for all symmetric stable Lévy processes.

Theorem 4.5 (Kahane's Problem). *Let X be a symmetric Lévy process in $\mathbb{R}^d$ with Lévy exponent Ψ. If the distribution of $X(t)$ is equivalent to λ_d for all $t > 0$, then for any disjoint compact sets $E_1, \ldots, E_k \subset \mathbb{R}_+$, $\mathbb{P}\{X(E_1) \cap \cdots \cap X(E_k) \neq \emptyset\} > 0$ if and only if $\mathcal{C}_f(E_1 \times E_2 \times \cdots \times E_k) > 0$, where*

$$f(x) := \int_{\mathbb{R}^{d(k-1)}} \exp\Big(- \sum_{j=1}^{k} |x_j| \Psi(\xi_{j-1} - \xi_j) \Big)\, d\xi, \qquad x \in \mathbb{R}^k,\ \xi_0 = \xi_k = 0.$$

To prove Theorem 4.5, it suffices to consider k independent Lévy processes $X_1, \ldots, X_k$ in $\mathbb{R}^d$ with exponent Ψ and define an additive Lévy process Z by (2.6). Then by Lemma 3.1, we need to find a necessary and sufficient condition for $\mathbb{E}\{\lambda_{d(k-1)}(Z(E_1 \times \cdots \times E_k))\} > 0$. This is done by Theorem 2.1 of Khoshnevisan, Xiao and Zhong (2003a) [see also Theorem 3.6, under some extra conditions].

Now we consider the fractal properties of the set $M_k^{(d)}$ of k-multiple points and the set $L_k^{(d)}$ of k-multiple times of X defined by

$$M_k^{(d)} = \{x \in \mathbb{R}^d :\ x = X(t_1) = \cdots = X(t_k) \text{ for distinct } t_1, \ldots, t_k \in \mathbb{R}_+\}$$

and

$$L_k^{(d)} = \big\{(t_1, \cdots, t_k) \in \mathbb{R}_+^k,\ t_1, \ldots, t_k \text{ are distinct and } X(t_1) = \cdots = X(t_k)\big\},$$

respectively.

The Hausdorff dimension $\dim_{\mathrm{H}} M_k^{(d)}$ has been determined by Taylor (1966), Fristedt (1967), Hendricks (1974), Hawkes (1978c), Shieh (1998) for stable Lévy processes, operator stable Lévy processes and certain isotropic Lévy processes, respectively. However, the following problem is still open:

Problem 4.6. Given a general d-dimensional Lévy process X, compute $\dim_{\mathrm{H}} M_k^{(d)}$ and $\dim_{\mathrm{P}} M_k^{(d)}$ explicitly in terms of the Lévy exponent of X, where $\dim_{\mathrm{P}}$ denotes packing dimension.

Note that Corollary 4.2 computes the Hausdorff dimension of M_1. In order to understand the fine structure of M_k for $k > 1$, we need to appeal to the potential theory for multiplicative Lévy processes $X = X_1 \otimes \cdots \otimes X_k$. Prefatory results, along these lines, have been developed earlier in Evans (1987a, b), Fitzsimmons and Salisbury (1989), Peres (1999), and Khoshnevisan (2002).

Very little is known about the packing dimension aspect of Problem 4.6, and completely new ideas are needed. For $k = 1$, Taylor (1986b) has proven that $\dim_{\mathrm{P}} X(\mathbb{R}_+)$ is equal to a certain exponent γ' a.s., but γ' seems, in general, incomputable. It would be desirable to represent γ' in terms of the Lévy exponent Ψ.

For the Hausdorff dimension of $L_k^{(d)}$, Rosen (1983) has shown that if X is a d-dimensional Brownian motion ($d = 2, 3$), then

$$\dim_{\mathrm{H}} L_2^{(3)} = \frac{1}{2} \qquad \text{and} \qquad \dim_{\mathrm{H}} L_k^{(2)} = 1 \quad \text{for all} \quad k \geq 2. \tag{4.14}$$

The following more general result is derived from Theorem 2.10 in Khoshnevisan and Xiao (2002) [see also Khoshnevisan and Xiao (2003a)]:

Theorem 4.7. *If X is a symmetric Lévy process in $\mathbb{R}^d$ with exponent Ψ such that $\xi \mapsto e^{-t\Psi(\xi)}$ is in $L^1(\mathbb{R}^d)$. Then*

$$\dim_{\mathrm{H}} L_k^{(d)} = \sup\left\{b > 0 : \int_{[0,1]^k} \frac{1}{|\mathbf{s}|^b}\, \Phi(\mathbf{s})\, d\mathbf{s} < \infty\right\},$$

where Φ is the gauge function on $\mathbb{R}^k$ defined by

$$\Phi(\mathbf{s}) = (2\pi)^{-d} \int_{\mathbb{R}^{(k-1)d}} \exp\left(-\sum_{j=1}^{k} |s_j| \Psi\left(\xi_j - \xi_{j-1}\right) \right) d\xi \qquad \textit{for } \mathbf{s} = \langle s_j \rangle \in \mathbb{R}^k.$$

In particular, if X is a symmetric stable Lévy processes in $\mathbb{R}^d$ with index $\alpha \in (0, 2]$ and such that $\alpha k > (k-1)d$ [i.e., $L_k^{(d)} \neq \emptyset$], then

$$\dim_{\mathrm{H}} L_k^{(d)} = k - \frac{(k-1)d}{\alpha}.$$

This extends Rosen's result (4.14).

The conditions of Theorem 4.7 are quite restrictive. It would be interesting to solve the following:

Problem 4.8. Find $\dim_{\rm H} L_k^{(d)}$ and $\dim_{\rm P} L_k^{(d)}$ for a general Lévy process X in $\mathbb{R}^d$.

Except for Brownian motion, there is no knowledge on the exact Hausdorff and packing measure functions of M_k and L_k. These problems have to be studied by using other methods than potential theory of additive Lévy processes, and hence will not be discussed here. We refer to Xiao (2003b, section 9) for more information.

References

[1] R. J. Adler (1981), *The Geometry of Random Fields.* Wiley, New York.

[2] R. J. Adler, D. Monrad, R. H. Scissors and R. Wilson (1981), Representation, decompositions and sample function continuity of random fields with independent increments. *Stoch. Proc. Appl.* **15**, 3–30.

[3] P. Becker-Kern, M. M. Meerschaert and H. P. Scheffler (2003), Hausdorff dimension of operator stable sample paths. *Monatsh. Math.* to appear.

[4] C. Berg and G. Forst (1975), *Potential Theory on Locally Compact Abelian Groups.* Springer-Verlag, New York-Heidelberg.

[5] J. Bertoin (1996), *Lévy Processes.* Cambridge Univ. Press.

[6] J. Bertoin (1999), *Subordinators: Examples and Applications.* In: *Lectures on Probability Theory and Statistics (Saint-Flour, 1997)*, pp. 1–91, Lecture Notes in Math., **1717**, Springer-Verlag, Berlin.

[7] R. M. Blumenthal and R. Getoor (1960), A dimension theorem for sample functions of stable processes. *Illinois J. Math.* **4**, 370–375.

[8] R. M. Blumenthal and R. Getoor (1961), Sample functions of stochastic processes with stationary independent increments. *J. Math. Mech.* **10**, 493–516.

[9] R. M. Blumenthal and R. Getoor (1968), *Markov Processes and Potential Theory.* Academic Press, New York.

[10] J. Bretagnolle (1971), Résultats de Kesten sur les processus à accroisements indépendants. *Sém. de Prob. V*, Lecture Notes in Math., **191**, 21–36, Springer-Verlag, Berlin.

[11] R. C. Dalang (2003), Level sets and excursions of the Brownian sheet. In: *Topics in spatial stochastic processes (Martina Franca, 2001)*, 167–208, *Lecture Notes in Math.*, **1802**, Springer, Berlin.

[12] R. C. Dalang and T. S. Mountford (1996), Nondifferentiability of curves on the Brownian sheet. *Ann. Probab.* **24**, 182–195.

[13] R. C. Dalang and T. S. Mountford (1997), Points of increase of the Brownian sheet, *Probab. Th. Rel. Fields*, **108**, 1–27.

[14] R. C. Dalang and T. Mountford (2001), Jordan curves in the level sets of additive Brownian motion. *Trans. Amer. Math. Soc.* **353**, 3531–3545.

[15] R. C. Dalang and T. Mountford (2002), Eccentric behaviors of the Brownian sheet along lines. *Ann. Probab.* **30**, 293–322.

[16] R. C. Dalang and T. Mountford (2003), Non-independence of excursions of the Brownian sheet and of additive Brownian motion. *Trans. Amer. Math. Soc.* **355**, 967–985

[17] R. C. Dalang and J. B. Walsh (1992), The sharp Markov property of Lévy sheets. *Ann. Probab.* **20**, 591–626.

[18] R. C. Dalang and J. B. Walsh (1993a), Geography of the level sets of the Brownian sheet. *Probab. Th. Rel. Fields*, **96**, 153–176.

[19] R. C. Dalang and J. B. Walsh (1993b), The structure of a Brownian bubble. *Probab. Th. Rel. Fields*, **96**, 475–501.

[20] A. Dembo, Y. Peres, J. Rosen and O. Zeitouni (2000), Thick points for spatial Brownian motion: Multifractal analysis of occupation measure. *Ann. Prob.*, **28**, 1–35.

[21] J. Doob (1984), *Classical Potential Theory and its Probabilistic Counterpart.* Springer-Verlag, Berlin.

[22] W. Ehm (1981), Sample function properties of multi-parameter stable processes. *Z. Wahrsch. Verw Gebiete* **56**, 195–228.

[23] S. N. Evans (1987a), Multiple points in the sample paths of a Lévy process. *Probab. Th. Rel. Fields* **76**, 359–367.

[24] S. N. Evans (1987b), Potential theory for a family of several Markov processes. *Ann. Inst. Henri Poincaré Probab. Statist.* **23**, 499–530.

[25] K. J. Falconer(1990), *Fractal Geometry – Mathematical Foundations and Applications.* Wiley & Sons, Chichester.

[26] P. J. Fitzsimmons and T. S. Salisbury (1989), Capacity and energy for multiparameter processes. *Ann. Inst. Henri Poincaré Probab. Statist.* **25**, 325–350.

[27] B. E. Fristedt (1967), An extension of a theorem of S. J. Taylor concerning the multiple points of the symmetric stable process. *Z. Wahrsch. Verw. Gebiete* **9**, 62–64.

[28] B. E. Fristedt (1974), Sample functions of stochastic processes with stationary, independent increments. *Adv. in Probab. III*, pp. 241–396, Dekker.

[29] B. E. Fristedt and W. E. Pruitt (1971), Lower functions for increasing random walks and subordinators. *Z. Wahrsch. Verw. Gebiete* **18**, 167–182.

[30] J. Hawkes (1977), Local properties of some Gaussian processes. *Z. Wahrsch. Verw. Gebiete* **40**, 309–315.

[31] J. Hawkes (1978a), Measures of Hausdorff type and stable processes. *Mathematika* **25**, 202–210.

[32] J. Hawkes (1978b), Image and intersection sets for subordinators. *J. London Math. Soc. (2)* **17**, 567–576.

[33] J. Hawkes (1978c), Multiple points for symmetric Lévy processes. *Math. Proc. Camb. Philos. Soc.* **83**, 83–90.

[34] J. Hawkes (1981), Trees generated by a simple branching process. *J. London Math. Soc. (2)* **24**, 373–384.

[35] J. Hawkes (1998), Exact capacity results for stable processes. *Probab. Th. Rel. Fields* **112**, 1–11.

[36] W. J. Hendricks (1973), A dimension theorem for sample functions of processes with stable components. *Ann. Probab.* **1**, 849–853.

[37] W. J. Hendricks (1974), Multiple points for a process in $\mathbb{R}^2$ with stable components. *Z. Wahrsch. Verw. Gebiete* **28**, 113–128.

[38] F. Hirsch (1995), Potential Theory related to some multiparameter processes. *Potential Analysis*, **4**, 245–267.

[39] F. Hirsch and S. Song (1995a), Symmetric Skorohod topology on n–variable functions and hierarchical Markov properties of n-parameter processes. *Prob. Th. Rel. Fields*, **103**(1), 25–43.

[40] F. Hirsch and S. Song (1995b), Markov properties of multiparameter processes and capacities. *Prob. Th. Rel. Fields*, **103**(1), 45–71.

[41] J.-P. Kahane (1972), Ensembles parfaits et processus de Lévy. *Period. Math. Hungar.* **2**, 49–59.

[42] J.-P. Kahane (1983), Points multiples des processus de Lévy symétriques stables restreints á un ensemble de valeurs du temps. *Seminar on Harmonic Analysis, 1981–1982,* pp. 74–105, *Publ. Math. Orsay* **83**, 2, Univ. Paris XI, Orsay.

[43] J.-P. Kahane (1985a), *Some Random Series of Functions.* 2nd edition. Cambridge Univ. Press.

[44] J.-P. Kahane (1985b), Ensembles aléatoires et dimensions. In: *Recent Progress in Fourier Analysis (El Escorial, 1983)*, pp. 65–121, North-Holland Math. Stud., **111**, North-Holland, Amsterdam.

[45] R. Kaufman (1972), Measures of Hausdorff-type, and Brownian motion. *Mathematika* **19**, 115–119.

[46] R. Kaufman (1975), Large increments of Brownian motion. *Nagoya Math. J.*, **56**, 139–145.

[47] R. Kaufman and J. M. Wu (1982), Parabolic potential theory. *J. Differential Equations* **43**, 204–234.

[48] H. Kesten (1969), Hitting probabilities of single points for processes with stationary independent increments. *Mem. Amer. Math. Soc.* **93**, 129 pp.

[49] D. Khoshnevisan (1995), On the distribution of bubbles of the Brownian sheet. *Ann. Probab.* **23**, 786–805.

[50] D. Khoshnevisan (1999), Brownian sheet images and Bessel-Riesz capacity. *Trans. Amer. Math. Soc.* **351**, 2607–2622.

[51] D. Khoshnevisan (2002), *Multi-Parameter Processes: An Introduction to Random Fields.* Springer-Verlag, Berlin.

[52] D. Khoshnevisan, Y. Peres and Y. Xiao (2000), Limsup random fractals. *Electron. J. Probab.* **5**, No.5, 24 pp.

[53] D. Khoshnevisan and Z. Shi (1999), Brownian sheet and capacity. *Ann. Probab.* **27**, 1135–1159.

[54] D. Khoshnevisan and Z. Shi (2000), Fast sets and points for fractional Brownian motion. *Séminaire de Probabilités* XXXIV, pp. 393–416, Lecture Notes in Math., **1729**, Springer-Verlag, Berlin.

[55] D. Khoshnevisan and Y. Xiao (2002), Level sets of additive Lévy processes. *Ann. Probab.* **30**, 62–100.

[56] D. Khoshnevisan and Y. Xiao (2003a), Weak unimodality of finite measures, and an application to potential theory of additive Lévy processes. *Proc. Amer. Math. Soc.* **131**, 2611–2616.

[57] D. Khoshnevisan and Y. Xiao (2003b), Lévy processes: capacity and Hausdorff dimension. *Submitted.*

[58] D. Khoshnevisan, Y. Xiao and Y. Zhong (2003a), Measuring the range of an additive Lévy processes. *Ann. Probab.* **31**, 1097–1141.

[59] D. Khoshnevisan, Y. Xiao and Y. Zhong (2003b), Local times of additive Lévy processes. *Stoch. Process. Appl.* **104**, 193–216.

[60] J.-F. Le Gall, J. Rosen and N.-R. Shieh (1989), Multiple points of Lévy processes. *Ann. Probab.* **17**, 503–515.

[61] R. Lyons (1990), Random walks and percolation on trees. *Ann. Probab.* **18**, 931–958.

[62] P. Mattila (1995), *Geometry of Sets and Measures in Euclidean Spaces.* Cambridge Univ. Press, Cambridge.

[63] H. P. McKean, Jr. (1955), Hausdorff–Besicovitch dimension of Brownian motion paths. *Duke Math. J.* **22**, 229–234.

[64] M. Meerschaert and Y. Xiao (2003), Dimension results for the sample paths of operator stable processes. *Submitted.*

[65] P. W. Millar (1971), Path behavior of processes with stationary independent increments. *Z. Wahrsch. Verw. Gebiete* **17**, 53–73.

[66] S. Orey and S. J. Taylor (1974), How often on a Brownian path does the law of the iterated logarithm fail? *Proc. London Math. Soc.*, **28**, 174–192.

[67] R. Pemantle, Y. Peres, and J. W. Shapiro (1996), The trace of spatial Brownian motion is capacity-equivalent to the unit square. *Probab. Th. Rel. Fields* **106**, 379–399.

[68] Y. Peres (1996), Remarks on intersection-equivalence and capacity-equivalence. *Ann. Inst. Henri Poincaré Phys. Théor.* **64**, 339–347.

[69] Y. Peres (1999), Probability on trees: an introductory climb. In: *Lectures on Probability Theory and Statistics (Saint-Flour, 1997)*, pp. 193–280, Lecture Notes in Math., **1717**, Springer-Verlag, Berlin.

[70] W. E. Pruitt (1969), The Hausdorff dimension of the range of a process with stationary independent increments. *J. Math. Mech.* **19**, 371–378.

[71] J. Rosen (1983), A local time approach to the self-intersections of Brownian paths in space. *Comm. Math. Phys.* **88**, 327–338.

[72] J. Rosen (1984), Self-intersections of random fields. *Ann. Probab.* **12**, 108–119.

[73] J. Rosen (2000), Capacitary moduli for Lévy processes and intersections. *Stoch. Process. Appl.* **89**, 269–285.

[74] K. Sato (1999), *Lévy Processes and Infinitely Divisible Distributions.* Cambridge Univ. Press, Cambridge.

[75] N. R. Shieh (1998), Multiple points of dilation-stable Lévy processes. *Ann. Probab.* **26**, 1341–1355.

[76] M. L. Straf (1972), Weak convergence of stochastic processes with several parameters. In: *Proc. Sixth Berkeley Sympos. Math. Statist. Probab. (Univ. California, Berkeley, Calif., 1970/1971)*, **2**, pp. 187–221. Univ. California Press, Berkeley.

[77] S. J. Taylor (1953), The Hausdorff α-dimensional measure of Brownian paths in n-space. *Proc. Camb. Philos. Soc.* **49**, 31–39.

[78] S. J. Taylor (1955), The α-dimensional measure of the graph and set of zeros of a Brownian path. *Proc. Camb. Philos. Soc.* **51**, 265–274.

[79] S. J. Taylor (1966), Multiple points for the sample paths of the symmetric stable process. Z. Wahrsch. Verw. Gebiete **5**, 247–264.

[80] S. J. Taylor (1986a), The measure theory of random fractals. *Math. Proc. Camb. Philos. Soc.* **100**, 383–406.

[81] S. J. Taylor (1986b), The use of packing measure in the analysis of random sets. *Lecture Notes in Math.*, **1203**, 214–222.

[82] S. J. Taylor and N. A. Watson (1985), A Hausdorff measure classification of polar sets for the heat equation. *Math. Proc. Camb. Philos. Soc.* **97**, 325–344.

[83] M. E. Vares (1983), Local times for two-parameter Levy processes. *Stochastic Process. Appl.*, **15**, 59–82.

[84] J. B. Walsh (1986). *Martingales with a Multidimensional Parameter and Stochastic Integrals in the Plane.* (Ed's: G. del Piño and R. Robodello), *Lecture Notes in Math.* **1215**, Springer, Berlin.

[85] N. A. Watson (1976), Green functions, potentials, and the Dirichlet problem for the heat equation. *Proc. London Math. Soc.* **33**, 251–298; Corrigendum, ibid., **37** (1978), 32–34.

[86] N. A. Watson (1978), Thermal capacity. *Proc. London Math. Soc.* **37**, 342–362.

[87] Y. Xiao (2003), Random fractals and Markov processes. *Proceedings of the Conference on Fractal Geometry and Applications: A Jubilee of Benoit Mandelbrot,* (to appear).

[88] Y. Zhong and Y. Xiao (1995), Self-intersection local times and multiple points of the stable sheet. *Acta Math. Sci. (Chinese)* **15**, 141–152.

Davar Khoshnevisan
Department of Mathematics
155 South 1400 East JWB 233
University of Utah
Salt Lake City, UT 84112–0090
USA
e-mail: `davar@math.utah.edu`
URL: `http://www.math.utah.edu/~davar`

Yimin Xiao
Department of Statistics and Probability
A-413 Wells Hall
Michigan State University
East Lansing, MI 48824
USA
e-mail: `xiao@stt.msu.edu`
URL: `http://www.stt.msu.edu/~xiaoyimi`

Part 4

Fractal Analysis in Euclidean Space

Progress in Probability, Vol. 57, 173–192

The Fractal Laplacian and Multifractal Quantities

Hans Triebel

Abstract. This survey deals with several quantities of Radon measures in the plane: *multifractal characteristics*, *Courant characteristics* related to distinguished eigenfunctions of fractal Laplacians, and *Besov characteristics* related to function spaces. It is the main aim to describe how closely these quantities are related to each other. In addition, spectral properties of fractal Laplacians are discussed resulting in *Weyl measures*.

Mathematics Subject Classification (2000). MSC 2000: Primary 46E35; Secondary 42B35, 28A80, 35P15.

Keywords. Multifractals, fractal Laplacian, function spaces.

1. Introduction

Let Ω be a bounded C^∞ domain in the plane $\mathbb{R}^2$ and let $(-\Delta)^{-1}$ be the inverse of the respective Dirichlet Laplacian. Let μ be a finite Radon measure in $\mathbb{R}^2$ such that

$$\operatorname{supp} \mu = \Gamma \subset \Omega.$$

In [17], Ch. V; [18], Ch. III, and in the survey [19] we dealt with the fractal Laplacian

$$B = (-\Delta)^{-1} \circ \mu, \tag{1.1}$$

preferably in case of d-sets, $0 < d < 2$,

$$\mu(B(\gamma, r)) \sim r^d, \quad \gamma \in \Gamma, \quad 0 < r \leq 1, \tag{1.2}$$

where $B(\gamma, r)$ is the circle centred at γ and of radius r. It comes out that under these circumstances B makes sense: it is a compact, self-adjoint, non-negative operator in the Hilbert space $\overset{\circ}{H}{}^1(\Omega)$, the largest eigenvalue $\varrho_1 = \varrho$ is simple (and positive), the respective eigenfunction u (called the *Courant function*) is positive in Ω (ignoring a constant complex factor), and the positive eigenvalues ϱ_l have the Weyl property,

$$\varrho_l \sim l^{-1}, \quad l \in \mathbb{N}. \tag{1.3}$$

In [17], [18], [19] and the underlying papers and reports we concentrated on the spectral property (1.3) not only for d-sets but also for more general measures μ which we called *Weyl measures.* The aim of the present survey is different and in some sense complementary to [19]. We deal now with the Courant function u and the symbiotic relationship between

- multifractal quantities of μ,
- smoothness properties of u,
- some characteristics of related Besov spaces.

As far as function spaces are concerned we restrict ourselves to the bare minimum: they are a tool not a subject. Most of the results are new, some will appear in forthcoming papers (we give the related references), others are formulated here for the first time and proofs will be published later on.
The plan of the paper is the following. In Section 2 we describe the general set-up and properties of the operator B in (1.1). Section 3 deals with the indicated relationship of diverse types of characteristics of μ: multifractal, Besov, Courant. At the end we formulated some problems.

2. The Dirichlet Laplacian

2.1. The classical theory

Let Ω be a bounded C^∞ domain in the plane $\mathbb{R}^2$ with the boundary $\partial\Omega$ and let $L_2(\Omega)$ be the usual Hilbert space of all complex-valued Lebesgue-measurable functions f such that

$$\|f\,|L_2(\Omega)\| = \left(\int_\Omega |f(x)|^2 \, \mu_L(dx) \right)^{\frac{1}{2}}$$

is finite. Here μ_L stands for the Lebesgue measure. Let $l \in \mathbb{N}$ (the collection of all natural numbers). Then $H^l(\Omega)$ is the usual Sobolev space

$$H^l(\Omega) = \{f : \ D^\alpha f \in L_2(\Omega), \ |\alpha| \le l\},$$

obviously normed. Of course, $D^\alpha f$ has the usual meaning of distributional derivatives. Let

$$H_0^2(\Omega) = \left\{f \in H^2(\Omega) : \ f|\partial\Omega = 0\right\}$$

and

$$\overset{\circ}{H}{}^1(\Omega) = \left\{f \in H^1(\Omega) : \ f|\partial\Omega = 0\right\},$$

where the latter space will be equipped with the scalar product

$$(f,g)_{\overset{\circ}{H}^1(\Omega)} = \sum_{m=1}^{2} \int_{\Omega} \frac{\partial f}{\partial x_m} \cdot \frac{\partial \overline{g}}{\partial x_m} \, \mu_L(dx). \tag{2.1}$$

Here the trace $f|\partial\Omega$ of $f \in H^l(\Omega)$ with $l \in \mathbb{N}$ makes sense and has the usual meaning. Recall that $D(\Omega) = C_0^\infty(\Omega)$ is dense in $\overset{\circ}{H}{}^1(\Omega)$ (but not in $H_0^2(\Omega)$, which explains the different way of writing). Then the Dirichlet Laplacian

$$-\Delta = -\frac{\partial^2}{\partial x_1^2} - \frac{\partial^2}{\partial x_2^2}, \qquad \text{dom } (-\Delta) = H_0^2(\Omega),$$

is a positive-definite, self-adjoint operator in $L_2(\Omega)$ with a pure point spectrum. We prefer here its inverse,

$$B = (-\Delta)^{-1} = (-\Delta)^{-1} \circ \mu_L, \tag{2.2}$$

where again μ_L is the Lebesgue measure (preparing notationally our later considerations). Then B is a positive, self-adjoint, compact operator in $L_2(\Omega)$. Let ϱ_k be its positive eigenvalues and u_k be its related eigenfunctions,

$$Bu_k = \varrho_k u_k, \qquad \varrho_1 > \varrho_2 \geq \cdots > 0, \qquad \varrho_k \to 0$$

if $k \to \infty$. Recall the following outstanding properties (again notationally adapted to our later needs):

(a) (H. Weyl, 1912, [24], [25]),

$$\varrho_k \sim k^{-1}, \qquad k \in \mathbb{N}.$$

(b) (R. Courant, 1924, [6]). The largest eigenvalue $\varrho = \varrho_1$ is simple and (ignoring a complex constant) the related eigenfunction $u_1(x) = u(x)$ is positive,

$$u(x) > 0, \qquad x \in \Omega. \tag{2.3}$$

(c) (Smoothness). In obvious notation,

$$u_k \in C^\infty(\overline{\Omega}), \qquad k \in \mathbb{N}.$$

Here and in the sequel we use the equivalence $\sim$ for two positive functions $a(x)$ and $b(x)$ or two sequences of positive numbers a_k and b_k (say with $k \in \mathbb{N}$) if there are two positive numbers c and C such that

$$ca(x) \leq b(x) \leq Ca(x) \quad \text{or} \quad ca_k \leq b_k \leq Ca_k$$

for all admitted variables x or k, respectively.

It is the main aim of this survey to describe what happens if the Lebesgue measure μ_L in (2.2) is replaced by appropriate Radon measures μ and to which extent the above properties (a) - (c) are preserved or how they must be modified. We concentrate in this survey on the *Courant function* u according to (2.3).

2.2. The fractal theory

Again Ω is a bounded C^∞ domain in the plane and $(-\Delta)^{-1}$ is the inverse of the respective Dirichlet Laplacian according to 2.1. Let μ be a Radon measure in $\mathbb{R}^2$ with

$$\text{supp}\,\mu = \Gamma \subset \Omega, \qquad 0 < \mu(\mathbb{R}^2) < \infty \quad \text{and} \quad |\Gamma| = 0, \tag{2.4}$$

where $|\Gamma|$ is the Lebesgue measure of Γ. In particular, Γ is compact in Ω. First one has to say what is meant by B in (1.1). It comes out that this question can be reduced to the trace problem asking for a constant $c > 0$ such that

$$\|\varphi\,|L_2(\Gamma,\mu)\| \le c\,\|\varphi\,|H^1(\Omega)\| \quad \text{for all } \varphi \in D(\Omega). \tag{2.5}$$

Here $L_2(\Gamma,\mu)$ is the naturally normed Hilbert space on Γ with respect to the measure μ. If one has an affirmative answer then one can define the linear and bounded trace operator tr_μ,

$$tr_\mu \;:\quad H^1(\Omega) \hookrightarrow L_2(\Gamma,\mu) \tag{2.6}$$

by completion. Since we have (2.4) one can replace Ω in (2.5) and (2.6) by $\mathbb{R}^2$. On the other hand, any $f \in L_2(\Gamma,\mu)$ can be interpreted in the usual way as a tempered distribution, denoted by $id_\mu f$. If we have (2.6) then it comes out that

$$id_\mu = tr'_\mu \;:\quad L_2(\Gamma,\mu) \hookrightarrow H^{-1}(\Omega), \tag{2.7}$$

is just the dual of tr_μ, where

$$H^{-1}(\Omega) = \left(\mathring{H}^1(\Omega)\right)' \tag{2.8}$$

according to the dual pairing $(D(\Omega), D'(\Omega))$, might be the shortest way to say what is meant by the Sobolev space $H^{-1}(\Omega)$ of smoothness -1. Then we have

$$id^\mu = id_\mu \circ tr_\mu \;:\quad \mathring{H}^1(\Omega) \hookrightarrow H^{-1}(\Omega) \tag{2.9}$$

and

$$B = (-\Delta)^{-1} \circ id^\mu = (-\Delta)^{-1} \circ \mu \;:\quad \mathring{H}^1(\Omega) \hookrightarrow \mathring{H}^1(\Omega) \tag{2.10}$$

is a linear and bounded operator in $\mathring{H}^1(\Omega)$, where the middle term in this equation is the precise definition of what is meant by the last expression in the equation. Details and further explanations concerning (2.5) - (2.10) may be found in [18], 9.2, pp. 122-124; 20.3, p. 297; 19.2, pp. 254, 255. We refer also to [19] where we discussed this background material in greater details. (As said the present survey might be considered as the continuation of the survey [19]). In other words, first of all one has to clarify under which circumstances the trace operator tr_μ according to (2.5), (2.6), exists.

We introduce a few notation. Let Q_{jm} with $j \in \mathbb{N}_0$ $(= \mathbb{N} \cup \{0\})$ and $m \in \mathbb{Z}^2$ (the lattice of all points in $\mathbb{R}^2$ with integer-valued coordinates) be the square centred at $2^{-j}m$ and with side-length 2^{-j+1}. (We use standard notation without further explanations; in case of doubt one may consult [18]). If μ is given by (2.4) then we put

$$\mu_j = \sup_{m \in \mathbb{Z}^2} \mu(Q_{jm}), \qquad j \in \mathbb{N}_0. \tag{2.11}$$

Furthermore let

$$f_{jm} = \int_{Q_{jm}} f(\gamma)\,\mu(d\gamma), \qquad j \in \mathbb{N}_0, \quad m \in \mathbb{Z}^2.$$

Finally, the measure μ according to (2.4) is called *isotropic* if there is a positive, continuous, monotonically increasing function h on the interval $(0,1]$ such that

$$\mu\left(B(\gamma,r)\right) \sim h(r) \quad \text{for all } \gamma \in \Gamma \text{ and } 0 < r \leq 1, \tag{2.12}$$

where $B(\gamma,r)$ is the circle centred at $\gamma \in \Gamma$ and of radius r. According to [1], [2], [3], for given h there is a Radon measure μ with (2.12) if, and only if, there is a positive, continuous, monotonically increasing function $\widetilde{h}$ on the interval $(0,1]$ such that

$$\widetilde{h}(r) \sim h(r) \quad \text{if} \quad 0 < r \leq 1 \quad \text{and} \quad \frac{\widetilde{h}(2^{-j-k})}{\widetilde{h}(2^{-j})} \geq 2^{-2k} \tag{2.13}$$

for $j \in \mathbb{N}_0$ and $k \in \mathbb{N}_0$. The final version, including standard and non-standard examples of isotropic measures may be found in [3]. (Recall that we are always in the plane $\mathbb{R}^2$, otherwise 2^{-2k} in (2.13) must be replaced in $\mathbb{R}^n$ by 2^{-nk}).

Theorem 2.1. *Let μ be given by* (2.4).

(i) *Then tr_μ according to* (2.5), (2.6) *exists if, and only if,*

$$\sup \sum_{j=0}^{\infty} \sum_{m \in \mathbb{Z}^2} f_{jm}^2 < \infty, \tag{2.14}$$

where the supremum is taken over all

$$f \in L_2(\Gamma,\mu) \quad \text{with} \quad f \geq 0 \quad \text{and} \quad \|f\,|L_2(\Gamma,\mu)\| \leq 1.$$

(ii) *If tr_μ exists, then*

$$\sum_{j=0}^{\infty} \sum_{m \in \mathbb{Z}^2} \mu(Q_{jm})^2 < \infty. \tag{2.15}$$

Conversely, if

$$\sum_{j=0}^{\infty} \mu_j < \infty, \tag{2.16}$$

then tr_μ exists and is compact.

(iii) *In addition, let μ be isotropic. Then tr_μ exists if, and only if,*

$$\sum_{j=0}^{\infty} h(2^{-j}) < \infty. \tag{2.17}$$

In this case, tr_μ is compact.

Remark 2.2. Part (i) is a special case of [18], Theorem 9.3, p. 125. Both the necessary condition (2.15) and the sufficient condition (2.16) are easy consequences of (2.14). If μ is isotropic then $\mu_j \sim h(2^{-j})$. Hence (2.16) and (2.17) coincide, and in case of isotropic measures this condition is not only sufficient but also necessary. It is remarkable that for isotropic measures tr_μ exists if, and only if, it is compact. Concerning the assertions in part (iii), but also in part (ii), we refer to [22], where one finds proofs also in more general situations (in $\mathbb{R}^n$, and with respect to more general spaces). In particular, [22], Proposition 3, covers also the compactness assertions of tr_μ in the parts (ii) and (iii), again in a more general setting.

In particular, (2.16) guarantees that B given by (2.10) is a linear and compact operator in $\mathring{H}^1(\Omega)$. However for our purpose it is desirable to strengthen this assertion as follows. Let $C(\overline{\Omega})$ be the naturally normed space of all complex-valued continuous functions in $\overline{\Omega}$.

Corollary 2.3. *Let*

$$\sum_{j=0}^{\infty} \mu_j^{\frac{1}{2}} < \infty,$$

where μ and μ_j are given by (2.4) *and* (2.11), *respectively. Then*

$$B = (-\Delta)^{-1} \circ \mu : \quad \mathring{H}^1(\Omega) \hookrightarrow \mathring{H}^1(\Omega) \cap C(\overline{\Omega}). \tag{2.18}$$

Remark 2.4. The stronger assertion (2.18), compared with (2.10) is based on

$$\left\| id_\mu \,:\, L_2(\Gamma,\mu) \hookrightarrow B^{-1}_{2,1}(\mathbb{R}^2) \right\| \le c \sum_{j=0}^{\infty} \mu_j^{\frac{1}{2}}, \tag{2.19}$$

where c is independent of μ with (2.4) and, as a consequence,

$$B : \quad \mathring{H}^1(\Omega) \hookrightarrow \mathring{B}^1_{2,1}(\Omega) \hookrightarrow \mathring{H}^1(\Omega) \cap C(\overline{\Omega}). \tag{2.20}$$

Here $B^{-1}_{2,1}$ and $B^1_{2,1}$ are special Besov spaces, which will be defined later on. In particular, $B^1_{2,1}$ is a smaller space than H^1 and the second inclusion in (2.20) is an embedding theorem of Sobolev type. Assertions of type (2.19) can be extended to $\mathbb{R}^n$, L_p-spaces, and $B^s_{p,q}$-spaces. But details must be shifted to a later occasion. They can be based on the so-called μ-property of function spaces as considered in [20].

After these preparations we are ready for the fractal counterpart of the classical theory as described in 2.1. Let Ω and Γ as in (2.4). We always assume that $\mathring{H}^1(\Omega)$ is normed according to (2.1). Furthermore $\mathring{H}^1(\Omega\backslash\Gamma)$ is the completion of $D(\Omega\backslash\Gamma)$ in $H^1(\Omega\backslash\Gamma)$ normed as in (2.1).

Theorem 2.5. *Let Ω be a bounded C^∞ domain in the plane $\mathbb{R}^2$ and let μ be a Radon measure in $\mathbb{R}^2$ with* (2.4) *and*

$$\sum_{j=0}^{\infty} \mu_j^{\frac{1}{2}} < \infty \quad \textit{where} \quad \mu_j = \sup_{m \in \mathbb{Z}^2} \mu(Q_{jm}) \tag{2.21}$$

as in (2.11). *Then*

$$B = (-\Delta)^{-1} \circ \mu$$

according to (2.10) *and the explanations given there is a self-adjoint, compact, non-negative operator in* $\overset{\circ}{H}{}^{1}(\Omega)$ *with null-space*

$$N(B) = \overset{\circ}{H}{}^{1}(\Omega \backslash \Gamma).$$

Let ϱ_k *be the positive eigenvalues of* B, *repeated according to multiplicity and ordered by magnitude and let* u_k *be the related eigenfunctions,*

$$Bu_k = \varrho_k u_k, \quad k \in \mathbb{N}.$$

(i) *The largest eigenvalue is simple,*

$$\varrho = \varrho_1 > \varrho_2 \geq \varrho_3 \cdots,$$

and the related eigenfunctions $u_1(x)$ *have no zeroes in* Ω,

$$u_1(x) = c\,u(x) \quad \textit{with} \quad u(x) > 0 \quad \textit{if} \quad x \in \Omega, \tag{2.22}$$

and a complex constant c.

(ii) *The eigenfunctions* u_k *are (classical) harmonic functions in* $\Omega \backslash \Gamma$,

$$\Delta u_k(x) = 0 \quad \textit{if} \quad x \in \Omega \backslash \Gamma, \quad \textit{and} \quad u_k(x) \in C(\overline{\Omega}). \tag{2.23}$$

Remark 2.6. By Theorem 2.1 and Corollary 2.3 it follows that B is compact in $\overset{\circ}{H}{}^{1}(\Omega)$ with (2.18). The self-adjointness depends on the special choice of the scalar product in $\overset{\circ}{H}{}^{1}(\Omega)$ according to (2.1). Afterwards both assertions in (2.23) can be obtained quite easily. Condition (2.21) covers both measures of power-type,

$$\mu(Q_{jm}) \leq c\,2^{-j\varepsilon} \quad \text{for some } c > 0 \text{ and } \varepsilon > 0,$$

and of log-type,

$$\mu(Q_{jm}) \leq c\,j^{-\varkappa} \quad \text{for some } c > 0 \text{ and } \varkappa > 2.$$

This applies in particular to suitable isotropic measures as described in (2.12), (2.13), making clear what can happen. The case of d-sets,

$$h(r) = r^d, \qquad 0 < r \leq 1, \qquad 0 < d < 2,$$

hence (1.2), has been considered in [18], Theorem 19.7, pp. 264-265, covering the above assertions (in an $\mathbb{R}^n$-setting), including the Weylian behaviour of the eigenvalues ϱ_k (in $\mathbb{R}^2$),

$$\varrho_k \sim k^{-1}, \qquad k \in \mathbb{N}. \tag{2.24}$$

To which extent this remarkable property remains valid for the more general measures in the above theorem is not so clear. However we dealt with this problem in detail in [18], Section 19, especially 19.17, 19.18; [19]; [23]; [22]. Here this problem will play only a marginal role. We are mostly interested to which extent the function u according to (2.22) reflects multifractal quantities of the underlying measure μ. But nevertheless we give a formal definition.

Definition 2.7. (i) A Radon measure μ with (2.4) is called a *Weyl measure* if B according to (2.10) is a self-adjoint non-negative compact operator and if one has (2.24) for its positive eigenvalues.

(ii) Let μ be a Radon measure in $\mathbb{R}^2$ with (2.4) and (2.21). Then $u = u^\Omega$ according to (2.22) is called the related *Courant function.*

3. Fractal characteristics of measures

3.1. Definitions

Let μ be a positive Radon measure in $\mathbb{R}^2$ with

$$\operatorname{supp} \mu = \Gamma \text{ compact}, \quad 0 < \mu(\mathbb{R}^2) < \infty, \quad |\Gamma| = 0,$$

where $|\Gamma|$ is the Lebesgue measure of Γ. Multifractal quantities of μ (in the context to the so-called multifractal formalism) attracted a lot of attention since the late eighties, but especially in the last few years, preferably asking for the behaviour of

$$\sum_{m\in\mathbb{Z}^2} \mu(Q_{jm})^p \quad \text{if} \quad j \to \infty,$$

where $0 < p < \infty$. The recent development from the point of view of fractal geometry may be found in [5], [12], [11], [8], [10]. Some aspects are surveyed in [7], Chapter 11. From the point of view of Fourier analysis and function spaces we refer to [13], [14], [20]. What follows might be considered as a continuation of [20] including some more recent results, partly published here for the first time. (Again we restrict ourselves to the plane $\mathbb{R}^2$, but most of the definitions and results can be extended to $\mathbb{R}^n$ with $n \in \mathbb{N}$). Recall that Q_{jm} with $j \in \mathbb{N}_0$ and $m \in \mathbb{Z}^2$ are the same squares as in connection with μ_j in (2.11) and in Theorem 2.1. Let $\lambda \in \mathbb{R}$ (real numbers). Then we put

$$\mu_p^\lambda = \sup_{j\in\mathbb{N}_0} 2^{j\lambda} \left(\sum_{m\in\mathbb{Z}^2} \mu(Q_{jm})^p \right)^{\frac{1}{p}} \quad \text{if} \quad 0 < p < \infty \tag{3.1}$$

and

$$\mu_\infty^\lambda = \sup_{j\in\mathbb{N}_0} 2^{j\lambda} \mu_j. \tag{3.2}$$

These quantities together with their generalisations

$$\mu_{pq}^\lambda = \left(\sum_{j=0}^\infty 2^{j\lambda q} \left(\sum_{m\in\mathbb{Z}^2} \mu(Q_{jm})^p \right)^{\frac{q}{p}} \right)^{\frac{1}{q}} \tag{3.3}$$

where $\lambda \in \mathbb{R}$, $0 < p \le \infty$, $0 < q \le \infty$ (obvious modifications if $p = \infty$ and/or $q = \infty$) played a decisive role both in multifractal geometry according to the above references and in the theory of function spaces, where in the latter case one may consult [18], Section 9, and more recently [20]. There one finds further references.

Next we describe briefly what is meant by Besov spaces $B^s_{pq}(\mathbb{R}^2)$ and $B^s_{pq}(\Omega)$ where the case $q=\infty$ is sufficient for us, hence

$$B^s_{p\infty}(\mathbb{R}^2), \qquad s\in\mathbb{R}, \qquad 0<p\le\infty.$$

We put $a_+ = \max(a,0)$ where $a\in\mathbb{R}$. Let

$$(\Delta_h f)(x) = f(x+h) - f(x) \quad \text{and} \quad \Delta_h^M = \Delta_h \Delta_h^{M-1}, \quad M-1\in\mathbb{N},$$

be the usual differences of functions, where $x\in\mathbb{R}^2$, $h\in\mathbb{R}^2$, and $\Delta_h^1 = \Delta_h$. Let

$$0<p\le\infty, \qquad 2\left(\frac{1}{p}-1\right)_+ < s < M \in \mathbb{N}. \tag{3.4}$$

Then $B^s_{p\infty}(\mathbb{R}^2)$ is the collection of all regular tempered distributions $f\in S'(\mathbb{R}^2)$ such that

$$\left\| f\,|L_p(\mathbb{R}^2)\right\| + \sup_{0<|h|\le 1} |h|^{-s}\left\|\Delta_h^M f\,|L_p(\mathbb{R}^2)\right\| < \infty. \tag{3.5}$$

Let

$$0<p\le\infty, \quad s\in\mathbb{R} \quad \text{and} \quad k\in\mathbb{N}_0 \text{ with } s+2k > 2\left(\frac{1}{p}-1\right)_+. \tag{3.6}$$

Then

$$B^s_{p\infty}(\mathbb{R}^2) = (id-\Delta)^k B^{s+2k}_{p\infty}(\mathbb{R}^2). \tag{3.7}$$

All the admitted quasi-norms (norms if $p\ge 1$) in (3.5) with (3.4) are equivalent to each other and $B^s_{p\infty}(\mathbb{R}^2)$ is independent of M. Similarly, $B^s_{p\infty}(\mathbb{R}^2)$ with (3.6) is independent of $k\in\mathbb{N}_0$. If one replaces in (3.5)

$$\sup_{0<|h|\le 1} |h|^{-s}\|\cdot\| \quad \text{by} \quad \left(\int_0^1 |h|^{-sq}\|\cdot\|^q \frac{dh}{|h|^2}\right)^{\frac{1}{q}}, \quad 0<q<\infty,$$

and applies the lift (3.7) then one gets the more general Besov spaces $B^s_{pq}(\mathbb{R}^2)$. Furthermore, $B^s_{p\infty}(\Omega)$ (and also $B^s_{pq}(\Omega)$) are the usual restrictions of the corresponding spaces on $\mathbb{R}^2$ to the given bounded C^∞ domain Ω in $\mathbb{R}^2$. If $l\in\mathbb{N}_0$ or $l=-1$, then

$$H^l(\Omega) = B^l_{2,2}(\Omega)$$

are the Sobolev spaces used in 2.1 and 2.2. Furthermore,

$$\mathcal{C}^s(\Omega) = B^s_{\infty,\infty}(\Omega), \qquad s>0,$$

are the classical Hölder-Zygmund spaces. Details about the theory of function spaces in general and about the above special assertions in particular may be found in [15], [16].

Definition 3.1. Let Ω be a bounded C^∞ domain in the plane $\mathbb{R}^2$ and let μ be a Radon measure in $\mathbb{R}^2$ with

$$\operatorname{supp}\mu = \Gamma\subset\Omega, \qquad 0<\mu(\mathbb{R}^2)<\infty, \qquad |\Gamma| = 0. \tag{3.8}$$

Let

$$0 \le t = \frac{1}{p} < \infty.$$

(i) Then

$$\lambda_\mu(t) = \sup\{\lambda \ : \ \mu_p^\lambda < \infty\} \tag{3.9}$$

are *multifractal characteristics* of μ.
(ii) Then

$$s_\mu(t) = \sup\{s \ : \ \mu \in B^s_{p\infty}(\mathbb{R}^2)\} \tag{3.10}$$

are *Besov characteristics* of μ.
(iii) Let, in addition,

$$\sum_{j=0}^{\infty} \mu_j^{\frac{1}{2}} < \infty \tag{3.11}$$

and let u be the Courant function according to Definition 2.7(ii). Then

$$\omega_\mu^\Omega(t) = \sup\{\omega \ : \ u \in B^\omega_{p\infty}(\Omega)\}$$

are *Courant characteristics* of μ.

Remark 3.2. We add a discussion about the above quantities. We always take log to base 2. It follows for $0 < t < \infty$ that

$$\lambda_\mu(t) = -t \limsup_{j\to\infty} \frac{1}{j} \log \sum_{m\in\mathbb{Z}^2} \mu(Q_{jm})^{\frac{1}{t}} \tag{3.12}$$

and for $t = 0$ that

$$\lambda_\mu(0) = -\limsup_{j\to\infty} \frac{1}{j} \log \sup_{m\in\mathbb{Z}^2} \mu(Q_{jm}). \tag{3.13}$$

These $\limsup_{j\to\infty}$ exist for all t with $0 \le t < \infty$. By the references given at the beginning of this subsection, (3.12), (3.13) are typical multifractal quantities as discussed nowadays in (multi)fractal geometry. We prefer here the version given in the above definition since the quantities μ_p^λ and, more generally μ_{pq}^λ according to (3.3), are intimately connected with norms in appropriate Besov spaces. A detailed discussion about this point may be found in [20]. Furthermore,

$$\lambda_\mu(1) = 0 \qquad \text{and} \qquad s_\mu(1) = 0.$$

The first assertion is obvious. The second assertion is well known for all Radon measures μ with (3.8) interpreted as tempered distributions. Furthermore, $s_\mu(t)$ is well defined for all $t \ge 0$ and looks typically as in Figure 1 below. In particular, $s_\mu(t)$ is in the triangle in Figure 1 with the corner points $(-2,0)$, $(0,0)$, $(1,0)$.

3.2. Main results

We say that a curve on $[0,\infty)$ is monotonically increasing if it is not decreasing.

Theorem 3.3. *Let Ω be a bounded C^∞ domain in $\mathbb{R}^2$ and let μ be a Radon measure in $\mathbb{R}^2$ with* (3.8).

(i) *Then $s = s_\mu(t)$ with $0 \le t < \infty$ according to* (3.10) *is a monotonically increasing concave function in the (t,s)-diagram in Figure* 1 *below with*

$$2(t-1) \le s_\mu(t) \le 0 \qquad \text{if} \quad 0 \le t \le 1.$$

In particular $s_\mu(1) = 0$.

(ii) *Then*

$$s_\mu(t) = \lambda_\mu(t) + 2(t-1) \qquad \text{if} \quad 0 \le t \le 1 \tag{3.14}$$

and

$$s_\mu(t) \ge \lambda_\mu(t) + 2(t-1) \qquad \text{if} \quad t > 1. \tag{3.15}$$

(iii) *Let, in addition,* (3.11) *be satisfied and let $u^\Omega = u$ be the related Courant function according to Definition* 2.7(ii). *Then $\mu_\Omega = u^\Omega \mu$ is a Radon measure satisfying* (3.8),

$$\omega_\mu^\Omega(t) = s_{\mu_\Omega}(t) + 2 \qquad \text{if} \quad 0 \le t < \infty \tag{3.16}$$

and

$$\omega_\mu^\Omega(t) = s_\mu(t) + 2 \qquad \text{if} \quad 0 \le t \le 1. \tag{3.17}$$

Remark 3.4. The parts (i) and (ii) are covered by [20]. By Theorem 2.5 the Courant function u^Ω is positive and continuous on Γ. Hence $\mu_\Omega = u^\Omega \mu$ is again a Radon measure satisfying the counterpart of (3.8). Then $s_{\mu_\Omega}(t)$ makes sense and (3.16) follows quite easily from the well-known smoothness theory of elliptic operators, in our case of the Laplacian. The independence of $\omega_\mu^\Omega(t)$ on Ω if $0 \le t \le 1$ and (3.17) follow from the considerations in [20]. But details must be shifted to a later occasion. However whether (3.17) holds also for $t > 1$ is not so clear. Later on we discuss a few examples where this is the case.

3.3. Tame measures

If $0 \le t \le 1$, then the relations between the multifractal, Besov, and Courant characteristics as described in Theorem 3.3 are perfect. This is not the case if $t > 1$. Then we have only (3.15), (3.16). Obviously,

$$0 \le s_\mu(t) \le 2(t-1) \qquad \text{if} \quad t > 1,$$

Figure 1, and, hence the curve $s_\mu(t)$ penetrates into the somewhat *gloomy region*

$$1 < t = \frac{1}{p} < \infty, \qquad 0 \le s \le 2(t-1),$$

of the otherwise *lucid function spaces universe,* $0 \le t < \infty$, $s \in \mathbb{R}$, and it requires surely additional efforts to clarify what happens there and whether the above characteristics of Radon measures are the *right* ones. We have no answers but we contribute to this apparently rather tricky problem by introducing a new notation and discussing it afterwards.

Definition 3.5. Let μ be a Radon measure in $\mathbb{R}^2$ with

$$\text{supp } \mu = \Gamma \text{ compact}, \quad 0 < \mu(\mathbb{R}^2) < \infty, \quad |\Gamma| = 0. \tag{3.18}$$

Let $\lambda_\mu(t)$ and $s_\mu(t)$ be given by (3.9) and (3.10), respectively. Then μ is called *tame* if

$$s_\mu(t) = \lambda_\mu(t) + 2(t-1) \quad \text{for all} \quad 0 \le t < \infty. \tag{3.19}$$

Remark 3.6. In other words, μ is called *tame* if one has not only in (3.14), but also in (3.15) equality. We have several examples of tame measures, which will be described later on, but so far no example of a *wild* (this means *not tame*) measure. However we have the somewhat unsupported impression that most of the measures are wild and that (3.19) reflects some type of regularity. In any case we offer here a modification of the multifractal characteristic $\lambda_\mu(t)$, denoted by $\lambda^\mu(t)$, for which we have the respective counterpart of (3.19) for all $t \ge 0$ and which coincides with $\lambda_\mu(t)$ if $0 \le t \le 1$. For this purpose we need some preparations.

We follow [21] restricting ourselves again to the plane $\mathbb{R}^2$, although everything works also in $\mathbb{R}^n$ with $n \in \mathbb{N}$. Let $e_1 = (1,0)$ and $e_2 = (0,1)$ be the unit vectors in $\mathbb{R}^2$ with respect to the axes of coordinates. Let

$$(\mathbb{D}^1 g)(x) = \frac{1}{4} \sum_{k=1}^{2} \left(g(x+e_k) - 2g(x) + g(x-e_k)\right), \quad x \in \mathbb{R}^2,$$

be the sum of the second symmetric differences, and let, by iteration,

$$\mathbb{D}^{M+1} g = \mathbb{D}^1 \mathbb{D}^M g, \qquad M \in \mathbb{N}.$$

Then

$$(\mathbb{D}^M g)(x) = 2^{-2M} \sum_{m \in \mathbb{Z}^2_M} d^M_m \, g(x+m), \qquad x \in \mathbb{R}^2, \quad M \in \mathbb{N}_0, \tag{3.20}$$

where

$$\mathbb{Z}^2_M = \{m \in \mathbb{Z}^2 \, : \, |m_1| + |m_2| \le M\}$$

is a rhombohedron and the coefficients $d^M_m \in \mathbb{Z}$ can be calculated explicitly. To extend this notation to distributions, in our case Radon measures, we use mollifications, or local means, which play a decisive role in the recent theory of function spaces, [16], 2.4.6, 2.5.3. Let K be a real C^∞ function in $\mathbb{R}^2$ with

$$K(x) > 0 \text{ if } |x| < c \quad \text{and} \quad K(y) = 0 \text{ if } |y| \ge c$$

for some $c > 0$. Then

$$K(\tau, f)(x) = \int_{\mathbb{R}^2} K(y) \, f(x - \tau y) \, dy = \tau^{-2} \int_{\mathbb{R}^2} K\left(\frac{x-y}{\tau}\right) f(y) \, dy \tag{3.21}$$

are the local means (or mollifications) of $f \in S'(\mathbb{R}^2)$ (tempered distributions), where $\tau > 0$. According to (3.20) we have for $\tau > 0$ and $f \in S'(\mathbb{R}^2)$,

$$\begin{aligned} K^M(\tau, f)(x) &= \int_{\mathbb{R}^2} (\mathbb{D}^M K)(y)\, f(x - \tau y)\, dy \\ &= 2^{-2M} \sum_{m \in \mathbb{Z}^2_M} d^M_m\, K(\tau, f)(x + \tau m). \end{aligned} \tag{3.22}$$

Again, let Q_{jm} be the squares as used in connection with (2.11) and later on. Let $M \in \mathbb{N}_0$ (natural number or 0) and $f \in S'(\mathbb{R}^2)$. Then, for $x \in \mathbb{R}^2$,

$$V^M_{jm} f = \begin{cases} \sup_{x \in Q_{jm}} |K^M(2^{-j}, f)(x)| & \text{if } j \in \mathbb{N},\ m \in \mathbb{Z}^2, \\ \sup_{x \in Q_{jm}} |K(1, f)(x)| & \text{if } j = 0,\ m \in \mathbb{Z}^2. \end{cases}$$

If $M = 0$ then $K^M = K$. We apply this notation to $f = \mu$, where μ is a Radon measure in $\mathbb{R}^2$ with (3.18), and generalise μ^λ_p in (3.1), (3.2) by

$$V\mu^{\lambda,M}_p = \sup_{j \in \mathbb{N}_0} 2^{j\lambda} \left(\sum_{m \in \mathbb{Z}^2} (V^M_{jm}\, \mu)^p \right)^{\frac{1}{p}}, \quad 0 < p < \infty,$$

and

$$V\mu^{\lambda,M}_\infty = \sup_{j \in \mathbb{N}_0, m \in \mathbb{Z}^2} 2^{j\lambda} V^M_{jm}\, \mu.$$

Here $\lambda \in \mathbb{R}$ and $M \in \mathbb{N}_0$. If $M = 0$ then it follows easily

$$V\mu^{\lambda,0}_p \sim \mu^\lambda_p, \qquad \lambda \in \mathbb{R}, \quad 0 < p \le \infty. \tag{3.23}$$

If $M \in \mathbb{N}$ then (3.22) with $f = \mu$ tries to capture wild irregular variations of μ at any scale.

Definition 3.7. Let μ be a Radon measure in $\mathbb{R}^2$ with (3.18). Let $0 \le t = \frac{1}{p} < \infty$. Then

$$\lambda^\mu(t) = \sup \left\{ \lambda \ :\ V\mu^{\lambda,M}_p < \infty \text{ for some } M \in \mathbb{N}_0 \right\}. \tag{3.24}$$

Remark 3.8. If for given p and $M \in \mathbb{N}_0$,

$$V\mu^{\lambda,M}_p < \infty, \quad \text{then} \quad V\mu^{\lambda,N}_p < \infty \quad \text{for} \quad N \ge M, \quad N \in \mathbb{N}_0. \tag{3.25}$$

Hence the situation is improving if $M \to \infty$. However beyond some $M \in \mathbb{N}_0$ there is no improvement: We shall give in Remark 3.10, (3.29), (3.30) an estimate. A possible larger value of $\lambda^\mu(t)$ compared with $\lambda_\mu(t)$ is not caused so much by the mollification (3.21), but by the variation in (3.22).

Theorem 3.9. *Let μ be a Radon measure in $\mathbb{R}^2$ with* (3.18). *Let $s_\mu(t)$, $\lambda_\mu(t)$, and $\lambda^\mu(t)$ be given by* (3.10), (3.9) *and* (3.24), *respectively.*

(i) *Then*

$$\lambda^\mu(t) \ge \lambda_\mu(t) \quad \text{if} \quad 0 \le t < \infty \tag{3.26}$$

and

$$\lambda^\mu(t) = \lambda_\mu(t) \quad \text{if} \quad 0 \le t \le 1. \tag{3.27}$$

(ii) *Then*

$$s_\mu(t) = \lambda^\mu(t) + 2(t-1) \quad \textit{if} \quad 0 \le t < \infty. \tag{3.28}$$

Remark 3.10. This is a special case of more general assertions in [21] (dealing with $\mathbb{R}^n$ and general compactly supported $f \in S'(\mathbb{R}^n)$). By (3.23) and Remark 3.8 one has (3.26). The proofs in [21] are based on characterisations of the spaces $B^s_{pq}(\mathbb{R}^n)$ in terms of the expressions in (3.22). As a consequence one gets the following complement of (3.25). If

$$(t-1)_+ \le M \in \mathbb{N}_0 \tag{3.29}$$

then

$$\lambda^\mu(t) = \sup\left\{\lambda \,:\, V\mu_p^{\lambda,M} < \infty\right\}, \qquad 0 \le t = \frac{1}{p} < \infty. \tag{3.30}$$

One may denote by $M_\mu(t)$ the smallest number M with (3.30) as the *degree of wildness* of μ (depending on $t \ge 0$). In particular,

$$M_\mu(t) = 0 \qquad \text{if} \quad 0 \le t \le 1,$$

what coincides with (3.27). Comparing (3.28) with (3.15), (3.14) we have now equality for all $t \ge 0$ at the expense of stepping from $\lambda_\mu(t)$ to $\lambda^\mu(t)$. It follows by Definition 3.5 that

$$\mu \text{ is tame if, and only if, } M_\mu(t) = 0 \text{ for } 0 \le t < \infty. \tag{3.31}$$

But as said it is not clear at this moment whether there are *wild* (this means not tame) measures.

3.4. Problems and examples

This survey complements the survey [19] and the relevant parts of the books [17], [18]. In [17], [18], especially Section 19, [19], and, more recently in [23], [22] we dealt in detail with the problem under which circumstances a Radon measure μ with (2.4) is a Weyl measure according to Definition 2.7(i). We did not stress this point in the recent survey but we wish to return to this question in the examples below. This may justify to repeat this problem here.

Problem 3.11. *Characterise all Weyl measures according to Definition* 2.7(i).

Remark 3.12. The most advanced assertion and a discussion afterwards may be found in [18], 19.17, 19.18. We returned to this question in [22], with [23] as a forerunner, including an extension of this problem from the plane $\mathbb{R}^2$ to $\mathbb{R}^n$ with $n \in \mathbb{N}$. One can convert the above problem into the following question:

Does there exist Radon measures μ according to (2.4),
such that B given by (2.10) is compact and
which are not Weyl measures?

No example is known so far. But a detailed discussion of this problem and further references may be found in [18], 19.18, 19.19, pp. 288-294.

In connection with the above considerations two further problems arise.

Problem 3.13. *Characterise all tame measures according to Definition* 3.5.

Remark 3.14. By Theorem 3.9 a measure μ is tame if, and only if,

$$\lambda_\mu(t) = \lambda^\mu(t) \quad \text{for all} \quad 0 \le t < \infty.$$

This can be reformulated by (3.31). But the situation is quite similar as in connection with Problem 3.11: No example of a wild (this means not tame) measure is known (to the author). If there exist wild measures then the *multfractal formalism* expressed by $\lambda_\mu(t)$ in (3.9) and, more explicitly by (3.12), (3.13), becomes questionable if $t > 1$. Then more sophisticated quantities are needed where $\lambda^\mu(t)$ according to Definition 3.7 is a proposal. Maybe the following question is closely related to these problems. Two Radon measures μ^1 and μ^2 with (3.18) are called equivalent if

$$\mu^1(E) \sim \mu^2(E), \qquad E \text{ Borel set in } \mathbb{R}^2.$$

(Hence they are related to each other by a Radon-Nikodym derivative, bounded from above and below by positive constants). Let $[\mu]$ be the equivalence class of the above measures which are equivalent to μ. Obviously, $\lambda_\mu(t)$ is the same for all measures of a given equivalence class. Is this also the case for $\lambda^\mu(t)$ with $t > 1$? (If $0 \le t \le 1$ then one has (3.27)). If not, then one may ask whether in a given equivalence class $[\mu]$ there is a (distinguished) tame representative.

Closely related to these speculative considerations is the question whether for given $t \ge 0$ the Courant characteristic $\omega_\mu^\Omega(t)$ according to Definition 3.1(iii) is independent of Ω. Based on (3.16), (3.17) one can strengthen this question as follows.

Definition 3.15. A Radon measure μ with (3.8), (3.11) is called a *Courant measure* if

$$\omega_\mu^\Omega(t) = s_\mu(t) + 2 \quad \text{for all} \quad 0 \le t < \infty. \tag{3.32}$$

Remark 3.16. The restriction (3.11) is somewhat disturbing. We needed it in Definition 3.1(iii) to ensure that u is continuous. But it is not clear whether this additional assumption is necessary or natural.

Problem 3.17. *Characterise all Courant measures according to Definition* 3.15.

Next we describe a few examples. To simplify the presentation the following notation seems to be reasonable.

Definition 3.18. A Radon measure μ in the plane $\mathbb{R}^2$ with (3.8) is called *distinguished* if it is simultaneously

- a Weyl measure according to Definition 2.7(i),
- a tame measure according to Definition 3.5,
- a Courant measure according to Definition 3.15.

Remark 3.19. By (3.19) and (3.32) a Weyl measure μ with (3.8) and (3.11) is distinguished according to the above definition if, and only if,

$$\omega_\mu^\Omega(t) = s_\mu(t) + 2 = \lambda_\mu(t) + 2t \quad \text{for all} \quad 0 \le t < \infty.$$

Example 3.20. Let Γ be a compact d-set in $\mathbb{R}^2$ with $0 < d < 2$. Hence there is Radon measure μ in $\mathbb{R}^2$ with $\Gamma = \operatorname{supp} \mu$ and

$$\mu(B(\gamma,r)) \sim r^d, \quad \gamma \in \Gamma, \quad 0 < r \le 1, \tag{3.33}$$

where the equivalence constants are independent of $\gamma \in \Gamma$ and r, and where $B(\gamma, r)$ is the circle centred at γ and of radius r. Then $\mu \sim \mathcal{H}^d|\Gamma$ where the latter is the restriction of the Hausdorff measure $\mathcal{H}^d$ in $\mathbb{R}^2$ to Γ. All these measures μ are distinguished measures according to Definition 3.18. We refer to [20], Remark 10, where we proved that μ is tame and

$$s_\mu(t) = (2-d)(t-1), \qquad 0 \le t < \infty. \tag{3.34}$$

By the same arguments it follows that μ is Courant. The Weyl property for these measures is one of the central assertions at the beginning of this theory in [17], Section 30, in connection with fractal drums (or better drums with fractal membranes).

Example 3.21. Let $\Gamma = \Gamma_1 \cup \Gamma_2$ where Γ_1 and Γ_2 are compact d_1-sets and d_2-sets, respectively, with $0 < d_1 < d_2 < 2$ and $\Gamma_1 \cap \Gamma_2 = \emptyset$. Let μ be a corresponding measure with $\operatorname{supp} \mu = \Gamma$, and (3.33) with d_1 on Γ_1 and d_2 on Γ_2. Then μ is a distinguished measure according to Definition 3.18 and

$$s_\mu(t) = \begin{cases} (2-d_1)(t-1) & \text{if } 0 \le t \le 1, \\ (2-d_2)(t-1) & \text{if } 1 \le t < \infty. \end{cases} \tag{3.35}$$

This follows essentially from Example 3.20. We refer again to [20], Remark 10, for details.

Example 3.22. So far, $s_\mu(t)$ in (3.34) and (3.35) is either a straight line or a broken line, respectively. Now we construct a distinguished probability measure where $s_\mu(t)$ is a genuine bent curve. Let Q be the closed square in $\mathbb{R}^2$, centred at the origin and of side-length 1. Let $\varrho > 0$ and let T_k be contractions in $\mathbb{R}^2$, where $k = 1, \cdots, N$, with $2 \le N \in \mathbb{N}$,

$$T_k : \quad \mathbb{R}^2 \ni x \mapsto \varrho x + x^k, \quad \text{where} \quad x^k \in \mathbb{R}^2,$$

such that

$$Q^k = T_k Q \subset Q, \qquad Q^k \cap Q^l = \emptyset \quad \text{if} \quad k \ne l.$$

Since Q is assumed to be closed and $N \ge 2$ it follows that $0 < \varrho < \frac{1}{2}$ and the squares Q^k have a positive distance from each other. Recall the standard procedure to create a self-similar fractal set Γ:

$$TQ = (TQ)^1 = \bigcup_{k=1}^N Q^k, \qquad (TQ)^0 = Q,$$

$$(TQ)^l = T\left((TQ)^{l-1}\right) = \bigcup_{1 \le j_r \le N} T_{j_1} \cdots T_{j_l} Q, \qquad l \in \mathbb{N},$$

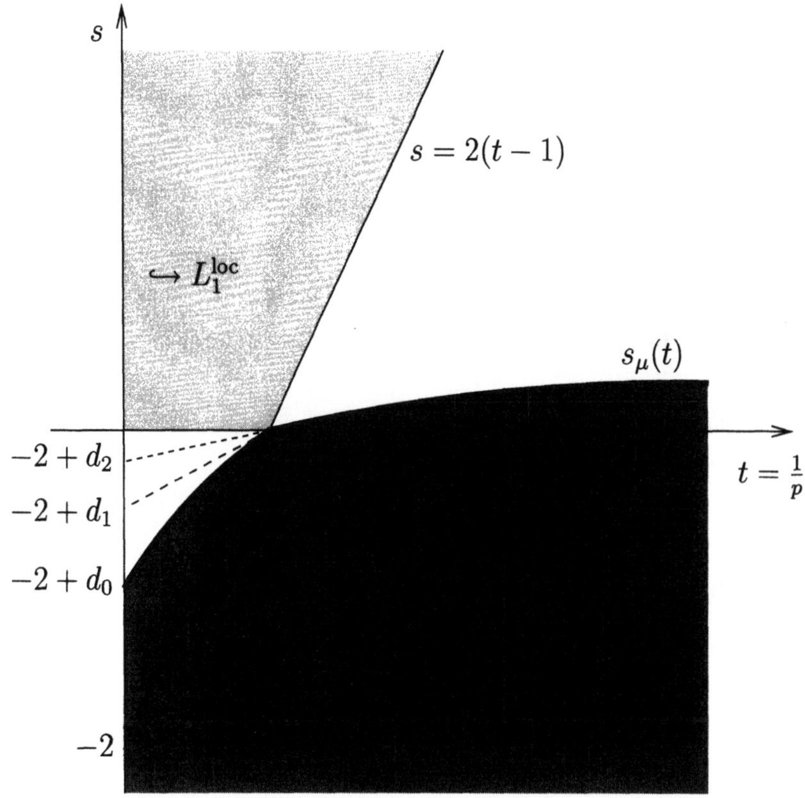

FIGURE 1.

and

$$\Gamma = (TQ)^\infty = \bigcap_{l \in \mathbb{N}} (TQ)^l = \lim_{l \to \infty} (TQ)^l .$$

By construction, Γ is disconnected, some type of *dust.* Let

$$w = (w_1, \dots, w_N), \quad w_1 \geq \cdots w_N > 0, \quad \sum_{l=1}^{N} w_l = 1,$$

be a probability. The unit mass in Q is distributed according to these probabilities and transforms. Then one gets a Radon measure μ with

$$\operatorname{supp} \mu = \Gamma, \qquad \mu \left(T_{j_1} \cdots T_{j_l} Q \right) = w_{j_1} \cdots w_{j_l},$$

$\mu(\mathbb{R}^2) = 1$ and $|\Gamma| = 0$. It comes out that μ is a distinguished measure according to Definition 3.18. Furthermore,

$$s_\mu(t) = 2(t-1) + \frac{t}{\log \varrho} \log \sum_{j=1}^{N} w_j^{\frac{1}{t}}, \qquad 0 < t < \infty,$$

and

$$s_\mu(0) = -2 + \frac{|\log w_1|}{|\log \varrho|}.$$

The curve $s_\mu(t)$ is shown in Figure 1. Let

$$d_0 = \frac{|\log w_1|}{\log \varrho|}, \qquad d_1 = \frac{\sum w_j |\log w_j|}{|\log \varrho|}, \qquad d_2 = \frac{\log N}{|\log \varrho|}.$$

Then we have $d_0 < d_1 < d_2$ in case of $N^{-1} < w_1 < 1$ and

$$s_\mu(0) = -2 + d_0, \quad s'_\mu(1) = 2 - d_1, \quad s'_\mu(t) \to 2 - d_2$$

if $t \to \infty$. In particular, $s_\mu(t)$ is a bent curve. Proofs will be shifted to a later occasion.

Example 3.23. Let μ be an isotropic Radon measure with the generating function h according to (2.12), (2.4). Let

$$h_- = \liminf_{j\to\infty} \frac{|\log h(2^{-j})|}{j}, \qquad h_+ = \limsup_{j\to\infty} \frac{|\log h(2^{-j})|}{j}.$$

Then one gets by (3.12), (3.13) that

$$\lambda_\mu(t) = \begin{cases} (1-t)\, h_- & \text{if } 0 \le t \le 1, \\ (1-t)\, h_+ & \text{if } t > 1, \end{cases}$$

and we have (3.14), (3.15) for $s_\mu(t)$. Furthermore,

$$\dim_H \Gamma = h_- \qquad \text{and} \qquad \dim_P \Gamma = h_+ \tag{3.36}$$

have a geometrical meaning as the Hausdorff dimension and the packing dimension of Γ, respectively. Background assertions concerning the lesser known packing dimension may be found in [9], Section 5. Concerning (3.36) we refer to [2], [3], [4] and the literature mentioned there. If, in addition,

$$h_- = h_+ = h_0, \tag{3.37}$$

then it comes out that μ is a tame measure and a Courant measure according to Definitions 3.5 and 3.15, respectively, and

$$s_\mu(t) = (t-1)(2-h_0) \qquad \text{for all} \quad 0 \le t < \infty.$$

This can be proved by the same arguments as in [20], Remark 10. According to [22], Definition 3.1(ii), the above isotropic measure μ with the generating function h is called *strongly isotropic* if there is a natural number k such that

$$h\left(2^{-j-k}\right) \le \frac{1}{2} h\left(2^{-j}\right) \qquad \text{for all} \quad j \in \mathbb{N}_0. \tag{3.38}$$

As proved in [22], Corollary 2, any strongly isotropic measure is a Weyl measure. Combining these results one gets the following assertion:

Any strongly isotropic Radon measure with the generating function h according to (2.12), (2.4), (3.38), *satisfying in addition* (3.37), *is a distinguished measure according to Definition* 3.18.

References

[1] M. Bricchi. *Existence and properties of h-sets.* Georgian Math. Journ. **9** (2002), 13–32.

[2] M. Bricchi. *Tailored Besov spaces and h-sets.* Math. Nachr. **263–264** (2004), 36–52.

[3] M. Bricchi. *Complements and results on h-sets.* In: Function Spaces, Differential Operators and Nonlinear Analysis, Birkhäuser, Basel, 2003, 219–229.

[4] M. Bricchi. *Compact embeddings between Besov spaces defined on h-sets.* Functiones Approximatio **30** (2002), 7–36.

[5] G. Brown, G. Michon, J. Peyrière. *On the multifractal analysis of measures.* Journ. Statistical Physics **66** (1992), 775–790.

[6] R. Courant, D. Hilbert. *Methoden der mathematischen Physik*, Springer, Berlin, 1993 (4th edition), 1st edition 1924.

[7] K.J. Falconer. *Techniques in fractal geometry.* Wiley, Chichester, 1997.

[8] Y. Heurteaux. *Estimations de la dimension intérieure et de la dimension supérieure des mesures.* Ann. Inst. H. Poincaré, Probab. Statist. **34** (1998), 309–338.

[9] P. Mattila. *Geometry of sets and measures in euclidean spaces.* Cambridge Univ. Press, Cambridge, 1995.

[10] F.B.Nasr, J. Bhouri, Y. Heurteaux. *The validity of the multifractal formalism: results and examples.* Adv. Math. **165** (2002), 264–284.

[11] S.-M. Ngai. *A dimension result arising from the L^q-spectrum of a measure.* Proc. Amer. Math. Soc. **125** (1997), 2943–2951.

[12] L. Olsen. *A multifractal formalism.* Adv. Math. **116** (1995), 82–196.

[13] R. S. Strichartz. *Self-similar measures and their Fourier transform III.* Indiana Univ. Math. Journ. **42** (1993), 367–411.

[14] R. S. Strichartz. *Self-similarity in harmonic analysis.* Journ. Fourier Analysis Appl. **1** (1994), 1–37.

[15] H. Triebel. *Theory of function spaces.* Birkhäuser, Basel, 1983.

[16] H. Triebel. *Theory of function spaces, II.* Birkhäuser, Basel, 1992.

[17] H. Triebel. *Fractals and Spectra.* Birkhäuser, Basel, 1997.

[18] H. Triebel. *The structure of functions.* Birkhäuser, Basel, 2001.

[19] H. Triebel. *Fraktale Analysis aus der Sicht der Funktionenräume.* Jahresbericht DMV **104** (2002), 171–199.

[20] H. Triebel. *Fractal characteristics of measures, an approach via function spaces.* Journ. Fourier Analysis Appl. **9** (2003), 411–431.

[21] H. Triebel. *Characterisation of function spaces via mollification; fractal quantities for distributions.* Journ. Function Spaces Appl. **1** (2003), 75–89.

[22] H. Triebel. *Approximation numbers in function spaces and the distribution of eigenvalues of some fractal elliptic operators.* Journ. Approximation Th. (to appear).

[23] H. Triebel. *The distribution of eigenvalues of some fractal elliptic operators and Weyl measures.* In: Operator Theory, Advances Appl. 147. The Erhard Meister Memorial Volume, Birkhäuser, Basel, 2004, 457–473.

[24] H. Weyl. *Über die Abhängigkeit der Eigenschwingungen einer Membran von deren Begrenzung.* Journ. Reine Angew. Mathematik **141** (1912), 1–11.

[25] H. Weyl. *Das asymptotische Verteilungsgesetz der Eigenwerte linearer partialler Differentialgleichungen.* Math. Ann. **71** (1912), 441–479.

Hans Triebel
Mathematisches Institut
Fakultät für Mathematik und Informatik
Friedrich-Schiller-Universität Jena
D-07740 Jena
Germany
e-mail: `triebel@minet.uni-jena.de`

Progress in Probability, Vol. 57, 193–204

Geometric Representations of Currents and Distributions

Jenny Harrison

Abstract. Currents are defined abstractly as dual spaces to differential forms and differential forms are defined abstractly as dual spaces to linear spaces of k-vectors. This double duality makes it sometimes difficult to recognize a specific example as a current or to give a current geometric meaning. Operators on currents are usually defined by dualizing analytic operators on differential forms. We give direct geometric definitions for a large class of currents and their operators. Chainlets are limits of polyhedral chains taken with respect to a norm. Integrals of differential forms over polyhedra have well defined limits to chainlets and a full exterior calculus has been established for chainlet domains. ([H2], [H3], [H4]) Every chainlet is thus a current. In this paper we specify which currents T correspond uniquely to chainlets, giving such T direct, geometric representation.

Mathematics Subject Classification (2000). Primary 53C65; Secondary 28A75.

Keywords. Currents, distributions, fractals, chainlets.

1. Introduction

De Rham's theory of currents [deR] is a calculus of abstract linear functionals, based on a generalized Stokes' theorem. The currents are functionals defined on a linear space of differential forms. Currents provide the most general expression for a domain of integration using integrands of smooth differential forms. Operations on currents are defined by duality via their effect on the differential forms. Using a family of norms defined on polyhedra there is a direct definition leading to a geometric interpretation of the currents and operations on them. Convergence is taken in the norm topology rather than the weak topology. Limits of polyhedra in these norms are called *chainlets.* In [H2] it is shown that every chainlet is a current. The geometric nature of chainlets makes it possible to identify currents that had heretofore been elusive including soap film surfaces [H5]. Many fractals are currents, including graphs of bounded measurable functions, the Smale horseshoe, the Van Koch snowflake and the toral solenoid [H6]. It is therefore possible to treat such fractals essentially as manifolds, complete with their operations and integrals. In this paper we identify subspaces of currents that are isomorphic to r-natural chainlets.

It follows from the representation theorem below that integrable currents can be approximated by a variety of geometric objects besides polyhedra, each choice giving a distinct flavor to the resulting exterior calculus. For example, self-similar fractals are dense in spaces of chainlets, and thus a fractal exterior calculus, equivalent to the smooth theory, can be developed based on self-similar fractals. (Related work may be seen in work such as in [K] and [L].) A major application of the representation theorem [H6] shows that point mass chains are dense in spaces of chainlets, leading to a theory of discrete exterior calculus.

In §2 we define r-natural norms on k-polyhedra leading to Banach spaces of chainlets $\mathcal{N}_k^r(\mathbb{R})$ and recall basic properties proved about them in [H1], [H2], [H3] and [H4]. In §2 we also define spaces of *integrable currents* $\mathcal{T}_k^r(\mathbb{R})$. In §3 and §4 we define two geometric operators on chainlets: Lie derivative and components. We prove the following representation theorem for integrable currents $T \in \mathcal{T}_k^r(\mathbb{R})$ in §5. (See §2 for definitions used in the statement.)

Theorem 1.1 (Geometric representation of currents). *For each $r \geq 0, k \geq 0$ there exists an isomorphism*

$$\theta : \mathcal{N}_k^r(\mathbb{R}) \to \mathcal{T}_k^r(\mathbb{R})$$

such that if $T = \theta(A)$ then

$$T[\omega] = \int_A \omega$$

for all $\omega \in \mathcal{L}_k^r(\mathbb{R})$. Furthermore, $spt(A) = spt(T)$ and $|A|^{\natural_r} = |T|^r$.

2. Normed spaces of polyhedra

The norms defined on polyhedra in this section are at the heart of chainlet geometry. Fortunately, they are easy to work with, both in proofs and in examples. It practice, it is rarely important to compute the precise norm of a polyhedron, but to find a reasonable upper bound. Good guesses usually work out.

It is helpful to recall the definition of a differentiable function in this section since the definition of the norms is quite similar. They both rely on limits of divided differences. Higher order differentiability relies on higher order divided differences. In this section we introduce multicells to give a geometric version of divided differences of functions.

A function $f : U \to V$ defined on bounded, open subsets $U \subset \mathbb{R}^n$ and $V \subset \mathbb{R}^m$ satisfies a *Lipschitz* condition if there exists a constant $C > 0$ such that $|f(x) - f(y)| \leq C|x - y|$ for all $x, y \in U$. A function is of class C^{r+Lip} if its rth order partial derivatives each satisfies some Lipschitz condition. The B^r norm of a differential form ω is defined as the sum of the $C^{r-1+Lip}$ norms of the local coefficient functions of ω and is denoted by $\|\omega\|_r$. We say ω is of class B^r

if $\|\omega\|_r < \infty$. Let $\mathcal{B}_k^r(G)$ denote the real linear space of k-forms ω in $\mathbb{R}^n$ of class B^r. Denote the Frechet space $\mathcal{B}_k^\infty(\mathbb{R}) = \bigcap_{r=0}^\infty \mathcal{B}_k^r(\mathbb{R})$. A discussion relating r-Lipschitz constants of functions to r^{th} derivatives is given in Whitney [W2]. His proofs easily generalize to the corresponding theorems for differential forms.

Polyhedral chains

For $k \geq 1$, a k-dimensional *cell* in $\mathbb{R}^k$ is defined to be the finite intersection of open half spaces. Examples of cells include simplexes and cubes. A k-cell in $\mathbb{R}^n$ is a subset of a k-dimensional subspace Π in $\mathbb{R}^n$ that is a k-cell in Π. All k-cells are assumed to be oriented. A 0-cell is a single point $\{x\}$ in $\mathbb{R}^n$. No orientation need be assigned to the 0-cell. A k-cellular chain $S = \sum a_i\sigma_i$ is a formal sum of k-cells σ_i with coefficients a_i in $\mathbb{R}$. Formal sums are too specific and restrictive for our purposes. We do not want to have to be concerned which simplicial subdivision or triangulation we are using for a given domain. We also wish cells with opposite orientation and with the same coefficient to cancel wherever they overlap. The standard way to treat this is to form equivalence classes of polyhedral chains $P = [S]$, saying that two formal sums are equivalent if integrals of smooth differential forms are the same over each of them. Let $\mathcal{P}_k(\mathbb{R})$ denote the linear space of polyhedral chains with coefficients in $\mathbb{R}$. Let ∂ denote the usual boundary operator on cells. For $k = 0$ we set $\partial P = 0$. The sequence of polyhedral chains of arbitrary dimension together with the operator ∂ form a chain complex $\mathcal{P}_*(\mathbb{R})$.

Mass of a polyhedral chain. For $k \geq 1$, let $M(\sigma)$ denote k-dimensional Hausdorff measure of a k-dimensional cell σ. Every 0-dimensional cell σ^0 takes the form $\sigma^0 = \{x\}$ and we set $M(\sigma^0) = 1$. The *mass* of $P = [\sum_{i=1}^m a_i\sigma_i]$ is defined by

$$M(P) = \sum_{i=1}^m |a_i| M(\sigma_i)$$

where the σ_i are non-overlapping.

Mass is a norm on the linear space $\mathcal{P}_k(\mathbb{R})$, but it does not lead to strong enough convergence to identify much of interest beyond polyhedral chains. We will make use of a family of finer norms, called *r-natural norms* defined on spaces of k-polyhedra with coefficients in $\mathbb{R}$ for each $r \geq 0$. The 0-natural norm of a polyhedron is just its mass. (We sometimes refer to mass in this way for consistency of notation.) Each norm can be used to complete $\mathcal{P}_k(\mathbb{R})$ to obtain a Banach space $\mathcal{N}_k^r(\mathbb{R})$. The norms are monotone decreasing in r, that is, $|P|^{\natural_0} \geq |P|^{\natural_1} \geq |P|^{\natural_2} \geq \cdots$ so the Banach spaces obtained on completion are monotone increasing. That is, each $\mathcal{N}_k^r$ is naturally included in $\mathcal{N}_k^{r+1}$ and we can write $\mathcal{N}_k^0 \subset \mathcal{N}_k^1 \subset \mathcal{N}_k^2 \subset \cdots$. The boundary operator $\partial : \mathcal{N}_k^r \to \mathcal{N}_{k-1}^{r+1}$ is well defined and since $\partial\partial = 0$

$$\mathcal{N}_n^r \xrightarrow{\partial} \mathcal{N}_{n-1}^{r+1} \xrightarrow{\partial} \cdots \xrightarrow{\partial} \mathcal{N}_1^{r+n+1} \xrightarrow{\partial} \mathcal{N}_0^{r+n}$$

is a chain complex for each $r \geq 0$. For $G = \mathbb{R}$, the direct limit of the $\mathcal{N}_k^r$ as $r \to \infty$, is a semi-normed space $\mathcal{N}_k^\infty$ dual to the Frechet space of C^∞ forms $\mathcal{B}_k^\infty$.

Multicellular chains

If σ is a cell and v a vector let $T_v\sigma$ denote the translation of σ through v.

A *k-dimensional* 1*-multicell* is a cellular chain of the form

$$\sigma^1 = \sigma^0 - T_{v_1}\sigma^0$$

where σ^0 is a k-cell and v_1 is a vector. We inductively define 2^j-multicells. Given a vector v_j and a 2^{j-1}-multicell σ^{j-1}, define the 2^j*-multicell* σ^j as the cellular chain

$$\sigma^j = \sigma^{j-1} - T_{v_j}\sigma^{j-1}.$$

Thus σ^j is generated by vectors $v_1, \ldots, v_j$, and a cell σ^0.

One can picture a 1-dimensional 2^j-multicell as the sum T of parallel oriented edges of a $(j+1)$-dimensional sheared parallelepiped in $\mathbb{R}^n$.

A 2^j*-multicellular chain in* $\mathbb{R}^n$ is a formal sum of 2^j-multicells, $S^j = \sum_{i=1}^n a_i\sigma_i^j$ with coefficients $a_i \in G$. Since it is also a cellular chain, it makes sense to form equivalence classes, or 2^j-multicellular polyhedra $P^j = [S^j]$.

Multicellular mass. Given a 2^j-multicell σ^j generated by a cell σ^0 and vectors $v_1, \cdots, v_j$, define $\|\sigma^0\|_0 = M(\sigma^0)$, $\|\sigma^1\|_1 = M(\sigma^0)|v_1|$ and for $j \geq 2$,

$$\|\sigma^j\|_j = M(\sigma^0)|v_1||v_2|\cdots|v_j|.$$

For $S^j = \sum a_i\sigma_i^j$ define

$$\|S^j\|_j = \sum_{i=1}^n |a_i|\|\sigma_i^j\|_j.$$

Example. Consider first a constant function $g(x) = c$ defined on $[a, b]$. The graph of g supports the oriented 1-cell with oriented vertices $-(a, c)$ and $+(b, c)$. Now suppose $f : [0, 1] \to \mathbb{R}$ is an L^1 function. Denote the L^1 norm of f by $|f|_{L^1}$. Approximate f with step functions g_k. The graph of each g_k is naturally represented as a polyhedral chain P_k by orienting each step of the graph as above. The difference $P_k - P_j$ is a 1-multicell.

Natural norms

Let $P \in \mathcal{P}_k(\mathbb{Z})$ be a polyhedral k-chain in $\mathbb{R}^n$ and $r \in \mathbb{R}^+$. For $r = 0$ define

$$|P|^{\natural_0} = M(P).$$

For $r \geq 1$ define the r*-natural* norm

$$|P|^{\natural_r} = \inf\left\{\sum_{s=0}^r \|S^j\|_j + |C|^{\natural_{r-1}}\right\} \tag{2.1}$$

where the infimum is taken over all decompositions

$$P = \sum_{s=0}^r [S^j] + \partial C$$

where S^j is a 2^j multicellular k-chain and C is a polyhedral $(k+1)$-chain. It is clear $|\quad|^{\natural_r}$ is a semi-norm. We shortly prove it is a norm.

Integration of differential forms over chainlets

The following basic integral inequality is proved in [H2].

Theorem 2.1. *Let P be a k-dimensional polyhedron and $\omega \in \mathcal{B}_k^r(\mathbb{R})$. Then*

$$\int_P \omega \leq |P|^{\natural_r} \|\omega\|_r.$$

Corollary 2.2. *$|P|^{\natural_r}$ is a norm on the space of polyhedral chains $\mathcal{P}_k(\mathbb{R})$.*

Proof. Suppose $P = \sum a_i \sigma_i$ is nonoverlapping and nonzero. We may assume $\sigma = \sigma_i \neq 0$ and parallel to a coordinate plane. Choose a coordinate n-cube Q meeting σ_1 and disjoint from each $\sigma_i, i > 1$. There exists a differential form ω supported in Q such that $\int_\sigma \omega = M(\sigma) \neq 0$. Then $0 < \int_P \omega \leq |P|^{\natural_r} \|\omega\|_r$ implies $|P|^{\natural_r} > 0$. □

The linear space of polyhedral k-chains $\mathcal{P}_k(\mathbb{R})$ completed with the norm $|\quad|^{\natural_r}$ is denoted $\mathcal{N}_k^r(\mathbb{R})$. The elements of $\mathcal{N}_k^r(\mathbb{R})$ are called *k-dimensional chainlets of class N^r with coefficients in* $\mathbb{R}$. It can be easily shown that that the r-natural norms satisfy the inequalities

$$|P|^{\natural_0} \geq |P|^{\natural_1} \geq |P|^{\natural_2} \geq \cdots$$

for polyhedra P.

Let $\mathcal{N}_k^\infty(\mathbb{R}) = \bigcup \mathcal{N}_k^r(\mathbb{R})$.

The integral of a differential form $\omega \in \mathcal{B}_k^r(\mathbb{R})$ over a chainlet $A \in \mathcal{N}_k^r(\mathbb{R})$ is defined by taking limits over polyhedra $A = \lim_{j\to\infty} P_j$ in the r-natural norm:

$$\int_A \omega = \lim_{j\to\infty} \int_{P_j} \omega. \tag{2.2}$$

The limit exists and is independent of the choice of P_j by Theorem 2.1. It follows that every chainlet is a current.

Suppose $P = \sum g_i \sigma_i$ where $g_i \in G$ and the $\{\sigma_i\}$ are nonoverlapping k-cells. The mass of P is defined by $M(P) = \sum |g_i| M(\sigma_i)$ where the $\{\sigma_i\}$ are nonoverlapping.

The support of a chainlet

The support *spt*σ of a cell σ is merely the points contained in its closure. If P is a polyhedron it can be written $P = \sum a_i \sigma_i$ where the σ_i are non-overlapping. The support of P is defined as $|P| = \cup spt\sigma_i$. We say a closed set E *supports* a chainlet A if for every open set U containing E there is a sequence $\{P_j\}$ of polyhedra tending to A such that each P_j lies in U. If there is a smallest set E which supports A then E is called the *support* of A and denoted *sptA*.

Examples

1. *An L^1 function f is represented by a chainlet supported in the graph of f* Continuing with the previous example, we can see
$$|P_k - P_j|^{\natural_1} \leq |g_k - g_j|_{L^1}$$
by writing $P_k - Pj$ as a 1-multicellular chain the norm of which is bounded by the area of the subgraph of $g_k - g_j$. Therefore, the sequence $\{P_k\}$ is Cauchy in the 1-natural norm and thus converges to a 1-natural chainlet Γ_f supported in the graph of f. This chainlet limit can be thought of as the oriented x-component of the graph of f.
2. *An L^1 function f is represented by a chainlet supported in the domain of f.* We can also represent a positive function $f : [a, b] \to \mathbb{R}$ by a chainlet of class N^1 this time supported in $[a, b]$. Again, let $g_k \to f$ be a monotone sequence of step functions converging to f. Let Q_k denote the cellular chain defined by weighting each 1-simplex σ_{k_i} of $[a, b]$ over which g_k is a constant c_{k_i} by the same constant. Thus $Q_k = \sum c_{k_i}\sigma_{k_i}$. We leave it to the reader to verify that Q_k forms a Cauchy sequence in the 1-natural norm and thus converges to a chainlet $A_f \in \mathcal{N}_1^1$. (See [H6], as well.) Furthermore, $|f|_{L^1} = |A_f|^{\natural_0}$.
3. *Mass cells.* Suppose τ is a $(k-1)$-cell in R^n, $0 \leq k - 1 \leq n$, and v is a unit vector in R^{n+1} orthogonal to τ. Let $\tau \times 2^{-j}v$ denote the oriented orbit of τ through the time-t map of the vector field $v(x) = v$ for $0 \leq t \leq 2^{-j}$. (Here $2^{-j}v$ refers to scalar multiplication of a vector v by 2^{-j}.)

   We show that the sequence of weighted cells $2^j(\tau \times 2^{-j}v)$ is Cauchy in the 1-natural norm.

Proof. Note that
$$\tau \times 2^{-j}v = (\tau \times 2^{-j-1}v) + (T_{2^{-j-1}v}\tau \times 2^{-j-1}v).$$
Since
$$\begin{aligned}\frac{\tau\times 2^{-j}v}{2^{-j}} - \frac{\tau\times 2^{-j-1}v}{2^{-j-1}} &= \frac{(T_{2^{-j-1}v}\tau\times 2^{-j-1}v)-(\tau\times 2^{-j-1}v)}{2^{-j}} \\ &= \frac{T_{2^{-j-1}v}(\tau\times 2^{-j-1}v)-(\tau\times 2^{-j-1}v)}{2^{-j}}\end{aligned}$$
we have
$$\begin{aligned}\left|\frac{\tau\times 2^{-j}v}{2^{-j}} - \frac{\tau\times 2^{-j-1}v}{2^{-j-1}}\right|^{\natural_1} &= \left|\frac{T_{2^{-j-1}v}(\tau\times 2^{-j-1}v)-(\tau\times 2^{-j-1}v)}{2^{-j}}\right|^{\natural_1} \\ &\leq \|v\|^2|\tau|^{\natural_0}2^{-j}/4.\end{aligned}$$
Thus $2^j(\tau \times 2^{-j}v)$ is Cauchy in the 1-natural norm. □

We denote the limit in the 1-natural norm by
$$i_v(\tau) = \lim_{j\to\infty} 2^j(\tau \times 2^{-j}v)$$
and call this a mass cell *a*. It is k-dimensional, but its support is τ which has dimension $k-1$. Mass cells were fundamental in [H5] where mathematical models of soap films were introduced.

Theorem 2.3. *If P is a polyhedral k-chain then*

$$|P|^{\natural_r} = \sup\left\{ \int_P \omega : \|\omega\|_r \leq 1 \right\}.$$

This follows from Theorem 2.1 and Theorem 4.7 of [H3].

A *current* $T \in \mathcal{T}_k^r(\mathbb{R})$ is a linear functional of k-forms $\omega \in \mathcal{B}_k^r(\mathbb{R})$. Define

$$|T|^r = sup\{T[\omega] : \omega \in \mathcal{B}_k^r(\mathbb{R}), \|\omega\|_r \leq 1\}.$$

3. Geometric Lie derivatives

Let $\omega \in \mathcal{B}_k^r(\mathbb{R})$ and v a unit vector in $\mathbb{R}^n$. Suppose U is an open subset of $spt(\omega)$.

Definition The *Lie derivative of ω at $x \in U$ in the direction v* is the k-covector defined by

$$\mathcal{L}_v\omega(x)(\beta) = \lim_{h\to 0}\left\{ \frac{\omega(x+hv)(\beta) - \omega(x)(\beta)}{h} \right\}$$

for all k-vectors β.

Note that ω is r-smooth in U if and only if for all unit vectors $v_1, \dots, v_r \in \mathbb{R}^n$, the differential form

$$D^{[r]}\omega = \mathcal{L}_{v_r} \cdots \mathcal{L}_{v_1}\omega(x)$$

exists and satisfies a Lipschitz condition in U. We call $D^{[r]}\omega$ an rth order *Lie derivative* of ω.

The definition of a mass cell $i_v(\sigma)$ was given in §2 for vectors v orthogonal to σ, but it can easily be extended to define the *interior product* of cell σ with respect to an arbitrary vector $v \in \mathbb{R}^n$, and is also denoted $i_v\sigma$. It follows from the definitions that $\int_{i_v\sigma}\omega = \int_\sigma i_v\omega$ for all Lipschitz differential forms ω.

Define $\mathcal{L}_v\sigma = \partial i_v\sigma + i_v\partial\sigma$. This is a chainlet of class N^2 satisfying

$$\int_{\mathcal{L}_v\sigma}\omega = \int_\sigma \mathcal{L}_v\omega$$

for all forms ω of class $\mathcal{B}^2$. It follows from the definitions that

$$\mathcal{L}_v\sigma = \lim_{h\to 0}\frac{T_{hv}\sigma - \sigma}{h}.$$

The following theorem is proved in [H6].

Theorem 3.1. *Given a polyhedral k-chain P and $v \in \mathbb{R}^n$, there exists a unique chainlet $\mathcal{L}_vP \in \mathcal{N}_k^1$ such that*

$$\int_{\mathcal{L}_vP}\omega = \int_P \mathcal{L}_v\omega$$

for all k-forms $\omega \in \mathcal{B}_k^1$. Furthermore,

$$|\mathcal{L}_vP|^{\natural_r} \leq |P|^{\natural_{r-1}}.$$

4. Components of chainlets

We next define the components of a k-simplex σ depending on a given coordinate system. Given an indexing set $I = \{x_{i_1}, \dots, x_{i_k}\}$, let S_I denote the k-dimensional plane of $\mathbb{R}^n$ defined by I. Let $\pi : \mathbb{R}^n \to S_I$ denote orthogonal projection. Fix an oriented k-simplex $\sigma \subset \mathbb{R}^n$. For each j, cover σ with n-cubes with vertices in the coordinate binary lattice and side length 2^{-j}. The projection π sends this cover onto a binary cover Q_i of $\pi(\sigma)$ in S_I. For each k-cube Q_i meeting $\pi(\sigma)$ choose a k-dimensional edge E_{j_i} in one of the n-cubes in the preimage $\pi^{-1}(Q_i)$. Orient each E_{j_i} according to the orientation of σ and define $A_j = \sum_i E_{j_i}$.

We prove that the sequence A_j is Cauchy in the 1-natural norm and that its limit σ_I in the 1-natural norm depends only on σ and I. Each edge E_{j_i} either projects to a Q_i contained entirely in $\pi(\sigma)$ or not. Those that do not project into $\pi(\sigma)$ have mass m_j tending to 0 as $j \to \infty$. Denote the chain of the E_{j_i} that project into $\pi(\sigma)$ by P_j. If $j \le \ell$ then $P_j - P_\ell$ is a dipole with 1-natural norm bounded above by $2^{-j}|\sigma|_0$. Since this bound tends to zero as $j \to \infty$, we conclude that P_j and therefore A_j forms a Cauchy sequence in the 1-natural norm. If different edges E_{j_i} are chosen we obtain the same limit since the error in the 1-natural norm is also bounded above by $2^{-j+1}|\sigma|_0$.

Lemma 4.1. *Given an oriented simplex $\sigma \subset \mathbb{R}^n$ and an indexing set I there exists a unique chainlet $\sigma_I \in \mathcal{N}_k^1$ such that*

$$\int_{\sigma_I} \omega = \int_\sigma \omega_I$$

for all $\omega \in \mathcal{B}_k^1$.

Proof. By Theorem 2.1

$$\int_{\sigma_I} \omega = \lim_{j\to\infty} \int_{A_j} \omega.$$

But $\int_{A_j} \omega = \int_{A_j} \omega_I$ which, in turn, limits to $\int_{\sigma_I} \omega_I$. The latter integral is the same as $\int_\sigma \omega_I$ because the projections $\pi(A_j)$ limit to $\pi(\sigma)$ and the supports of A_j tend to the support of σ. □

In [H3] (Corollary 5.4) it is shown that a chainlet is determined by the integrals of forms over the chain. That is

Theorem 4.2. *Let A and B be chainlets in $\mathcal{N}_k^{\natural r}$. Then*

$$\int_A \omega = \int_B \omega$$

for all $\omega \in \mathcal{B}_k^r$ if and only if $A = B$.

Given a simplicial chain $S = \sum_i a_i \sigma_i$, define S_I to be the chainlet $S_I = \sum_i a_i \sigma_{iI}$.

Lemma 4.3. *If $S \sim T$ then $S_I = T_I$.*

Proof. It suffices to show that $\int_{S_I} \omega = \int_{T_I} \omega$ for all $\omega \in \mathcal{B}_k^1$ by Theorem 4.2. By Lemma 4.1 and since $S \sim T$, we have

$$\int_{S_I} \omega = \int_S \omega_I = \int_T \omega_I = \int_{T_I} \omega. \qquad \square$$

We may now define the *component* of a polyhedral chain P by

$$P_I = S_I$$

where S is any simplicial representative of P.

Theorem 4.4. *Given any polyhedral k-chain P and indexing set I, there exists a unique chainlet $P_I \in \mathcal{N}_k^1$ such that*

$$\int_{P_I} \omega = \int_P \omega_I$$

for all $\omega \in \mathcal{B}_k^1$. Furthermore

$$|P_I|^{\natural_r} \le |P|^{\natural_r}.$$

Proof. According to Theorems 2.3, 2.1 and Lemma 4.1

$$\begin{aligned} |P_I|^{\natural_r} &= \sup\left\{\textstyle\int_{P_I} \omega : \|\omega\|_{C^r} \le 1\right\} \\ &= \sup\left\{\textstyle\int_P \omega_I : \|\omega_I\|_{C^r} \le 1\right\} \\ &\le \sup\left\{\textstyle\int_P \omega : \|\omega\|_{C^r} \le 1\right\} \\ &= |P|^{\natural_r}. \end{aligned} \qquad \square$$

As a consequence of this theorem we may now define the I-component operator $I : \mathcal{N}_k^r \to \mathcal{N}_k^r$ sending a chainlet A to its component A_I with

$$\int_{A_I} \omega = \int_A \omega_I$$

for all $\omega \in \mathcal{B}_k^r$.

5. Correspondence of currents and chainlets

Let $L^1\left(\mathbb{R}^n, V^{[k]}\right)$ denote the space of all k-vector valued L^1 functions on $\mathbb{R}^n$ and let m denote n-dimensional Lebesgue measure in $\mathbb{R}^n$.

Every function $f \in L^1\left(\mathbb{R}^n, V^{[k]}\right)$ determines a current T_f defined by

$$T_f[\omega] = \int_{\mathbb{R}^n} \omega(x; f(x))dm$$

for all $\omega \in \mathcal{B}_k^1$.

The *flat norm* of a k-dimensional polyhedral chain is defined as

$$|P|^\flat = \inf\{M(B) + M(C) : P = B + \partial C : B \in \mathcal{P}_k(\mathbb{R}), C \in \mathcal{P}_{k+1}(G\}$$

Flat k-chains are naturally included in the space of chainlets $\mathcal{N}_k^1$ since $|P|^{\natural_1} \le |P|^\flat$ for all polyhedra P.

Whitney [W] proved that T_f corresponds to a unique flat chain A_f for every $f \in L^1\left(\mathbb{R}^n, V^{[k]}\right)$.

Theorem 5.1. *For every $f \in L^1\left(\mathbb{R}^n, V^{[k]}\right)$ there exists a unique flat k-chain A_f such that $|f|_{L^1} = |A_f|_0$ and*

$$\int_{A_f} \omega = \int_{\mathbb{R}^n} \omega(x; f(x))dm$$

for all Lipschitz k-forms ω defined on $\mathbb{R}^n$.

The basic idea is similar to the example given earlier demonstrating that an L^1 function $f : [a,b] \to \mathbb{R}$ is represented by a chainlet supported in $[a,b]$. Given a k-vector valued function f on $\mathbb{R}^n$ there is a sequence of polyhedral k-chains P_i in $\mathbb{R}^n$ constructed by forming Riemann sums associated with f. Since f is an L^1 function, it follows that the sequence limits in the 1-norm to a unique chainlet $A_f \in \mathcal{N}_k^1$ satisfying the integral equation in Theorem 5.1.

If T is a linear form on $\mathcal{B}_k^r$ continuous in the weak * topology then T is called an *integrable k-current* of order $\leq r$. The subspace of all integrable k-currents of order $\leq r$ is denoted $\mathcal{T}_k^r$. Observe that $\mathcal{T}_k^r \subset (\mathcal{B}_k^r)^*$.

Representations of integrable distributions

The following theorem is due to Schwartz ([Sch], Théoreme XXV; see also [B-W] Theorem 1.26)

Theorem 5.2. *A distribution c is integrable if and only if $c = \sum_{s\leq r} D^s f_s$ for some r, where each $f_s \in L^1$.*

Proposition 5.3. *If c is an integrable distribution there exists an integer r and a unique chainlet $A \in \mathcal{N}_0^r$ such that*

$$c[\phi] = \int_A \phi$$

for all $\phi \in \mathcal{B}_0^r$.

Proof. Existence follows from Theorems 3.1, 5.1 and 5.2: Suppose that $c = \sum_{s\leq r-1} D^s f_s$. Then there exists a chainlet $A_s \in \mathcal{N}_0^1$ such that

$$\begin{aligned} c[\phi] &= \sum \int D^s\phi(x) f_s(x) dm \\ &= \sum \int_{A_s} D^s\phi \\ &= \sum \int_{D^s A_s} \phi \\ &= \int_{\sum D^s A_s} \phi. \end{aligned}$$

Each $D^s A_s$ is an element of $\mathcal{N}_0^{s+1}$. Therefore it is an element of $\mathcal{N}_0^r$. Set $A = \sum D^s A_s$.

Uniqueness: Suppose $\int_A \phi = \int_B \phi$ for all $\phi \in \mathcal{B}_0^r$. By Theorem 4.2 it follows that $A = B$. □

We use components and Lie derivatives of chains to translate this into a result for currents. If c_I is a distribution defined for each indexing set $I = \{x_{i_1}, \ldots x_{i_k}\}$ then $\sum_I c_I dx_I$ uniquely determines a current T defined by $T[\omega] = \sum_I c_I \omega_I$ where $\omega = \sum_I \omega_I dx_I$. This correspondence is $1-1$, that is, given a current T, and indexing set I define the distribution $c_I[\phi] = T[\phi dx_I]$. Then $\sum_I c_I dx_I$ determines T as above. The expression $\sum_I c_I dx_I$ is called a *differential form with distributional coefficients* with respect to Euclidean coordinates. Observe that T is integrable if each c_I is an integrable distribution.

Theorem 5.4. *Let $r > 0$. There exists an isomorphism*

$$\theta : \mathcal{N}_k^r \to \mathcal{T}_k^r$$

such that if $T = \theta(A)$ then

$$T[\omega] = \int_A \omega$$

for all $\omega \in \mathcal{B}_k^r$. Furthermore, spt$(A)$ = spt(T), and $|A|^{\natural_r} = |T|^{\natural_r}$.

Proof. Let $A \in \mathcal{N}_k^r$. Define the current $\theta(A)[\omega] = \int_A \omega$. We show $\theta(A)$ is a linear form continuous in the weak * topology of $\mathcal{B}_k^r$. Let D_{X_j} be a sequence of k-forms in $\mathcal{B}_k^r$ which converges to zero in the weak * topology of $\mathcal{B}_k^r$. Then $X_j \to 0$ in the weak * topology of $(\mathcal{N}_k^r)^*$. It follows that $\theta(A)[D_{X_j}] = A \cdot X_j \to 0$ as $j \to \infty$.

We show θ is onto. Let $T \in \mathcal{T}_k^r$. For each indexing set $I = \{x_{i_1}, \ldots, x_{i_k}\}$, let x_I denote the unit k-vector $x_{i_1} \wedge \cdots \wedge x_{i_k}$.

Since $T[\omega] = \sum T_I[\omega] = \sum T[\omega_I]$ we can write $T = \sum_I c_I[\omega]$ where each c_I is an integrable distribution defined by $c_I[\omega] = T[\omega_I]$.

Define the k-vector valued function f_I by $f_I(x) = c_I[\omega]x_I$. According to Theorem 5.1 there exist chainlets $A_I \in \mathcal{N}_k^1$ such that

$$c_I[\omega] = \int_{\mathbb{R}^n} \omega(x, f_I(x))dm = \int_{A_I} \omega.$$

Let $A = \sum A_I$. Then

$$T[\omega] = \sum c_I[\omega] = \sum \int_{A_I} \omega = \int_A \omega$$

for all $\omega \in \mathcal{B}_k^r$. It follows immediately that spt(A) = spt(T) and $T \in \mathcal{T}_k^r$.

By Theorem 2.1 and 2.3

$$|A|^{\natural_r} = \sup\left\{\int_A \omega : \|\omega\|_{C^r} \le 1\right\} = \sup\{T[\omega] : \|\omega\|_{C^r} \le 1\} = |T|^r. \qquad \square$$

Nonreflexivity of the natural norms

The natural norms are not reflexive since the spaces $\mathcal{N}_k^{\natural_r}$ are separable, but their conjugate spaces are not ([H2]). The conjugate spaces of $\mathcal{N}_k^{\natural_r}$ are isomorphic to spaces of differential forms of class $\mathcal{B}_k^r$ ([H3]). We have

$$\mathcal{N}_k^{\natural_r} \cong \mathcal{T}_k^r \subset (\mathcal{B}_k^r)^* \cong (\mathcal{N}_k^{\natural_r})^{**}.$$

Operators on currents

The following operators have been constructed geometrically for chainlets in ([H2], [H3], [H4], [H6]): boundary, star, coboundary, pushforward, contraction, Laplace, and Dirac. The isomorphism commutes with the analytic versions of these operators defined for forms. Therefore the operators on currents are represented by geometrically defined operators on chainlets.

References

[deR] G. de Rham, *Variétés différentiables*, Herman, Paris, 1955.

[Sch] L. Schwartz, *Théorie des distributions*, Hermann, 1966.

[B-W] E. Beltrami and M.R. Wohlers, *Distributions and the Boundary Values of Analytic Functions*, Academic Press, New York and London, 1966.

[H1] J. Harrison, Stokes' theorem on nonsmooth chains, *Bulletin AMS*, October 1993.

[H2] J. Harrison, Continuity of the integral as a function of the domain, *Journal of Geometric Analysis* **8** no. 5 (1998), 769–795.

[H3] J. Harrison, Isomorphisms of differential forms and cochains, *Journal of Geometric Analysis* **8** no. 5 (1998), 797–807.

[H4] J. Harrison, Geometric dual to the Hodge star operator with applications to the theorems of Gauss and Green, preprint

[H5] J. Harrison, Cartan's magic formula and soap film structures, to appear, *Journal of Geometric Analysis*, 2004.

[H6] J. Harrison, Discrete exterior calculus, preprint in preparation.

[K] J. Kigami, Harmonic calculus on p.c.f. self–similar sets, *Trans. Amer. Math. Soc.* **335** (1993), 721–755.

[L] M.L. Lapidus, Fractal drum, inverse spectral problems of elliptic operators and a partial resolution of the Weyl-Berry conjecture, *Transactions of the American Mathematical Society* **325** (1991), 465–529.

[W] H. Whitney, *Geometric integration theory*, Princeton University Press, 1957.

[W2] H. Whitney, Derivatives, difference quotients, and Taylor's formula II, *Annals of Mathematics* **35** (1934), 476–481.

Jenny Harrison
Department of Mathematics
University of California, Berkeley
Berkeley, CA
94705
USA
e-mail: `harrison@math.berkeley.edu`

Progress in Probability, Vol. 57, 205–217

Variational Principles and Transmission Conditions for Fractal Layers

Maria Agostina Vivaldi

Abstract. We review some recent results for second order transmission problems with fractal layers.

Mathematics Subject Classification (2000). Primary 35J25; Secondary 35D10.
Keywords. Transmission problems, highly conductive, fractal layers.

1. Introduction

The aim of this talk is to describe second order transmission problems involving a layer of fractal type, which is imbedded in an Euclidean domain. Boundary value problems, in which boundaries are "large" and volumes are "small" emerge naturally in transmission problems of absorption or irrigation type, where surface effects are enhanced. In this context, fractal boundaries and fractal layers provide new interesting setting. Fractals are geometric objects with highly non-Euclidean characteristics: despite their tricky geometry, there are however large families of fractals which possess a very rich analytic structure. So we are able to study fractals both as intrinsic bodies, in which it is possible to give a suitable notion of Laplacian and as *boundaries* of Euclidean domains supporting traces of functions belonging to classic spaces like Sobolev spaces or possibly as *bodies and boundaries* at the same time, which is the situation we focus on in this presentation.

We are particulary interested in the case of *highly conductive* fractal layers: these layers enjoy higher conductivity with respect to the surrounding space and, because of that, they absorb energy and convey intrinsic diffusion more efficiently. From the analytical point of view one deals with *transmission conditions* involving at the same time *traces* of functions from classic Sobolev spaces and intrinsic *Laplacians* within the layer.

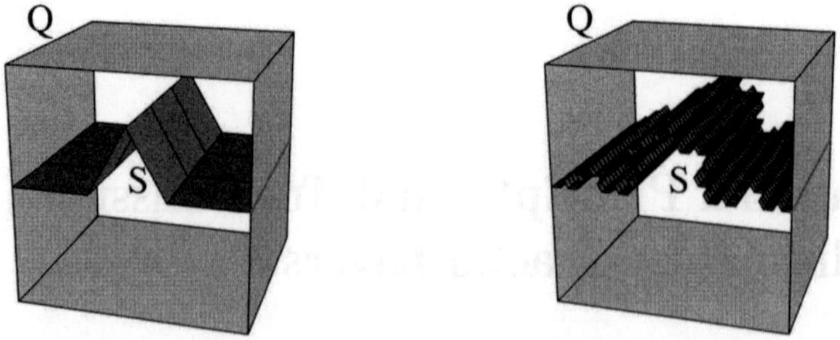

FIGURE 1. Koch roof layer

A broader definition of layer brings into the picture more general *absorption* models, in which a "thin" material inclusion acts as a spongy boundary that absorbs energy or heat from the surrounding space, the material inclusion is a subset with empty interior but positive capacity of an open Euclidean domain possibly, a fractal set, see for example the *Sierpinski layer* in Figure 2

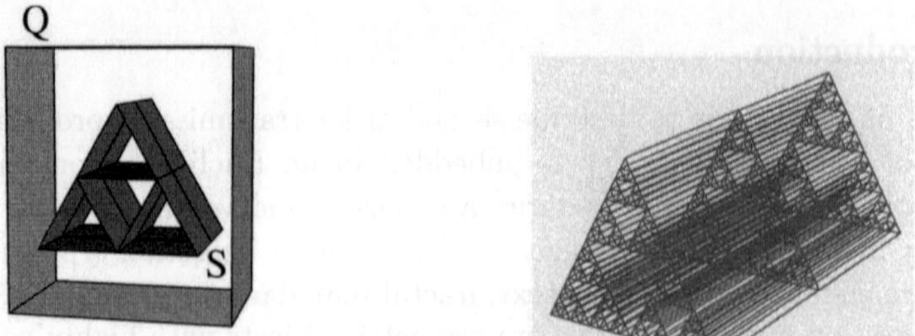

FIGURE 2. Sierpinski layer

Let me note incidentally, that conductive layers, as those considered in this paper, have an interesting counterpart in the so-called *insulating* thin layers: "thin" material inclusions with vanishing conductivity that lead to other kind of layer conditions, see for example Mosco [27].

Boundary value problems with unusual boundary conditions can be traced back to the early work of Venttsel in the late fifties [35]. Second order transmission conditions have been considered in the early seventies in connection with various applications: we only mention here the early work of Cannon-Meyer [3] on the flow of oil in a fractured medium, the contributions of Pham Huy-Sanchez Palencia [33] in electro-static and magneto-statics. Variety of applications leading to transmission problems, not necessarily of second order, are described by Dautray-Lions in [4].

Transmission problems with *fractal*, or *pre-fractal*, highly conductive layers are new: to our knowledge the first examples have been given by Lancia [19]. Let us once more stress the fact that these problems belong to the family of boundary value problems with large boundaries and they provide an interesting crossing of Euclidean and fractal concepts and techniques.

The approach is based on suitable *variational principles*, so we will first describe the variational formulation.

2. Variational formulations

We describe the *variational formulation* of transmission problems with highly conductive layers, by considering examples both of *pre-fractal* and *fractal* layers S in a domain Q of $\mathbf{R}^3$. By Q we denote a (bounded open) domain in $\mathbf{R}^3$; in our basic models Q denotes the cube $Q =]0,1[^3$ in $\mathbf{R}^3$ and S denotes a layer in Q either of the type $S = K \times I$, where $I = [0,1]$ and K is the von Koch curve in $\mathbf{R}^2$ or of the type $S = K_h \times I$ where K_h is one of the *pre-fractal* sets associated with K. We assume that S is located in a median position inside Q, as illustrated in Figure 3.

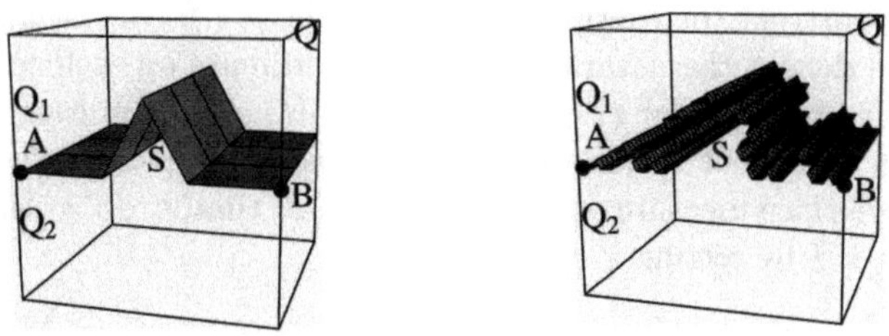

FIGURE 3. Basic model

We consider the *total energy* functional

$$E = E_Q + E_S \tag{2.1}$$

which is the sum of a *volume energy* E_Q and a *layer energy* E_S. We define the volume energy E_Q to be the *Dirichlet integral*

$$E_Q = E_Q[u] = \int_Q |\nabla u|^2 \, dQ, \qquad u \in H_0^1(Q) \tag{2.2}$$

where $dQ = dx_1 dx_2 dy$ is the volume element.

Here $H_0^1(Q)$ denotes the usual Sobolev space of functions of finite energy vanishing on the boundary ∂Q. Every function $u \in H^1(Q)$ possesses a well defined quasi-everywhere *trace* $u_{|S}$ on S, in the sense of the H^1-capacity.

In order to define layer energies, we give a point $P \in S$ the Cartesian *coordinates* $P = (x, y)$, where $x = (x_1, x_2)$ are the coordinates of the orthogonal projection of P on the plane containing K and y is the coordinate of the orthogonal projection

FIGURE 4. Layer coordinates

of P on the y–line containing the interval I: $P = (x, y) \in S$, $x = (x_1, x_2) \in K$, $y \in I$. Similar coordinates we give to points belonging to the *pre-fractal* piece-wise affine layer $S = K_h \times I$, $h = 1, 2, \ldots$; K_h is the *piece-wise linear* pre-fractal approximation of K at the step h, here occurring as the section of the pre-fractal layers S, obtained by h iterations of the similarity mappings of K, with initial set $K_0 = segment[A, B]$. By V_h we denote the discrete set obtained after h iterations of the similarity mappings of K, with initial set $\Gamma = \{A, B\}$ (the set of *essential fixed points* of the family of similarities).

We first construct the energy forms E_S on the *pre-fractal* layers $S = K_h \times I$, $h \geq 1$. By ℓ we denote the natural arc-length coordinate on each edge of K_h and we introduce the coordinates $x_1 = x_1(\ell)$, $x_2 = x_2(\ell)$, $y = y$ on every affine "face" S_j of S. By $d\ell$ we denote the one–dimensional measure given by the arc-length ℓ and by dS the surface measure on each face S_j of S, that is, $dS = d\ell dy$. We define E_S on $S = K_h \times I$ by setting

$$E_S = E_S[u] = \sum_j \left(\int_{S_j} (\sigma_h^1 |D_\ell u|^2 + \sigma_h^2 |D_y u|^2)\, dS \right) \tag{2.3}$$

where σ_h^1, σ_h^2 are positive constants that must be chosen conveniently. By Fubini theorem, we can write this functional in the form

$$E_S[u] = \sigma_h^1 \int_I \left(\int_{K_h} |D_\ell u|^2 d\ell \right) dy + \sigma_h^2 \int_{K_h} \left(\int_I |D_y u|^2 dy \right) d\ell. \tag{2.4}$$

We now go back to the total energy E in (2.1), where now we take E_Q given by (2.2) and E_S given by (2.4). The functional E is well defined on the domain

$$D_0(E) = \{u \in H_0^1(Q) : u_{|S} \in H_0^1(S)\} \tag{2.5}$$

where $H_0^1(S)$ denotes the Sobolev space of functions on the piece-wise affine set S vanishing on ∂S. This space is defined, for instance, according to Nečas [31].

From now on, we denote by the same letter both the quadratic energy forms and the associated bilinear forms defined on the space of functions of finite energy.

Theorem 2.1. *In the previous notations the space $D_0(E)$ defined in (2.5) is a Hilbert space under the norm $\| u \| = (E[u])^{1/2}$ and the form E with domain $D_0(E)$ is a regular Dirichlet form in $L^2(Q)$.*

This result is a direct consequence of trace and density results for Sobolev spaces in polyhedra-like domains, as in Grisvard [5], [6], Nečas [31], Buffa-Ciarlet [2]. See [22] and [29] for details.

At this stage, since h is fixed, the value of the constants in (2.4) does not play any significant role. A good choice of the constants σ_h^1 and σ_h^2, on the other hand, is essential in the asymptotic theory that sees the pre-fractals converging to the limit fractal.

As a corollary of Theorem 2.1, for any choice of f in $L^2(Q)$ there exists a unique function $u_h \in D_0(E)$ that minimizes the total energy:

$$\frac{1}{2}E[u] - \int_Q f u dQ. \tag{2.6}$$

This gives the *variational solution* of the transmission problem in Q with pre-fractal layer $S = K_h \times I$ and Dirichlet boundary conditions $u_h = 0$ on ∂Q and $u_{h|S} = 0$ on ∂S.

We now define the energy form E_S on the *fractal* layer $S = K \times I$ by setting

$$E_S(u,v) = \sigma^1 \int_I \int_K \mathcal{L}_x(u,v)(dx)dy + \sigma^2 \int_K \int_I D_y u D_y v dy \mathcal{H}^{d_f}(dx). \tag{2.7}$$

where $\sigma^1$and σ^2 are positive constants. Here, $\mathcal{L}_x(\cdot,\cdot)(dx)$ denotes the (measure-valued) Lagrangian of the energy form E_K of K with domain $D_0(K)$, now acting on $u(x,y)$ and $v(x,y)$ as functions of $x \in K$ for a.e. $y \in I$; $\mathcal{H}^{d_f}(dx)$ is the Hausdorff measure acting on each section K of S for a.e. $y \in I$ with $d_f = \frac{\log 4}{\log 3}$. We note that the "volume measure" dS in S is the product measure $dS = d\mathcal{H}^{d_f} \times dy$. We recall that the energy form E_K is constructed on the Koch curve (and more generally, on the so-called *nested fractals* in the sense of [24]) taking limits of *discrete* energy functionals E_h on the discrete pre-fractal sets V_h. These discrete functionals can be written in terms of a *graph gradient*, based on differential quotients with respect to a suitably re-normalized *quasi-distance* d of the kind $d(P,P') = |P-P'|^\delta$ for vertices $P, P' \in V_h$. Here δ is $= \frac{\log 4}{\log 3}$ (a parameter greater than 1 on all nested fractals), and has the role of transforming the Euclidean distance into an intrinsic *effective length* within the fractal and, accordingly, Euclidean differential quotients into *effective* differential quotients, see [28].

For nested fractals the convergence of the sequence of energies E_h, obtained by similarity, is studied by Kusuoka in [18]. In the present (simpler) situation, the form E_K on the fractal K is the limit of the forms E_h as $h \to \infty$ and the limit form E_K has the integral representation

$$E_K[u] = \int_K d\mathcal{L}[u] \tag{2.8}$$

where the local energy $\mathcal{L}[\cdot]$ is a *(measure-valued) Lagrangian* on K, see [27], [26] and [28]. E_K is a Dirichlet form in the Hilbert space $L^2(K, \mathcal{H}^{d_f}(dx))$. The *Laplace operator* $\triangle_K$ on K, with *Dirichlet* boundary condition $u_{|\Gamma} = 0$ on Γ, is obtained from the bilinear form

$$E_K(u,v) = \int_K d\mathcal{L}(u,v) \tag{2.9}$$

taken with domain

$$D_0(K) := \{u \in L^2(K, \mathcal{H}^{d_f}(dx)) : \int_K d\mathcal{L}[u] < \infty, u_{|\Gamma} = 0\} \tag{2.10}$$

by the identity

$$E_K(u,v) = \int_K -(\Delta_K u) v \mathcal{H}^{d_f}(dx) .$$

The Laplacian Δ_K is a non-positive *self-adjoint operator*, with a domain D_{Δ_K} dense in $L^2(K, \mathcal{H}^{d_f})$, and the previous identity holds for every $u \in D_{\Delta_K}$ and every $v \in D_0(K)$. We shall denote by the same symbol "Δ_K" the Laplace operator as *variational operator* from $D_0(K) \to (D_0(K))'$ defined by the identity

$$E_K(u,v) = - < \Delta_K u, v >, \qquad u, v \in D_0(K) . \tag{2.11}$$

In the following we will also make use of the non homogeneous domain

$$D(K) = \{u \in L^2(K, \mathcal{H}^{d_f}(dx)) : \int_K d\mathcal{L}[u] < \infty .\} \tag{2.12}$$

We then write the total energy E for the problem with the fractal layer S, as in (2.1), where the volume energy E_Q is still given by (2.2),while the layer energy E_S is now given by (2.7). The functional E is well defined on the domain

$$D_0(E) = \{u \in H_0^1(Q) : u_{|S} \in D_0(S)\}. \tag{2.13}$$

where $D_0(S)$ is the closure in the intrinsic norm $\|u\| = (E_S[u])^{\frac{1}{2}}$ of the set

$$C_0(S) \cap L^2(I; D_0(K)) \cap H_0^1(I; L^2(K)) \tag{2.14}$$
$$\text{where} \quad L^2(K) := L^2(K, \mathcal{H}^{d_f}).$$

Theorem 2.2. *In the previous notations and assumptions the space* $D_0(E)$ *in* (2.13) *is a Hilbert space under the intrinsic norm* $\| u \| = (E[u])^{1/2}$ *and the form* E *with domain* $D_0(E)$ *is a regular Dirichlet form in* $L^2(Q)$.

The proof makes use of the theory of *Besov spaces* on d-sets developed by Jonsson-Wallin [13], Triebel [34] and of the theory of the Besov spaces on closed sets (that are no d–sets) developed by Jonsson [10]. It is to be pointed out that interesting relations between Dirichlet forms and Brownian motion penetrating fractals have been addressed by Lindstrøm [25], Jonsson [12] and Kumagai [17].

As in the pre-fractal case, for every $f \in L^2(Q)$ there exists a unique function $u \in D_0(E)$ that minimizes the total energy functional (2.6). This function provides the *variational*, or *weak*, solution to the *transmission problem* in Q, with fractal layer S and Dirichlet conditions $u = 0$ on ∂Q and $u_{|S} = 0$ on ∂S.

3. Transmission conditions

The (closed) layer S divides the (open) domain Q into two adjacent (open) domains, Q^1 and Q^2, $Q = Q^1 \cup Q^2 \cup (S \cap Q)$, (see Figure 3). In this case, the variational solution of the transmission problem in Q with layer S, obtained in the previous section, satisfies a *second order transmission condition*, which comes to light once we write the (weak) *Euler equation* satisfied by the variational minimum. This leads to the *strong formulation* of the transmission problem, obtained by integration by parts through Green formulas in each domain Q^i. As already mentioned, highly conductive layers are characterized with respect to the surrounding space for having a much greater conductivity, or permeability: *heat* or *flow* in the space is absorbed by the layer and starts diffusing within it much more efficiently than in the surrounding volume. The normal derivative from each side of the layer has a *jump* across the layer which acts as a *source term* for the Laplace operator generating the layer diffusion. The resulting boundary transmission condition is thus of *second order*, what is in some sense unusual for second order elliptic b.v.p.'s. Moreover, the condition has an *implicit* character, since the source term of the layer equation — the jump of normal derivatives — is not among the data of the problem, but depends on the solution itself.

Formally, the equations in Q are

$$-\Delta u^i = f \quad \text{on} \quad Q^i \tag{3.1}$$

$$u = 0 \quad \text{on} \quad \partial Q \tag{3.2}$$

$$u^1 = u^2 \quad \text{on} \quad S, \tag{3.3}$$

where $u^i = u_{|Q^i}, i = 1, 2$, and these equations must be coupled with the equations on the layer S

$$u_{|S} = 0 \qquad \text{on} \quad \partial S \tag{3.4}$$

$$\frac{\partial u^1}{\partial \nu^1} + \frac{\partial u^2}{\partial \nu^2} = -\text{“}\Delta_S u\text{”} \quad \text{on} \quad S \tag{3.5}$$

where $\frac{\partial u^i}{\partial \nu^i}$ denotes the (formal) exterior normal derivative to the boundary of Q^i and "$\Delta_S u$" the Laplacian associated to the Dirichlet form E_S. The rigorous definition of the operators and the spaces is one of the main technical difficulties of this kind of problems.

We now summarize the main results for the pre–fractal transmission.

Theorem 3.1. *In the assumptions and notations of Theorem* 2.1 *we have that the variational solution* u_h *satisfies*

$$u_h \in C(\overline{Q}) \tag{3.6}$$

$$u_h^1 \in H^{\frac{8}{5}-\varepsilon}(Q^1), \quad u_h^2 \in H^{\frac{7}{4}-\varepsilon}(Q^2) \tag{3.7}$$

$$\frac{\partial u_h^i}{\partial \nu^i} \in L^2(S) \quad i = 1, 2. \tag{3.8}$$

Where $u_h^i = u_h|_{Q^i}$ and $\frac{\partial u_h^i}{\partial \nu^i}$ denote the exterior normal derivatives to the boundary of Q^i, $i = 1, 2$. Let us note that the norms of the functions u_h^i in the fractional Sobolev spaces (in (3.7)) depend on h.

As a consequence the variational solution u_h is indeed the strong solution: equality (3.1) holds a.e. in Q^i, equalities (3.2), (3.3) and (3.4) are satisfied pointwise and the transmission condition (3.5) holds in $L^2(S)$. Here the Laplacian "Δ_S" in (3.5) is the linear combination of the "piece–wise" second order tangential derivative along the sides of K_h and the "usual" second order partial derivative in y:

$$``\Delta_S" = \sigma_h^1 D_\ell^2 + \sigma_h^2 D_y^2. \tag{3.9}$$

We point out that the regularity result of the preceding theorem is new even in the Euclidean case of flat or smooth layers of Pham Huy-Sanchez Palencia. Both in the Euclidean and in the fractal case, the difficulty in proving regularity properties of solutions stems from the implicit character of the volume equations in Q^1, Q^2 and of the Poisson equation within the layer. Actually we proved in the "flat" case that both the restrictions of the (variational) solution to Q^i belong to the Sobolev space $H^2(Q^i)$ see Theorem 3.2 in [22].

As to the different Sobolev regularity exponents for u_h^1 and u_h^2, this discrepancy is due to the geometry of the polyhedra Q^1 and Q^2, which have different (largest) dihedral angles, $(5/3)\pi$ and $(4/3)\pi$ respectively. As it is known from the regularity theory, the regularity of the solutions improves if the opening of the inner dihedral angles becomes smaller. This effect holds on despite the implicit character of the equations and the dependence of the regularity exponent on the angle remains unperturbed. Just to present the main tools we sketch the proof of Theorem 3.1.

- **Step 1.** Let us denote by $\hat{w}^i$ the solution of the Dirichlet problem

$$\begin{cases} \Delta \hat{w}^i = 0 & \text{in } Q^i \\ \hat{w}^i = u_h^i & \text{on } \partial Q^i \end{cases}$$

Then, according to the results of Jerison and Kenig (see [9]), we prove

$$\| \frac{\partial \hat{w}^i}{\partial \nu_i} \|_{L^2(S)} \leq c(h) \|f\|_{L^2(Q)}. \tag{3.10}$$

On the other hand by using Kondrat'ev type results (see [16] and Theorem 10.2.3 in [30]), we obtain the following weighted estimates for the second order derivatives of $w^i = u_h^i - \hat{w}_i$:

$$\sum_{|\alpha|=2} \int_{Q^1} r^{2(\sigma_1 - \mu_1)} \rho^{2\mu_1} |D^\alpha w^1|^2 dxdy \leq c(\mu_1) \|f\|_{L^2(Q)}^2 \tag{3.11}$$

$$\sum_{|\alpha|=2} \int_{Q^2} r^{2(\sigma_2 - \mu_2)} \rho^{2\mu_2} |D^\alpha w^2|^2 dxdy \leq c(\mu_2) \|f\|_{L^2(Q)}^2 \tag{3.12}$$

$$\text{for any} \quad 1 > \mu_1, \sigma_1 > \frac{2}{5}, \quad 1 > \mu_2, \sigma_2 > \frac{1}{4} \tag{3.13}$$

where $\rho = \rho(x, y)$ denotes the distance of the point (x, y) from the nearest edge and $r = r(x, y)$ the distance of the point (x, y) from the nearest vertex of the polyhedron.

Now we choose $\mu_1 = \sigma_1$, $\mu_2 = \sigma_2$ and interpolation techniques, according to [8], allow us to show that w^i belongs to usual (fractional) Sobolev spaces: more precisely, we have

$$\|w^i\|_{H^{2-\mu_i}(Q^i)} \leq c(\mu_i, h)\|f\|_{L^2(Q)}, \quad i = 1, 2. \tag{3.14}$$

$$\mu_1 \text{ and } \mu_2 \text{ as in (3.13)}$$

• **Step 2.** Starting from (3.10) and (3.14) and using trace results (see [5] and [2]) we obtain (3.8). Green formula and the use of suitable test functions allow us to prove that $\Delta_S u_h$ belongs to $L^2(S)$ and that u_h is continuous on S. Making use of extension theorems in polyhedra and the before mentioned Kondrat'ev type results we prove (3.7). Finally the Morrey–Sobolev embedding results and the fact the functions u_h^i have the same (continuous) trace on S allow us to show the continuity of u_h on the closure of Q and to conclude the proof of Theorem 3.1.

Let us consider the fractal transmission

Theorem 3.2. *In the assumption and notation of Theorem 2.2 we have that the variational solution u satisfies:*

$$\frac{\partial u^1}{\partial \nu^1}, \frac{\partial u^2}{\partial \nu^2} \quad \in \quad (B^{2,2}_{\beta,0}(S))' \tag{3.15}$$

where $(B^{2,2}_{\beta,0}(S))'$ *is the dual space of* $B^{2,2}_{\beta,0}(S)$, $\quad \beta = d_f/2$

$$\frac{\partial u^1}{\partial \nu^1} + \frac{\partial u^2}{\partial \nu^2} \quad = \quad -\Delta_S u \quad \text{in the dual space } (D_0(S))' \text{of} D_0(S). \tag{3.16}$$

Here $u^i = u_{|Q^i}$, $i = 1, 2$ and Δ_S denotes the *variational* Laplace operator from $D_0(S)$ to the dual $(D_0(S))'$ associated with the Dirichlet forms E_S as in (2.11). Hence formula (3.5) has to be intended in the sense of the natural dual space of the right hand side.

The proof of Theorem 3.2 is delicate and makes use of complicated results from sophisticated subspaces of the Besov spaces on d–sets S and K. The space $B^{2,2}_\beta(S)$, is the space of functions on S for which is finite the norm

$$\| u \|_{L^2(S,dS)} + \left(\int\int_{|x-y|<1} \frac{|u(x) - u(y)|^2}{|x - y|^{d+2\beta}} dS(x)dS(y) \right)^{1/2} . \tag{3.17}$$

here $d = 1 + d_f$, and $\beta = d_f/2$. Let us recall that the definition of Besov spaces on (d-sets) with smoothness index greater or equal to 1 is much more complicated than the previous one, see [13] and [34].

The space $B^{2,2}_{\beta,0}(S)$ is a subspace of $B^{2,2}_\beta(S)$ and is the fractal analogue of the Lions-Magenes space $H^{\frac{1}{2}}_{0,0}(S)$, namely

$$B^{2,2}_{\beta,0}(S) = \{u \in L^2(S, dS) : \exists w \in H^1_0(Q), w_{|S} = u \,\text{on}\, S\}$$

taken with the norm

$$\| u \|_{B^{2,2}_{\beta,0}(S)} = \inf\{\| w \|_{H^1_0(Q)} : \exists w \in H^1_0(Q), w_{|S} = u \,\mathrm{on}\, S\}.$$

The space $D_0(S)$ introduced in Section 2 (see (2.14)) turns out to be a subspace of the Besov space $B^{2,2}_{\beta,0}(S)$, the proof can be done as in Remark 6.1 of [19] where the embedding of $D_0(K)$ in $B^{2,2}_{\beta,0}(K)$ is shown. The Besov spaces on K: $B^{2,2}_{\beta}(K)$ and $B^{2,2}_{\beta,0}(K)$ are defined in an analogous way where now in (3.17) d is equal to d_f. An important role, in proving Theorem 3.2 is played by the characterization of the domain $D(K)$ (see Section 2 formula (2.12)) in terms of the "tricky" Lipschitz spaces $Lip(\cdot, 2, \infty, K)$, obtained in [21], Theorem 1. These Lipschitz spaces $Lip(\cdot, 2, \infty, \cdot)$ have been firstly introduced by Jonsson to characterize the domain of the Dirichlet form of the Brownian motion on the Sierpinski gasket (see [11]). Their "collocation" in the "Besov scale" and several interesting properties are stated (see [11] and, for further application, [30] and [17]).

4. Asymptotics

An asymptotic "constructive" theory for the prefractal approximation is an important step toward the numerical analysis of the problem.

For sake of simplicity, we summarize the main results in the 2-dimensional case, the 3-dimensional case being more involved and requiring some additional work. So from now on Q denotes the unit square $Q =]0,1[^2$ in $\mathbf{R}^2$. The analogues of previous Theorems 2.1, 2.2, 3.1 and 3.2 hold and we address the question whether the pre-fractal solutions u_h do converge to the fractal solution u, as the pre-fractal layer $S = K_h$ converges to the fractal layer $S = K$ for $h \to \infty$.

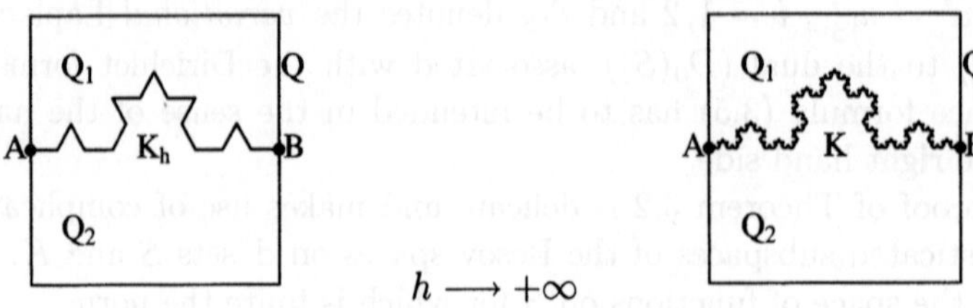

FIGURE 5. Basic model

As h increases the prefractal layers develop increasing length up to reach the infinite length of the limit fractal curve. The results of formula (3.7) hold with same smoothness exponents (see Theorem 2.2 of [22]) but the norm in the fractional Sobolev spaces (with smoothness index greater than 1) blows up as $h \to +\infty$. So we are forced to choose a different notion of convergence, actually an energy convergence.

The Euclidean pre-fractal energies must be re-normalized in order to keep the energy finite in the limit. This amounts to choose the constant σ^1_h in (2.4) conveniently. We note that in the 2-dimensional case the layer energy functional is given by

$$E_S[u] = \sigma_h^1 \int_{K_h} |D_\ell u|^2 d\ell.$$

The good choice of the re-normalization factor is obtained by taking into account the effect of the d_f-dimensional length intrinsic to the fractal curve.

Theorem 4.1. *Let u be the variational solution of the fractal transmission problem in the domain Q of $\mathbf{R}^2$, with layer $S = K$ the fractal Koch curve. For every integer $h \geq 1$, let u_h be the variational solution of the transmission problems in Q with pre-fractal layer $S = K_h$. If we scale the energy functionals (2.4), where $S = K_h$, by taking $\sigma_h^1 = \sigma^1(3^{d_f-1})^h$ then as $h \to \infty$ we find:*

$$u_h \quad \to \quad u \text{ strongly in } H_0^1(Q) \tag{4.1}$$

$$\int_{K_h} \frac{\partial u_h^i}{\partial \nu^i} \phi \, d\ell \quad \to \quad \langle \frac{\partial u^i}{\partial \nu^i}, \phi \rangle_{(B_{\beta,0}^{2,2}(K))', B_{\beta,0}^{2,2}(K)} \quad \forall \phi \in H_0^1(Q) \tag{4.2}$$

$$\beta = d_f/2, \; i = 1, 2$$

$$\sigma_h^1 \int_{K_h} \Delta_{K_h} u_h \phi d\ell \quad \to \quad \sigma^1 \langle \Delta_K u, \phi \rangle_{(D_0(K))', D_0(K)} \quad \forall \phi \in D_0(E). \tag{4.3}$$

where Δ_{K_h} is the piecewise second order tangential derivative D_ℓ^2 along the polygonal arc K_h and Δ_K the variational Laplacian defined by (2.11).

A stronger result is indeed proved in [23]: the so-called *M-convergence* of the prefractal energies to the fractal energy. We recall that M-convergence implies the convergence of several objects as the strong convergence of *resolvent operators* and of *semigroups* and leads also to convergence of *Brownian processes*, *spectra* and *spectral measures*, see Mosco [27].

Remark 4.2. A quantitative estimate of the rate of convergence in (4.1) would be very useful for numerical solutions of the fractal problem.

Remark 4.3. From the probabilistic point of view, Brownian motions penetrating fractal sets – a probabilistic counterpart of the analytic variational approach adopted here – have been constructed by Lindstrøm [25] and Kumagai [17], however without reference to transmission problems and related transmission conditions.

Remark 4.4. We could also consider perturbations of the volume energy E_Q in (2.2) caused by variable volume conductivities $a^{(\varepsilon)}(x)$ in Q, with $a^{(\varepsilon)}(x)$ some function on Q that becomes singular in small neighboring strips of size ϵ around each pre-fractal layer K_h, ϵ depending on h and tending to 0 as $h \to \infty$. This would give perturbed volume energy functionals

$$E_Q^{(\epsilon)}[u] = \int_Q | \nabla u|^2 a^{(\epsilon)}(x) dQ \tag{4.4}$$

and we should study the convergence of $E_Q^{(\epsilon)}$ as $\varepsilon \to 0$. For Euclidean flat layers S, this problem has already been faced by Pham Huy-Sanchez Palencia, [33], who proved in this case the variational convergence of (4.4) to the *singular* energy functional (2.1), see also [27]. The theory described before should help to obtain (weak) convergence results for piecewise affine layers.

References

[1] F. Brezzi, G. Gilardi, *Finite Elements Mathematics.*, Finite Element Handbook, Kardenstucer H., Norrie D.H. Eds., McGraw-Hill Books Co., New York, 1987.

[2] A. Buffa, P. Ciarlet, *On traces for functional spaces related to Maxwell's Equations, Part I: An integration by parts formula in Lipschitz Polyhedra*, Math. Meth. Appl. Sci., **24** 1 (2001), 9–30.

[3] J. R. Cannon, G. H. Meyer, *On a diffusion in a fractured medium*, SIAM J. Appl. Math. **3** (1971), 434–448.

[4] R. Dautray, J. L. Lions, *Mathematical Analysis and Numerical Methods for Science and technology*, Vol.2, Springer, Berlin, 1988.

[5] P. Grisvard, *Elliptic problems in nonsmooth domains*, Pitman, Boston, 1985.

[6] P. Grisvard, *Théorèmes de traces relatifs à un polyèdre*, C.R. Acad. Sci. Paris, Série A **278** (1974), 1581–1583.

[7] H. Yserantant, *On the Multi-Level splitting of Finite Element Spaces*, Numer. Math. **49** (1986), 379–412.

[8] D. Jerison, C. E. Kenig, *The inhomogeneous Dirichlet problem in Lipschitz domains*, J. Funct. Anal. **130** (1995), 161–219.

[9] D. Jerison, C. E. Kenig *The Neumann Problem in Lipschitz domains*, Bull. Amer. Math. Soc. **4** (1981), 203–207.

[10] A. Jonsson, *Besov spaces on closed subsets of* $\mathbf{R}^n$, Trans. Amer. Math. Soc. **341** (1994), 355–378.

[11] A. Jonsson, *Brownian motion on fractals and function spaces*, Math. Z. **222** (1996), 495–504.

[12] A. Jonsson, *Dirichlet forms and Brownian motion penetrating fractals*, Potential Analysis **13** (2000), 69–80.

[13] A. Jonsson, H. Wallin, *Function Spaces on Subsets of* $\mathbf{R}^n$, Part 1 (Math. Reports Vol. 2). London: Harwood Acad. Publ. 1984.

[14] A. Jonsson, H. Wallin, *The dual of Besov spaces on fractals*, Studia Math. **112** (1995), 285–300.

[15] J. Kigami, *Harmonic calculus on p.c.f. selfsimilar sets*, Trans. Amer. Math. Soc. **335** (1993), 721–755.

[16] V. A. Kondrat'ev, *Boundary-value problems for elliptic equations in domains with conical or angular point*, Trans. Moscow Math. Soc. **16** (1967), 227–314.

[17] T. Kumagai, *Brownian motion penetrating fractals and application of the trace theorem of Besov spaces*, J. Funct. Anal. **170** (2000), 69–92.

[18] S. Kusuoka, *Diffusion processes on nested fractals*, Lecture Notes in Math., 1567, Springer New York (1993).

[19] M. R. Lancia, *A transmission problem with a fractal interface*, Z. Anal. und Ihre Anwend. **21** (2002), 113–133.

[20] M. R. Lancia, *Second order transmission problems across cylindrical fractal surfaces*, (In preparation).

[21] M. R. Lancia , M. A. Vivaldi, *Lipschitz spaces and Besov traces on selfsimilar fractals*, Rend. Accad. Naz. XL, Mem. Mat. Appl. **23** (1999), 101–106.

[22] M. R. Lancia , M. A. Vivaldi, *On the regularity of the solutions for transmission problems*, Adv. Math. Sc. Appl. **12** (2002), 455–466.

[23] M. R. Lancia , M. A. Vivaldi, *Asymptotic convergence of transmission energy forms*, Adv. Math. Sc. Appl. **13** (2003), 315–341.

[24] T. Lindstrøm, *Brownian motion on nested fractals*, Mem. Am. Math. Soc. n. 20 (1990).

[25] T. Lindstrøm, *Brownian motion penetrating the Siepinski gasket*, *Asymptotic problems in probability theory: stochastics models and diffusions on fractals*, Elworthy K.D., Ikeda N. Eds., Longman, 1993.

[26] J. Malý, U. Mosco, *Remarks on measure-valued Lagrangians on homogeneous spaces*, Ric. Mat. Napoli, **48** suppl. (1999), 217–231.

[27] U. Mosco, *Composite media and asymptotic Dirichlet forms*, J. Funct. Anal. **123**(2) (1994), 368–421.

[28] U. Mosco, *Energy functionals on certain fractal structures*, J. Convex Anal. **9**(1) (2002), 1–20.

[29] U. Mosco, M. A. Vivaldi, *Variational problems with fractal layers*, Preprint Me. Mo. Mat. n. 23 (2003).

[30] S. A. Nazarov, B. A. Plamenovsky, *Elliptic problems in domains with piece-wise smooth boundaries*, De Gruyter expocitions in Mathematics, Berlin-New York (1994).

[31] J. Nečas, *Les méthodes directes en théorie des équationes elliptiques*, Masson, Paris, 1967.

[32] K. P. Paluba, *Some function space related to the Brownian motion on simple nested fractals*, Stoch. and Stoch. Rep. **67** (1999), 267–285.

[33] H. Pham Huy, E. Sanchez–Palencia, *Phénomènes des transmission à travers des couches minces de conductivité élevée*, J. Math. Anal. Appl. **47** (1974), 284–309.

[34] H. Triebel, *Fractals and Spectra. Related to Fourier Analysis and Function Spaces*, Monographs in Mathematics, Vol. 91, Birkhäuser, Basel, 1997.

[35] A. D. Venttsel, *On boundary conditions for multidimensional diffusion processes*, Teor. Veroyatnost. i Primenen. **4** (1959), 172–185; English translation, Theor. Probability Appl. **4** (1959), 164–177.

Maria Agostina Vivaldi
Dipartimento di Metodi e Modelli Matematici per le Scienze Applicate
Università degli Studi di Roma “La Sapienza”
Via A. Scarpa 16
00161 Roma
Italy
e-mail: `vivaldi@dmmm.uniroma1.it`

Part 5

Harmonic Analysis on Fractals

Progress in Probability, Vol. 57, 221–234

Function Spaces and Stochastic Processes on Fractals

Takashi Kumagai

Abstract. We summarize recent work on function spaces and stochastic processes on fractals. We discuss relations between various non-local Dirichlet forms on fractals whose domains are Besov spaces. The corresponding stochastic processes are jump-type processes. Results on heat kernel estimates for the processes are introduced. We will also discuss how jump processes and diffusion processes are related by observing their function spaces.

Mathematics Subject Classification (2000). Primary 46E35; Secondary 60J35; 60J75.

Keywords. Besov-Lipschitz spaces, heat kernels, jump processes, sub-Gaussian estimates, stable-like processes.

1. Introduction

In this article, we summarize recent work on function spaces and stochastic processes on fractals. Mathematical study of stochastic processes and the corresponding operators on fractals has its original motivation in physics to understand the heat transfer on such disordered media ([17]). Diffusion processes on fractals typically have sub-Gaussian behaviour and detailed properties such as estimates of the heat kernels (fundamental solutions of the heat equations) are obtained for typical fractals (see, for instance, [2, 23, 24, 28, 33]). On the other hand, there has been deep study of Besov spaces (fractional extensions of Sobolev spaces) on the so called d-sets (see [22, 36, 37]). Recently, it is getting clear that the domains of Dirichlet forms (quadratic forms that determine stochastic processes and the corresponding operators) on fractals are Besov-Lipschitz spaces. This motivates further study of Besov-type function spaces using probability and potential theory. For instance, detailed estimates of heat kernels can be obtained for operators naturally defined from norms of Besov spaces ([8, 9]). The purpose of this article is

This research is supported in part by the Inamori Foundation.

to make it clearer how the recent results on this direction are connected. Because of the limitation of pages, we do not give any proof except that of Theorem 3.2. The proof and further details are given in the references cited.

The organization of the paper is as follows. In Section 2, we first review basic facts on the Dirichlet form theory which connect function spaces and stochastic processes. We will also give a definition of d-sets, the state space we work on. In Section 3, we discuss about function spaces appear as domains of local regular Dirichlet forms on fractals, whose corresponding generators are so called Laplacians on fractals. Characterization of the domains is possible through heat kernel estimates (3.1) of the Laplacians. In Section 4, we introduce other types of function spaces appear as domains of non-local Dirichlet forms on d-sets, whose corresponding processes are stable-like jump processes. Three natural jump-type processes are introduced on d-sets and the corresponding three forms are shown to be equivalent. It turns out that one of the forms (and the corresponding operators) corresponds to the form (and the operator) studied by Triebel in [36, 37] and by Zähle in [39, 40, 16]. In Section 5, we will summarize results on heat kernel estimates for the stable-like jump processes. Finally in Section 6, we will give some sufficient conditions to construct diffusion processes on d-sets. The relation to stable-like jump processes is also discussed.

Throughout the paper, we will mainly consider compact fractals, but most of the results hold for unbounded fractals with suitable modifications of the statements.

For readers who are interested in the topics, we note that function spaces on fractals are also discussed in [14, 34]. Surveys and references on jump-type process on fractals may be found in [26, 38].

2. Dirichlet forms and d-sets

In this section, we will briefly review the definition of Dirichlet forms and the correspondence to processes following [12]. We will also introduce d-sets.

Let X be a locally compact separable metric space and ν be a positive Radon measure on X whose support is X. Let $\mathcal{E}$ be a symmetric bilinear closed form on $\mathbb{L}^2(X,\nu)$ with domain $\mathcal{F}$. $(\mathcal{E},\mathcal{F})$ is called a Dirichlet form if it is Markovian, i.e., for each $u \in \mathcal{F}$, $v := (0 \vee u) \wedge 1 \in \mathcal{F}$ and $\mathcal{E}(v,v) \le \mathcal{E}(u,u)$. A Dirichlet form $(\mathcal{E},\mathcal{F})$ is regular if there exists $C \subset \mathcal{F} \cap C_0(X)$ such that C is dense in $\mathcal{F}$ under $\mathcal{E}_1$-norm and C is dense in $C_0(X)$ under the uniform norm, where $C_0(X)$ is a space of continuous compact supported functions on X and $\mathcal{E}_1(\cdot,\cdot) = \mathcal{E}(\cdot,\cdot) + \|\cdot\|^2_{\mathbb{L}^2}$. $(\mathcal{E},\mathcal{F})$ is local if for each $u, v \in \mathcal{F}$ whose supports are disjoint compact sets, $\mathcal{E}(u,v) = 0$. There is a one to one correspondence between a regular Dirichlet form on $\mathbb{L}^2(X,\nu)$ and a ν-symmetric Hunt process (i.e. a strong Markov process whose paths are right continuous and quasi-left continuous w.r.t. some filtration) on X except for some exceptional set of starting points. Further, if the regular Dirichlet form is local, then the corresponding process is a diffusion process (i.e. Hunt process with continuous paths).

We next introduce our state space. Let G be a compact d-set in $\mathbb{R}^n$ ($n \geq 2, 0 < d \leq n$). That is, $G \subset \mathbb{R}^n$ and there exist a positive Radon measure μ on G and $c_{2.1}, c_{2.2} > 0$ such that

$$c_{2.1} r^d \leq \mu(B(x,r)) \leq c_{2.2} r^d \qquad \text{for all } x \in G,\ 0 \leq r \leq 1, \tag{2.1}$$

where $B(x,r)$ is a ball centered at x and radius r w.r.t. the Euclidean metric. Thus d is the Hausdorff dimension of G and μ the Hausdorff measure on G. We normalize the size of G so that the diameter of G is 1.

3. Lipschitz spaces and domains of Dirichlet forms

For several sub-classes of d-sets, diffusion processes, corresponding Laplace operators and Dirichlet forms have been studied extensively (see [2, 23, 24, 28, 33] etc). The most typical example is the Sierpinski gasket which we will define later. In this section, we will consider a class of d-sets which has a *fractional diffusion* in the sense of Barlow [2] and show that the domain of the Dirichlet form is a Lipschitz space. We first give a definition of the fractional diffusion.

Definition 3.1. Let (G,ρ) be a complete compact metric space. (G,ρ) is called a fractional metric space and $\{B_t^G\}_{t\geq 0}$ is called a fractional diffusion if the following holds.
1) ρ has the midpoint property; for each $x, y \in G$, there exists $z \in G$ such that $\rho(x,y) = \rho(x,z)/2 = \rho(z,y)/2$. Further, there exists a Borel measure μ which satisfies (2.1) w.r.t. ρ.
2) $\{B_t^G\}_{t\geq 0}$ is a μ-symmetric conservative Feller diffusion on G which has a symmetric jointly continuous heat kernel (fundamental solution of the heat equation) $p_t(x,y)$ $(t > 0, x, y \in G)$ satisfying the following estimate,

$$\begin{aligned} & c_{3.1} t^{-d_s/2} \exp(-c_{3.2}(\rho(x,y)^{d_w} t^{-1})^{1/(d_w-1)}) \leq p_t(x,y) \qquad (3.1) \\ \leq\ & c_{3.3} t^{-d_s/2} \exp(-c_{3.4}(\rho(x,y)^{d_w} t^{-1})^{1/(d_w-1)}) \ \text{ for all } 0 < t < 1,\ x, y \in G, \end{aligned}$$

with some constants $d_s > 0, d_w \geq 2$.

We note that in the original definition of the fractional diffusions in [2], G is not necessarily compact. For simplicity, we further assume that $\rho(\cdot,\cdot)$ is equivalent to the Euclidean metric, i.e.,

$$c_{3.5}|x-y| \leq \rho(x,y) \leq c_{3.6}|x-y| \qquad \text{for all } x, y \in G. \tag{3.2}$$

In this case, $d_s/2 = d/d_w$ holds.

Example: Sierpinski gasket. Let D_n be a n-dimensional simplex whose vertices are $\{p_0, p_1, \cdots, p_n\}$. For $i = 1, 2, \cdots, n+1$, let $F_i(z) = (z - p_i)/2 + p_i$, $z \in \mathbb{R}^n$. Then, there exists a unique non-void compact set G such that $G = \cup_{i=1}^{n+1} F_i(G)$. This G is called a (*n-dimensional*) *Sierpinski gasket.* It is known that the Sierpinski gasket has a fractional diffusion with $d = \log(n+1)/\log 2$, $d_s = 2\log(n+1)/\log(n+3)$, $d_w = \log(n+3)/\log 2$. (Note that $d_s < 2$ for this example.)

In general, (affine) nested fractals, (which is a class of finitely ramified fractals including Sierpinski gaskets) and Sierpinski carpets have fractional diffusions with $c_1|x-y|^{d_c} \le \rho(x,y) \le c_2|x-y|^{d_c}$ for some $d_c \ge 1$ instead of (3.2).

We now introduce Lipschitz spaces. For $1 \le p < \infty$, $1 \le q \le \infty$, $\beta \ge 0$ and $m \in \mathbb{N} \cup \{0\}$, set

$$a_m(\beta, f) := L^{m\beta}(L^{md}\int\int_{|x-y|<c_0 L^{-m}} |f(x) - f(y)|^p d\mu(x) d\mu(y))^{1/p}, \; f \in \mathbb{L}^p(G,\mu),$$

where $1 < L < \infty$, $0 < c_0 < \infty$. Define a *Lipschitz space* $\mathrm{Lip}(\beta,p,q)(G)$ as a set of $f \in \mathbb{L}^p(G,\mu)$ such that $\bar{a}(\beta,f) := \{a_m(\beta,f)\}_{m=0}^{\infty} \in l^q$. $\mathrm{Lip}(\beta,p,q)(G)$ is a Banach space with the norm $\|f\|_{\mathrm{Lip}} := \|f\|_{\mathbb{L}^p} + \|\bar{a}(\beta,f)\|_{l^q}$. Note that the Lipschitz space is determined independently of the choice of L and c_0 as long as the former is greater than 1 and the latter is positive.

Theorem 3.2. ([20, 25, 30, 15, 14, 27]) *Let $(\mathcal{E},\mathcal{F})$ be a local regular Dirichlet form on G which corresponds to a fractional diffusion. Then, the following holds.*

$$\mathcal{F} = Lip(\frac{d_w}{2}, 2, \infty)(G).$$

Proof. We will follow the argument in [30]. We first prove $\mathcal{F} \subset \mathrm{Lip}$. For $f \in \mathbb{L}^2(G,\mu)$, let $\mathcal{E}_t(f,f) := (f - P_t f, f)_{\mathbb{L}^2}/t$, where P_t is a semigroup corresponding to $(\mathcal{E},\mathcal{F})$. Then,

$$\begin{aligned} \mathcal{E}_t(f,f) &= \frac{1}{2t}\int\int_{G\times G} (f(x)-f(y))^2 p_t(x,y)\mu(dx)\mu(dy) \\ &\ge \frac{1}{2t}\int\int_{|x-y|\le c_0 t^{1/d_w}} (f(x)-f(y))^2 p_t(x,y)\mu(dx)\mu(dy) \\ &\ge \frac{c_1}{2t}\int\int_{|x-y|\le c_0 t^{1/d_w}} t^{-d_s/2}(f(x)-f(y))^2\mu(dx)\mu(dy), \qquad (3.3) \end{aligned}$$

where we use the lower bound of (3.1) in the last inequality. Taking $t = L^{-md_w}$ and using the fact $d_s/2 = d/d_w$, we see that (3.3) is equal to $c_1 a_m(d_w/2, f)^2$. It is well known that $\mathcal{E}_t(f,f) \nearrow \mathcal{E}(f,f)$ as $t \downarrow 0$ ([12], Lemma 1.3.4). We thus obtain $\sup_m a_m(d_w/2, f) \le c_2\sqrt{\mathcal{E}(f,f)}$ and the result holds.

We next prove $\mathcal{F} \supset \mathrm{Lip}$. Set $\gamma = 1/(d_w - 1)$. Since the diameter of G is 1, we have for each $g \in \mathrm{Lip}$,

$$\begin{aligned}
\mathcal{E}_t(g,g) &= \frac{1}{2t}\int\int_{\substack{x,y\in G\\ |x-y|\le 1}} (g(x)-g(y))^2 p_t(x,y)\mu(dx)\mu(dy)\\
&\le \frac{1}{2t}\sum_{m=1}^{\infty} c_3 t^{-d_s/2} e^{-c_4(tL^{md_w})^{-\gamma}} \int\int_{L^{-m}<|x-y|\le L^{-m+1}} (g(x)-g(y))^2\mu(dx)\mu(dy)\\
&\le c_3 t^{-(1+d_s/2)} \sum_{m=1}^{\infty} e^{-c_4(tL^{md_w})^{-\gamma}} L^{-m(d_w+d)} a_{m-1}(d_w/2,g)^2, \qquad (3.4)
\end{aligned}$$

where we use the upper bound of (3.1) in the first inequality. For $0 < t$ and $0 \le x$, let $\Phi_t(x) = e^{-c_4(tL^{xd_w})^{-\gamma}} L^{-x(d_w+d)}$. By elementary calculations, we see that $\Phi_t(0) > 0$, $\lim_{x\to\infty}\Phi_t(x) = 0$ and $\int_0^\infty \Phi_t(x)dx = c_5 t^{1+d_s/2}$. Further, there exists $x_t > 0$ such that $\Phi_t(x)$ is increasing for $0 \le x < x_t$, decreasing for $x_t < x < \infty$ and $\Phi_t(x_t) = c_6 t^{1+d_s/2}$. Thus, $\sum_{m=1}^\infty \Phi_t(m) \le \int_0^\infty \Phi_t(x)dx + 2\Phi_t(x_t) \le c_7 t^{1+d_s/2}$. Since (3.4) is less than or equal to $c_3 t^{-(1+d_s/2)}\|g\|^2_{\mathrm{Lip}}\sum_{m=1}^\infty \Phi_t(m)$, we conclude that $\sup_{t>0}\mathcal{E}_t(g,g) = \lim_{t\to 0}\mathcal{E}_t(g,g) \le c_8\|g\|^2_{\mathrm{Lip}}$ and the result holds. □

Remark 3.3. 1) The argument here can be extended to a class of diffusions wider than fractional diffusions. Indeed, by checking the proof of [30, 15] carefully, we see that the same results hold for diffusions whose heat kernels satisfy estimates similar to (3.1), but with different orders $0 < \gamma_1, \gamma_2 < \infty$ on the shoulders of $\rho(x,y)^{d_w}t^{-1}$ (instead of $1/(d_w-1)$). If we weaken the condition in this way, then we can include all diffusions on p.c.f. self-similar sets (which roughly corresponds to finitely ramified fractals) as mentioned in [27].
2) In [14], characterization of domains of Dirichlet forms is given using heat kernel estimates under a more general and abstract setting.

4. Dirichlet forms and jump type processes on d-sets

In [26], three natural non-local regular Dirichlet forms are introduced, whose corresponding processes are stable-like jump type processes on compact d-sets. Here we will survey them and give relations to other results.

4.1. Jump process as a Besov space on a d-set

We first introduce Besov spaces on G and their trace theory within the scope of our use (see [22, 36] etc. for details).

For $0 < \alpha < 1$, we introduce a Besov space $B^{2,2}_\alpha(G)$ as follows,

$$\|u|B^{2,2}_\alpha(G)\| = \|u\|_{\mathbb{L}^2(G,\mu)} + \Big(\int\int_{G\times G}\frac{|u(x)-u(y)|^2}{|x-y|^{d+2\alpha}}\mu(dx)\mu(dy)\Big)^{1/2}, \qquad (4.1)$$

$$B^{2,2}_\alpha(G) = \{u : u \text{ is measurable}, \|u|B^{2,2}_\alpha(G)\| < \infty\}. \qquad (4.2)$$

In [22], it is shown that for $0 < \alpha < 1$, $B_\alpha^{2,2}(G) = \mathrm{Lip}(\alpha, 2, 2)(G)$ and the two norms are equivalent (Chapter V, Proposition 3).

For each $f \in \mathbb{L}^1_{loc}(\mathbb{R}^n)$ and $x \in \mathbb{R}^n$, define

$$Rf(x) = \lim_{r \downarrow 0} \frac{1}{m(B(x,r))} \int_{B(x,r)} f(y)dy,$$

if the limit exists, where m is the Lebesgue measure in $\mathbb{R}^n$. It is well-known that the limit exists quasi-everywhere (i.e. except a set of zero capacity) in $\mathbb{R}^n$ with respect to the Newtonian capacity if $n \geq 3$ or logarithmic capacity if $n = 2$ and coincides with $f(x)$ almost everywhere in $\mathbb{R}^n$. For each $\beta > 0$, denote by $B_\beta^{2,2}(\mathbb{R}^n)$ the classical Besov space on $\mathbb{R}^n$ (see Remark 4.2 below for its definition). The following trace theorem plays an important role in the study of Besov spaces on d-sets (see, for instance, Chapters V and VI in [22]).

Proposition 4.1. *For $0 < s < 1$, the trace operator $Tr_G : f \mapsto Rf$ is a bounded linear surjection from $B^{2,2}_{s+(n-d)/2}(\mathbb{R}^n)$ onto $B_s^{2,2}(G)$ and it has a bounded linear right inverse operator E_G (which is called the extension operator in literature) so that $Tr_G \circ E_G$ is the identity map on $B_s^{2,2}(G)$.*

Remark 4.2. Note that for $\beta > 0$ with integer $k < \beta \leq k+1$, the classical Besov space $B_\beta^{2,2}(\mathbb{R}^n)$ is defined to be

$$B_\beta^{2,2}(\mathbb{R}^n) = \left\{ u \in C^k(\mathbb{R}^n) : \ \|u\|_{B_\beta^{2,2}} := \sum_{0 \leq |j| \leq k} \|D^j u\|_2 + \sum_{|j|=k} \left(\int_{\mathbb{R}^n} \frac{\|\Delta_h D^j f\|_2^2}{|h|^{n+2(\beta-k)}} dh \right)^{1/2} < \infty \right\},$$

where for $j = (j_1, j_2, \cdots, j_n) \in \mathbb{Z}^n_+$, $|j| = \sum_{k=1}^n j_k$ and $D^j = \frac{\partial^{|j|}}{\partial x_1^{j_1} \cdots \partial x_n^{j_n}}$, Δ_h is a difference operator so that for $h \in \mathbb{R}^n$, $(\Delta_h f)(x) = f(x+h) - f(x)$, and $\|\cdot\|_2$ denotes the $\mathbb{L}^2$-norm in $\mathbb{L}^2(\mathbb{R}^n, m)$ (see, for instance, section I.1.5 in [22]). It is known (cf. Section V.1.1 in [22]) that when $0 < \beta < 1$, the norm $\|u\|_{B_\beta^{2,2}}$ is equivalent to $\left\| u \mid B_\beta^{2,2}(\mathbb{R}^n) \right\|$ defined by (4.2) with $G = \mathbb{R}^n$, and therefore $B_\beta^{2,2}(\mathbb{R}^n)$ is the same as the space defined by (4.1) with $G = \mathbb{R}^n$. Furthermore, the space $B_\beta^{2,2}(\mathbb{R}^n)$ coincides with the classical Bessel potential space on $\mathbb{R}^n$ (also called the fractional Sobolev space or the Liouville space); see, for instance, p. 8 in Section I.1.5 of [22].

Now, for $0 < \alpha < 1$ and $u, v \in B_\alpha^{2,2}(G)$, define

$$\mathcal{E}_{Y^{(\alpha)}}(u,v) = \int\int_{G \times G} \frac{(u(x)-u(y))(v(x)-v(y))}{|x-y|^{d+2\alpha}} \mu(dx)\mu(dy).$$

By standard properties of Besov spaces, it is easy to check that $(\mathcal{E}_{Y^{(\alpha)}}, B_\alpha^{2,2}(G))$ is a regular Dirichlet space on $\mathbb{L}^2(G, \mu)$ (a detailed proof is given, for instance, in Theorem 3 of [32]). We denote $\{Y_t^{(\alpha)}\}_{t \geq 0}$ the corresponding Hunt process on G.

We note that when $G = \mathbb{R}^n$, this is a (2α)-stable process on $\mathbb{R}^n$. When G is an open n-set, it is the so called reflected (2α)-stable process on G proposed in [5].

4.2. Jump process as a subordination of a diffusion

In this subsection, we assume that there exists a fractional diffusion $\{B_t^G\}_{t\geq 0}$ on G.

For $0 < \alpha < 1$, let $\{\xi_t\}_{t>0}$ be the strictly α-stable subordinator, i.e., it is a one dimensional non-negative Lévy process independent of $\{B_t^G\}_{t\geq 0}$ with the generating function $E[\exp(-u\xi_t)] = \exp(-tu^\alpha)$. Let $\{\eta_t(u) : t > 0, u \geq 0\}$ be the distribution density of $\{\xi_t\}_{t>0}$. Using $p_t(x,y)$ in (3.1), we define

$$q_t(x,y) = \int_0^\infty p_u(x,y)\eta_t(u)du \qquad \text{for all } t > 0,\ x,y \in G.$$

Then, by a general theory, $q_t(x,y)$ is a heat kernel of some Markov process which we denote by $\{X_t^{(\alpha)}\}_{t\geq 0}$, called the subordinate process (see [4, 31]). In our case, $\{X_t^{(\alpha)}\}_{t\geq 0}$ is a μ-symmetric Hunt process and we denote the corresponding Dirichlet form on $\mathbb{L}^2(G,\mu)$ as $(\mathcal{E}_{X^{(\alpha)}}, \mathcal{F}_{X^{(\alpha)}})$.

4.3. Jump process as a time change of a stable process on $\mathbb{R}^n$

We first briefly give a result by Triebel. In [36], Triebel defines a Besov space on d-set G as follows. (In order to distinguish it from the previous one, we put $\hat{\cdot}$.)

$$\|u|\hat{B}_\alpha^{2,2}(G)\| \;=\; \inf_{Tr_G g = u} \|g|B^{2,2}_{\alpha+(n-d)/2}(\mathbb{R}^n)\|, \tag{4.3}$$

$$\hat{B}_\alpha^{2,2}(G) \;=\; Tr_G\, B^{2,2}_{\alpha+(n-d)/2}(\mathbb{R}^n), \tag{4.4}$$

where $\alpha > 0$. Here Tr_G is the trace operator given in Proposition 4.1 and the norm of $B^{2,2}_{\alpha+(n-d)/2}(\mathbb{R}^n)$ is the one given in Remark 4.2. (To be precise, the definition of the trace operator in [36] is different from the one given here, but it turns out to be the same; see p 261 in [37].) In general, this Besov space is different from the one defined by Jonsson-Wallin ([22]), but it is known that for $0 < \alpha < 1$, the two spaces coincide and the two norms are equivalent. Let H_α be the corresponding self-adjoint operator on G so that

$$(H_\alpha^{1/2}u, H_\alpha^{1/2}u)_{\mathbb{L}^2(G,\mu)} = \|u|\hat{B}_\alpha^{2,2}(G)\|^2,\ \text{Dom}\ (H_\alpha^{1/2}) = \hat{B}_\alpha^{2,2}(G). \tag{4.5}$$

Theorem 4.3. ([36]; Theorem 25.2) *H_α is a positive definite self-adjoint operator on $\mathbb{L}^2(G,\mu)$ with pure point spectrum. Let μ_k be its k-th eigenvalue (including multiplicities). Then there exist $c_{4.1}, c_{4.2} > 0$ such that the following holds.*

$$c_{4.1}k^{2\alpha/d} \leq \mu_k \leq c_{4.2}k^{2\alpha/d} \text{ for all } k \in \mathbb{N}.$$

We now give the third jump process on d-sets ([26]). It will be shown that the corresponding operator of the process is the one given by Triebel when $0 < \alpha < 1$.

We construct a jump process though a time change of (2α)-stable process on $\mathbb{R}^n$. Let G be a d-set on $\mathbb{R}^n$ with a Radon measure μ which satisfies (2.1). Also, let $\{B_t^{(\alpha)}\}_{t\geq 0}$ $(0 < \alpha \leq 1)$ be a rotation invariant (2α)-stable process on $\mathbb{R}^n$

(when $\alpha = 1$ it is a Brownian motion) and let $(\mathcal{E}^{(\alpha)}, H^\alpha(\mathbb{R}^n))$ be the corresponding Dirichlet form. (Note that $H^\alpha(\mathbb{R}^n) = B^{2,2}_\alpha(\mathbb{R}^n)$ when $0 < \alpha < 1$.) Then, it is proved in [26] that when $2\alpha > n - d$, then μ is a smooth measure w.r.t. $\{B_t^{(\alpha)}\}_t$, i.e. μ charges no set of zero capacity w.r.t. the form corresponding to $\{B_t^{(\alpha)}\}_t$. Thus, by a general theory (see [12]), there exists a unique positive continuous additive functional $\{A_t^{(\alpha)}\}_{t\geq 0}$ which is in Revuz correspondence with μ (thus, $A_t^{(\alpha)}$ increases only when $B_t^{(\alpha)} \in G$). Set $\tau_t = \inf\{s > 0 : A_s^{(\alpha)} > t\}$ and define $Z_t^{(\alpha)} = B_{\tau_t}^{(\alpha)}$. Then, again by a general theory, $\{Z_t^{(\alpha)}\}_{t\geq 0}$ is a μ-symmetric jump process and the corresponding regular Dirichlet form can be expressed as

$$\begin{aligned}\mathcal{F}_Z &= \{\phi \in \mathbb{L}^2(G,\mu) : \phi = u \ \mu - \text{a.e. for some } u \in H_e^\alpha(\mathbb{R}^n)\},\\ \mathcal{E}_Z(\phi,\phi) &= \mathcal{E}^{(\alpha)}(H_{\mathcal{S}}u, H_{\mathcal{S}}u),\ \phi \in \mathcal{F}_Z, \phi = u \ \mu - \text{a.e.}, u \in H_e^\alpha(\mathbb{R}^n).\end{aligned}$$

Here $\mathcal{S}$ is a quasi-support of μ (see p168 of [12]), $H_{\mathcal{S}}u(x) = E^x[u(B_{\sigma_{\mathcal{S}}}^{(\alpha)})]$ where $\sigma_{\mathcal{S}} = \inf\{t \geq 0 : B_t^{(\alpha)} \in \mathcal{S}\}$ and $H_e^\alpha(\mathbb{R}^n)$ is the extended Dirichlet space. See Theorem 6.2.1 in [12] for details.

In subsection 2.3 of [26], it is proved that for $2\alpha > n - d$,

$$P^x(\sigma_{\mathcal{S}} = 0) = 1 \qquad \text{for all } x \in G,$$

so $H_{\mathcal{S}}u(x) = u(x)$ for μ-a.e. $x \in G$. As μ is the smooth measure, we can prove $Tr_G H_{\mathcal{S}} u = u$, μ-a.e.. Since $H_{\mathcal{S}}u$ is harmonic outside G, we obtain

$$\mathcal{E}^{(\alpha)}(H_{\mathcal{S}}u, H_{\mathcal{S}}u) = \inf_{Tr_G g = u} \|g|B^{2,2}_\alpha(\mathbb{R}^n)\|.$$

Therefore, $\mathcal{E}_Z$, the trace of the Dirichlet form of the (2α)-stable process on $\mathbb{R}^n$, is equal to $\|\cdot|\hat{B}^{2,2}_{\alpha-(n-d)/2}(G)\|$ in (4.3) so that the corresponding operator is $H_{\alpha-(n-d)/2}$ given above.

Remark 4.4. When $A \subset \mathbb{R}^n$ is a bounded domain with smooth boundary ∂A, a similar time changed form (with exponential killing) is studied in [18], where the original form lies on A and the time changed form lies on ∂A. In [11], a similar problem is treated when A is a bounded (ϵ, δ)-domain with boundary ∂A which is a d-set with $n - 1 \leq d < n$.

There is an alternative analytic approach by Zähle using Riesz potentials ([39, 40, 16]). For $0 < 2\alpha < d$, define a Riesz potential of order 2α as follows

$$I_\mu^{2\alpha} f(x) := c_{n,2\alpha+n-d} \int \frac{f(y)d\mu(y)}{|x-y|^{d-2\alpha}}, \qquad \forall f \in \mathbb{L}^2(G,\mu),$$

where $c_{n,s} := \Gamma((n-s)/2)/(2^s\pi^{n/2}\Gamma(s/2))$. In [39], it is proved that $I_\mu^{2\alpha}$ is invertible. Define $D_\mu^{2\alpha} := (I_\mu^{2\alpha})^{-1}$; the domain of this operator is $\hat{B}^{2,2}_{2\alpha}(G)$ ([40]). Let

$$\mathcal{E}_\mu^{2\alpha}(f,g) = \int_G \sqrt{D_\mu^{2\alpha}} f(x) \sqrt{D_\mu^{2\alpha}} g(x) d\mu(x), \qquad \forall f,g \in \hat{B}^{2,2}_\alpha(G).$$

It is proved in [40] that $(\mathcal{E}_\mu^{2\alpha}, \hat{B}_\alpha^{2,2}(G))$ is a regular Dirichlet form on $\mathbb{L}^2(G,\mu)$ for $0 < \alpha + (n-d)/2 \leq 1$. In [16], it is shown that $(\mathcal{E}_\mu^{2\alpha}, \hat{B}_\alpha^{2,2}(G))$ coincides with $(\mathcal{E}_{Z^{(s)}}, \mathcal{F}_{Z^{(s)}})$ where $s = \alpha + (n-d)/2$. Thus, $\sqrt{D_\mu^{2\alpha}} = H_\alpha^{1/2}$ given in (4.5).

4.4. Comparison of the forms and heat kernel bounds

Define $\bar{\alpha} = \alpha d_w/2$ and $\hat{\alpha} = \alpha - (n-d)/2$. We then have the following.

Proposition 4.5. ([26, 32])
Let G be a d-set. For $(n-d)/2 < \alpha < 1$ or $\alpha = 1, n-2 < d < n$,

$$c_{4.3}\mathcal{E}_{Y^{(\hat{\alpha})}}(f,f) \leq \mathcal{E}_{Z^{(\alpha)}}(f,f) \leq c_{4.4}\mathcal{E}_{Y^{(\hat{\alpha})}}(f,f) \text{ for all } f \in \mathbb{L}^2(G,\mu). \tag{4.6}$$

Assume further that there exists a fractional diffusion on G. For $0 < \alpha < 1$,

$$c_{4.5}\mathcal{E}_{Y^{(\bar{\alpha})}}(f,f) \leq \mathcal{E}_{X^{(\alpha)}}(f,f) \leq c_{4.6}\mathcal{E}_{Y^{(\bar{\alpha})}}(f,f) \text{ for all } f \in \mathbb{L}^2(G,\mu). \tag{4.7}$$

In particular, under the conditions,

$$\mathcal{F}_{Z^{(\alpha)}} = B_{\hat{\alpha}}^{2,2}(G), \ \mathcal{F}_{X^{(\alpha)}} = B_{\bar{\alpha}}^{2,2}(G).$$

Remark 4.6. 1) In [26], (4.7) is stated for $0 < \alpha < 2/d_w$. But the proof there shows that it actually holds for $0 < \alpha < 1$. From this fact, we see that when G is a d-set on which there exists a fractional diffusion, $(\mathcal{E}_{Y^{(\alpha)}}, B_\alpha^{2,2}(G))$ is a regular Dirichlet form for $0 < \alpha < d_w/2$.
2) Note that in general the three-type Dirichlet forms introduced are different and the corresponding processes cannot be obtained by time changes of others by positive continuous additive functionals (see [26]).

5. Heat kernel estimates for stable-like processes on d-sets

In [8], detailed estimates of heat kernels for $\{Y^{(\alpha)}\}$ are obtained. There exists a non-negative bounded heat kernel $p_t(x,y)$ on $(t,x,y) \in (0,\infty) \times G \times G$ with

$$P_t^{Y^{(\alpha)}} f(x) = \int_G p_t(x,y) f(y)\mu(dy) \qquad \text{for all} \quad x \in G, \ f \in \mathbb{L}^2(G,\mu),$$

(where $P_t^{Y^{(\alpha)}}$ is the heat semigroup w.r.t. $\mathcal{E}_{Y^{(\alpha)}}$), satisfying the following.

Theorem 5.1. ([8]) *For $0 < \alpha < 1$, the following holds.*

1) *There are constants $c_{5.1}, c_{5.2} > 0$ such that for all $x, y \in G$, $0 < t < 1$,*

$$c_{5.1}(t^{-\frac{d}{2\alpha}} \wedge \frac{t}{|x-y|^{d+2\alpha}}) \leq p_t(x,y) \leq c_{5.2}(t^{-\frac{d}{2\alpha}} \wedge \frac{t}{|x-y|^{d+2\alpha}}).$$

2) *There are constants $c_{5.3} > 0$ and $\beta > 0$ such that for any $0 < t, s < 1$ and $(x_i, y_i) \in G \times G$ with $i = 1, 2$,*

$$|p_s(x_1,y_1) - p_t(x_2,y_2)| \leq c_{5.3}\,(t \wedge s)^{-\frac{d+\beta}{2\alpha}} \left(|t-s|^{\frac{1}{2\alpha}} + |x_1 - x_2| + |y_1 - y_2|\right)^\beta.$$

Theorem 5.2. ([8]) *For every $x \in F$, $\mathbb{P}^x$-a.s., the Hausdorff dimension of $Y[0,1] := \{Y_t : 0 \leq t \leq 1\}$ is $d \wedge (2\alpha)$.*

Note that above theorems hold under a wider framework, i.e. when the form is expressed as

$$\mathcal{E}(f,f) = \int\int_{G\times G} (f(x) - f(y))^2 n(x,y) d\mu(x) d\mu(y),$$

where $n(x,y)$, $x, y \in G$ is a jointly measurable function such that $n(x,y) = n(y,x)$ for all $x, y \in G$ and satisfies

$$\frac{c_{5.4}}{|x-y|^{d+2\alpha}} \le n(x,y) \le \frac{c_{5.5}}{|x-y|^{d+2\alpha}}.$$

In [8], it is also proved that all the parabolic functions (two variable functions which satisfy the heat equation) satisfy the parabolic Harnack inequality.

Remark 5.3. 1) When G is a global d-set (i.e. an unbounded closed d-set satisfying $\mu(B(x,r)) \le c_1 r^d$ for all $x \in G$ and $r > 0$), Theorem 5.1 holds for all $0 < t, s < \infty$ ([9]). Thus $Y^{(\alpha)}$ is transient if and only if $d > 2\alpha$, it is point recurrent if and only if $d < 2\alpha$.
2) In [7, 6], similar heat kernel estimates for the subordinate process $\{X^{(\alpha)}\}$ $(0 < \alpha < 1)$ of the fractional diffusion (see subsection 4.2) are obtained. There the original diffusion processes *is assumed* to satisfy the properties in Definition 3.1.
3) In relation to Theorem 5.2, results on ranges of stochastic processes are surveyed in [38]. In [3], among others, Hausdorff dimensions of boundary traces of reflecting Brownian motions and stable-like processes are computed on various Euclidean domains including the von Koch snowflake domain.

6. From jump-type processes to diffusion processes

6.1. Construction of diffusion processes on fractals

In [27], some sufficient conditions for constructing diffusion processes on metric measure spaces are given. In this subsection, we will briefly explain some of the results in the paper. (The results in the paper are more general, but we simplify to avoid complicated settings.)

Let

$$\mathcal{E}^{\beta,r}(u,u) = \frac{1}{r^{2\beta}} \int_G \fint_{B(x,r)} [u(x) - u(y)]^2 \mu(dy)\, \mu(dx) \tag{6.1}$$

for all $u \in \mathbb{L}^2(G,\mu)$. Here $\fint_A \cdots \mu(dy) := \mu(A)^{-1} \int_A \cdots \mu(dy)$ denotes the normalized integral. We fix a sequence $(r_n)_n$ of positive numbers decreasing to 0 and put

$$\begin{aligned} \mathcal{E}^{\beta,*}(u,u) &:= \limsup_{n\to\infty} \mathcal{E}^{\beta,r_n}(u,u), \\ \mathcal{C}^{\beta,*} &:= \left\{ u \in \mathcal{C}_0(G) : \ \mathcal{E}^{\beta,*}(u,u) < \infty \right\}. \end{aligned}$$

Let

$$\bar{\beta} := \sup\{\beta : \{u \in \mathbb{L}^2(G,\mu) : \mathcal{E}^{\beta,*}(u,u) < \infty\} \text{ is dense in } \mathbb{L}^2(G,\mu).\}.$$

In the following, we assume

- $\{u \in \mathbb{L}^2(G,\mu) : \mathcal{E}^{\bar{\beta},*}(u,u) = 0\}$ is not dense in $\mathbb{L}^2(G,\mu)$,
- $\{u \in \mathbb{L}^2(G,\mu) : \mathcal{E}^{\bar{\beta},*}(u,u) < \infty\}$ is dense in $\mathbb{L}^2(G,\mu)$,

and denote $d_w = 2\bar{\beta}$.

Theorem 6.1. ([27]) (i) *Assume that $\mathcal{C}^{d_w/2,*}$ is dense in $\mathcal{C}_0(G)$. Then for a suitable subsequence (r_n') of (r_n) the Γ-limit*

$$\mathcal{E}^0 := \Gamma\text{-}\lim_{n\to\infty} \mathcal{E}^{d_w/2,r_n'}$$

exists, is dominated by $\mathcal{E}^*$ on $\mathbb{L}^2(G,\mu)$ and is the extension of a regular Dirichlet form $(\mathcal{E},\mathcal{F})$ on $\mathbb{L}^2(G,\mu)$ with core $\mathcal{C}^{d_w/2,*}$.
Hence, there exists a μ-reversible strong Markov process associated with $(\mathcal{E},\mathcal{F})$.
(ii) *Assume in addition the following.*
There exists $\delta > 0$ such that

$$\liminf_{n\to\infty} \mathcal{E}^{d_w/2,r_n}(u_n,u_n) \geq \delta \cdot \mathcal{E}^{d_w/2,*}(u,u) \tag{6.2}$$

for all $u \in \mathbb{L}^2(G,\mu)$ and all $(u_n)_n \subset \mathbb{L}^2(G,\mu)$ with $u_n \to u$ in $\mathbb{L}^2(G,\mu)$.
Let $\mathcal{E}^{(c)}$ denote the diffusion part of $\mathcal{E}$. Then $\mathcal{C}^{d_w/2,*} = \mathcal{F} \cap \mathcal{C}_0(G)$,

$$\mathcal{E}^{d_w/2,*} \geq \mathcal{E} \geq \mathcal{E}^{(c)} \geq \delta \cdot \mathcal{E}^{d_w/2,*}$$

on $\mathcal{F}$ and $(\mathcal{E}^{(c)},\mathcal{F})$ is a strongly local regular Dirichlet form.
Hence, there exists a diffusion process associated with $(\mathcal{E}^{(c)},\mathcal{F})$. It is μ-reversible and strongly Markovian.

Let

$$\mathrm{Lip}(\frac{d_w}{2},2,\infty)(G) = \left\{u \in \mathbb{L}^2(G,\mu) : \sup_n \mathcal{E}^{d_w/2,r_n}(u,u) < \infty\right\}.$$

If we assume the following instead of (6.2),

$$\liminf_{n\to\infty} \mathcal{E}^{d_w/2,r_n}(u_n,u_n) \geq \delta \cdot \sup_n \mathcal{E}^{d_w/2,r_n}(u,u), \tag{6.3}$$

then $\mathcal{C}^{d_w/2,*} = \mathrm{Lip}(d_w/2,2,\infty)(G) \cap \mathcal{C}_0(G)$ holds. Note that when $r_n = L^{-n}$, then $\mathrm{Lip}(d_w/2,2,\infty)(G)$ coincides with the Lipschitz space discussed in Section 3. When there is a fractional diffusions on G, then the assumptions given in this section always hold.

6.2. Jump-type processes for $1 \leq \alpha < d_w/2$

It is natural to ask whether the results in Section 5 can be extended to $\alpha \geq 1$. The problem is that in this case Davies methods cannot be applied, likewise the case of diffusions on fractals. Let

$$\bar{d}_w := \sup\{2\alpha : (\mathcal{E}_{Y^{(\alpha)}}, B_\alpha^{2,2}(G)) \text{ is regular in } \mathbb{L}^2(G,\mu)\}.$$

It is easy to show $\bar{d}_w \leq d_w$. We do not know whether the equality holds in general or not (clearly the equality holds when there is a fractional diffusion on G).

We believe we can prove the theorems in Section 5 for all $\alpha < \bar{d}_w/2$ when $\bar{d}_w > d$, i.e. when the process is point recurrent ([9]). It is still currently under investigation whether or not these results hold in general for all $\alpha < \bar{d}_w/2$ or not.

References

[1] D.R. Adams and L.I. Hedberg, *Function spaces and potential theory,* 1996, Springer, Berlin-Heidelberg.

[2] M.T. Barlow, *Diffusions on fractals,* Lectures in Probability Theory and Statistics: Ecole d'été de probabilités de Saint-Flour XXV, Lect. Notes Math., **1690** (1998), Springer, New York.

[3] I. Benjamini, Z.-Q. Chen and S. Rohde, *Boundary Trace of Reflecting Brownian Motions,* Probab. Theory Relat. Fields, **129** (2004), 1–17.

[4] J. Bertoin, *Lévy processes,* 1996, Cambridge Univ. Press, Cambridge.

[5] K. Bogdan, K. Burdzy and Z.-Q. Chen, *Censored stable processes,* Probab. Theory Relat. Fields, **127** (2003), 89–152.

[6] K. Bogdan, A. Stós and P. Sztonyk, *Harnack inequality for symmetric stable processes on fractals,* C. R. Acad. Sci. Paris, Ser. I, **335** (2002), 59–63.

[7] K. Bogdan, A. Stós and P. Sztonyk, *Harnack inequality for stable processes on d-sets,* Studia Math., **158** (2003), 163–198.

[8] Z.-Q. Chen and T. Kumagai, *Heat kernel estimates for stable-like processes on d-sets,* Stochastic Process. Appl., **108** (2003), 27–62.

[9] Z.-Q. Chen and T. Kumagai, in preparation.

[10] E.B. Davies, *Heat kernels and spectral theory,* 1989, Cambridge Univ. Press, Cambridge.

[11] W. Farkas and N. Jacob, *Sobolev spaces on non-smooth domains and Dirichlet forms related to subordinate reflecting diffusions,* Math. Nachr., **224** (2001), 75–104.

[12] M. Fukushima, Y. Oshima, and M. Takeda, *Dirichlet forms and symmetric Markov processes,* 1994, de Gruyter, Berlin.

[13] M. Fukushima and T. Uemura, *On Sobolev and capacitary inequalities for contractive Besov spaces over d-sets,* Potential Anal., **18** (2003), 59–77.

[14] A. Grigoryan, *Heat kernels and function theory on metric measure spaces,* Contemporary Math., **338** (2003), 143–172.

[15] A. Grigoryan, J. Hu and K.S. Lau, *Heat kernels on metric-measure spaces and an application to semi-linear elliptic equations,* Trans. Amer. Math. Soc., **355** (2003), 2065–2095.

[16] W. Hansen and M. Zähle, *Restricting isotropic α-stable Lévy processes from $\mathbb{R}^n$ to fractal sets,* Preprint 2003.

[17] S. Havlin and D. Ben-Avraham, *Diffusion in disordered media,* Adv. Phys., **36** (1987), 695–798.

[18] N. Jacob and R.L. Schilling, *Some Dirichlet spaces obtained by subordinate reflected diffusions,* Rev. Mat. Iberoamericana, **15** (1999), 59–91.

[19] N. Jacob and R.L. Schilling, *Lévy-type processes and pseudodifferential operators,* In: O.E. Barndorff-Nielsen et al. (eds.), Lévy Processes, Birkhäuser, Boston, pp. 139–168, 2001.

[20] A. Jonsson, *Brownian motion on fractals and function spaces,* Math. Z., **222** (1996), 496–504.

[21] A. Jonsson, *Dirichlet forms and Brownian motion penetrating fractals,* Potential Anal., **13** (2000), 69–80.

[22] A. Jonsson and H. Wallin, *Function spaces on subsets of* $\mathbb{R}^n$, Mathematical Reports, Vol. **2**, Part 1, 1984, Acad. Publ., Harwood.

[23] J. Kigami, *Analysis on fractals,* 2001, Cambridge Univ. Press, Cambridge.

[24] T. Kumagai, *Stochastic processes on fractals and related topics,* (in Japanese) Sugaku, **49** (1997), 158–172, (English transl.) Sugaku Expositions, Amer. Math. Soc., **13** (2000), 55–71.

[25] T. Kumagai, *Brownian motion penetrating fractals -An application of the trace theorem of Besov spaces-,* J. Func. Anal., **170** (2000), 69–92.

[26] T. Kumagai, *Some remarks for jump processes on fractals,* In: P. Grabner and W. Woess (eds.), Trends in Math.: Fractals in Graz 2001, Birkhäuser, pp. 185–196, 2002.

[27] T. Kumagai and K.T. Sturm, *Construction of diffusion processes on fractals, d-sets, and general metric measure spaces,* Preprint 2003 (available at http://www.kurims.kyoto-u.ac.jp/~kumagai/).

[28] S. Kusuoka, *Diffusion processes on nested fractals,* In: R.L. Dobrushin and S. Kusuoka: Statistical Mechanics and Fractals, Lect. Notes Math., **1567** (1993), Springer, New York.

[29] T. Lindstrøm, *Brownian motion on nested fractals,* Memoirs Amer. Math. Soc., 420 **83** (1990).

[30] K. Pietruska-Paluba, *On function spaces related to fractional diffusions on d-sets,* Stochastics Rep., **70** (2000), 153–164.

[31] K. Sato, *Lévy processes and infinitely divisible distributions,* Cambridge Studies in Advanced Math., Vol. **68** (1999), Cambridge Univ. Press, Cambridge.

[32] A. Stós, *Symmetric α-stable processes on d-sets,* Bull. Polish Acad. Sci. Math., **48** (2000), 237–245.

[33] R.S. Strichartz, *Analysis on fractals,* Notices Amer. Math. Soc., **46** (1999), 1199–1208.

[34] R.S. Strichartz, *Function spaces on fractals,* J. Funct. Anal., **198** (2003), 43–83.

[35] H. Triebel, *Interpolation Theory, Function Spaces, Differential Operators,* 1995, Johann Ambrosius Barth, Heidelberg.

[36] H. Triebel, *Fractals and spectra -related to Fourier analysis and function spaces-,* Monographs in Math., Vol. **91** (1997), Birkhäuser, Basel-Boston-Berlin.

[37] H. Triebel, *The structure of functions,* Monographs in Math., Vol. **97** (2001), Birkhäuser, Basel-Boston-Berlin.

[38] Y. Xiao, *Random fractals and Markov processes,* Proc. of Symposia in Pure Math., to appear.

[39] M. Zähle, *Harmonic calculus on fractals - a measure geometric approach II,* Preprint 2000.

[40] M. Zähle, *Riesz potentials and Liouville operators on fractals,* Potential Anal., to appear.

Acknowledgment

The author expresses his gratitude to Zhenqing Chen and Martina Zähle for valuable comments on a preliminary version of this paper.

Takashi Kumagai
Research Institute for Mathematical Sciences
Kyoto University
Kyoto 606-8502
Japan
e-mail: `kumagai@kurims.kyoto-u.ac.jp`
Current address:
Department of Mathematics
Malott Hall, Cornell University
Ithaca, NY 14853
USA
e-mail: `kumagai@math.cornell.edu`

Progress in Probability, Vol. 57, 235–244

A Dirichlet Form on the Sierpinski Gasket, Related Function Spaces, and Traces

Alf Jonsson

Abstract. A trace result for the domain of a Dirichlet form on the Sierpinski gasket is given, and it is explained how the result relates to various results for for Sobolev, Besov, and Lipschitz spaces.

Mathematics Subject Classification (2000). 46E35, 31C25.

Keywords. Function spaces, Dirichlet form, traces.

1. Introduction

A Dirichlet form on the Sierpinski gasket has been constructed and studied in connection with the theory for Brownian motion on fractals, see e.g. [16] or [1] for an introduction to these topics. The domain $\mathcal{D}$ of this form is a function space, which can also be looked upon as a function space in the spirit of the theory of function spaces on closed sets developed in [9]. The purpose of this note is to describe this relation, with emphasis on trace results. In particular we present, in Section 6, a new result on the trace to the sides of the gasket of functions in $\mathcal{D}$. This result uses a characterization of $\mathcal{D}$ in terms of a piecewise harmonic basis given in Section 3. Possible applications are indicated in Section 6.

2. A Dirichlet form on the Sierpinski gasket. Harmonic functions

Let Δ be the closed simplex in $\mathbb{R}^2$ with vertices $P_1 = (0,0)$, $P_2 = (1,0)$, and $P_3 = (1/2, \sqrt{3}/2)$, and put $\mathcal{T}_0 = \{\Delta\}$. Next, for $i = 1,2,3$, consider the simplices which have P_i as one vertex and the points $(P_i + P_j)/2$, $j \neq i$, as the others, and denote the set of these three simplices by $\mathcal{T}_1$. From each simplex in $\mathcal{T}_1$ construct three new simplices in the same fashion, and denote the set of 9 simplices then obtained by $\mathcal{T}_2$, and so on. The intersection of the sets $F_n = \cup_{\Delta\in\mathcal{T}_n}\Delta$, $n \geq 0$, is *the Sierpinski gasket* F. The set of vertices of a simplex Δ is denoted by $V(\Delta)$ and the set of all vertices on level n is denoted by $\mathcal{U}_n$, i.e. $\mathcal{U}_n = \{p : p \in V(\Delta) \text{ for some } \Delta \in \mathcal{T}_n\}$, and we let $\mathcal{U}_{-1} = \emptyset$. For $n \geq 0$, put $\mathcal{V}_n = \mathcal{U}_n \setminus \mathcal{U}_{n-1}$. In other words, $\mathcal{V}_n$ is the set of new vertices added when one passes from $\mathcal{T}_{n-1}$ to $\mathcal{T}_n$.

In connection with the theory for Brownian motion on fractals, a Dirichlet form on the Sierpinski gasket F was introduced, see [2]. Our interest here is the domain $\mathcal{D}$ of this form, which is defined as follows. Let f be defined on F, and let

$$\mathcal{E}^n(f,f) = \frac{1}{2}\left(\frac{5}{3}\right)^n \sum_{\Delta \in \mathcal{T}_n} \sum_{p,q \in V(\Delta)} (f(p) - f(q))^2 \quad \text{for } n \geq 0.$$

It can be shown that $\mathcal{E}^n(f,f)$ is non-decreasing in n. Let $\mathcal{E}(f,f) = \lim_{n\to\infty} \mathcal{E}^n(f,f)$ and

$$\mathcal{D} = \{f : f \text{ is continuous on } F \text{ and } \mathcal{E}(f,f) < \infty\}.$$

As a *norm* in $\mathcal{D}$ we take $(\sum_{\xi \in \mathcal{U}_0} f(\xi)^2 + \mathcal{E}(f,f))^{1/2}$.

Another concept from the theory of analysis on fractals is harmonic functions. A continuous function f defined on F is *harmonic* if, for $n > 0$ and $p \in \mathcal{U}_n \setminus \mathcal{U}_0$, holds $f(p) - (1/4)\sum f(q) = 0$, where the sum is taken over the four vertices q in $\mathcal{U}_n$ with $q \neq p$ such that q and p belong to a common simplex in $\mathcal{T}_n$. Harmonic functions are determined by their value at the vertices in $\mathcal{U}_0$, and the value at a point $p \in \mathcal{V}_n$, $n > 0$, satisfies

$$f(p) = (2/5)f(p_1) + (2/5)f(p_2) + (1/5)f(p_3),$$

where p_1, p_2, and p_3, are the vertices of the simplex $\Delta \in \mathcal{T}_{n-1}$ such that $p \in \Delta$, and p_3 is the vertex which is not in the face of Δ which contains p. For $\Delta \in \mathcal{T}_{n_0}$, where $n_0 \geq 0$, we say that a function f is *harmonic in* Δ if f is defined and harmonic in $F \cap \Delta$, where the latter statement means that the above definition of a harmonic function holds when in the natural way adapted to a function defined on $F \cap \Delta$. The maximum and the minimum of a function defined and harmonic in a simplex Δ are both attained at a vertex of Δ.

3. A basis in the domain of the Dirichlet form

We shall now exhibit a basis in $\mathcal{D}$ (see also [12], Section 3.2) which will be a useful tool when studying traces in Section 6. Denote by H_n the set of piecewise harmonic functions on level n, i.e. $f \in H_n$ iff f is continuous on F and harmonic in each $\Delta \in \mathcal{T}_n$. For $\xi \in \mathcal{V}_n$, where $n \geq 0$, we let ψ_ξ denote the piecewise harmonic function in H_n which equals 1 at ξ and 0 at every other vertex of $\mathcal{U}_n$.

Let f be continuous on F, for short $f \in C(F)$, and let $S_n f$ for $n \geq 0$ be the unique piecewise harmonic function in H_n which interpolates f at all points in $\mathcal{U}_n$, also, put $S_{-1}f = 0$. Then it is easy to see that $S_n f$ has the representation

$$S_n f = \sum_{i=0}^{n} \sum_{\xi \in \mathcal{V}_i} c_\xi \psi_\xi,$$

where for $\xi \in \mathcal{V}_i, 0 \leq i \leq n$, the constant c_ξ is given, inductively, by

$$c_\xi = f(\xi) - S_{i-1}f(\xi).$$

It is also easy to show, using the above mentioned maximum and minimum principle for harmonic functions, that $S_n f$ tends to f uniformly on F as $n \to \infty$. Thus any $f \in C(F)$ has a unique representation in terms of the functions ψ_ξ

$$f = \sum_{i=0}^{\infty} \sum_{\xi \in \mathcal{V}_i} c_\xi \psi_\xi \, . \tag{1}$$

Theorem 3.1. *Let F be the Sierpinski gasket. Then holds, if $f \in C(F)$ with the representation* (1),

$$f \in \mathcal{D} \iff \sum_{i=0}^{\infty} \sum_{\xi \in \mathcal{V}_i} c_\xi^2 (5/3)^i < \infty.$$

Moreover, the norm of f in $\mathcal{D}$ is equivalent to $(\sum_{i=0}^{\infty} \sum_{\xi \in \mathcal{V}_i} c_\xi^2 (5/3)^i)^{1/2}$.

Moreover, if $\{c_\xi\}$ is any sequence satisfying $\sum_{i=0}^{\infty} \sum_{\xi \in \mathcal{V}_i} c_\xi^2 (5/3)^i < \infty$, then the sum $\sum_{i=0}^{\infty} \sum_{\xi \in \mathcal{V}_i} c_\xi \psi_\xi$ converges uniformly to a function f which, consequently, is a continuous function which belongs to $\mathcal{D}$. In other words, the theorem exhibits an equivalence between $\mathcal{D}$ and a certain sequence space.

A proof of Theorem 3.1 can be found in [4]. The result should be compared with some results on bases in classical function spaces, see Remark 5.5 below.

4. Besov spaces on closed subsets of $\mathbb{R}^n$. Traces

Besov spaces were introduced in the 1950's, although there were, of course, predecessors. For example, the Besov spaces $B_\alpha^{\infty,\infty}(\mathbb{R}^n)$ coincide with ordinary Lipschitz spaces if α is noninteger and with Zygmund spaces if α is an integer. They have since then become a useful tool in many branches of mathematical analysis. There are many equivalent ways to define Besov spaces on $\mathbb{R}^n$, and we recall here the one using differences. For an overview over Besov spaces and other function spaces on $\mathbb{R}^n$, see e.g.[18]. Denote by $D^j f$ the derivatives of f in the distribution sense, where j is a nonnegative multiindex $(j_1, j_2, \ldots, j_n)$ of length $|j| = j_1 + j_2 + \ldots, j_n$, and by $\| \cdot \|_p$ the L^p-norm in $\mathbb{R}^n$.

Definition 4.1. *Let $k < \alpha \leq k+1$, where k is a nonnegative integer. If $k < \alpha < k+1$, a function f belongs to* the Besov space $B_\alpha^{p,q}(\mathbb{R}^n)$ *iff the norm in $B_\alpha^{p,q}(\mathbb{R}^n)$ given by*

$$\sum_{|j| \leq k} \|D^j f\|_p + \sum_{|j|=k} \Big(\int_{|h| \leq 1} |h|^{-n-(\alpha-k)q} \Big(\int |D^j f(x+h) - D^j f(x)|^p dx \Big)^{q/p} dh \Big)^{1/q}$$

is finite. If $\alpha = k+1$, the first difference in the integrand shall be replaced by the second difference $D^j f(x+2h) - 2D^j f(x+h) + D^j f(x)$.

Besov spaces $B_\alpha^{p,q}(F)$ of functions defined on closed subsets F of $\mathbb{R}^n$ were studied in [9]. As we shall see below, they are defined in such a way that natural generalizations of classical trace theorems hold, see also [19]. In [17] the approach is somewhat different, which gives other spaces for higher smoothness.

Before defining these Besov spaces we define two classes of closed sets F which play an important role in the theory. Let $0 < d \leq n$, and denote by $B(x,r)$ the closed ball with center x and radius r.

Definition 4.2. *A closed set $F \subset \mathbb{R}^n$ is a* d-set *if there is a positive Borel measure μ with support F such that, for some positive constants c_1 and c_2, holds*

$$c_1 r^d \leq \mu(B(x,r)) \leq c_2 r^d, x \in F, 0 < r \leq 1. \tag{2}$$

If F is a d-set, then the restriction to F of the d-dimensional Hausdorff measure satisfies (2), and we call it *the d-measure* μ on F; any measure on F satisfying (2) is equivalent to this measure μ.

Examples. The closure of a Lipschitz domain in $\mathbb{R}^n$ is an n-set. A domain in $\mathbb{R}^2$ bounded by von Koch curves is a 2-set. A self-similar fractal F satisfying the open set condition is a d-set, with d equal to the Hausdorff dimension of the F. For our purpose, we add the condition that F is not a subset of an $n-1$-dimensional affine subspace. In particular, the Sierpinski gasket is a d-set with $d = \ln 3/\ln 2$.

Denote by $\mathcal{P}_m$ the set of all polynomials in n variables of total degree less than or equal to m.

Definition 4.3. *A closed set $F \subset \mathbb{R}^n$* preserves Markov's inequality *if for every fixed positive integer m there exists a constant c, such that for all polynomials $P \in \mathcal{P}_m$ and all balls $B = B(x_0, r), x_0 \in F, 0 < r \leq 1$, holds*

$$\max_{F\cap B} |\nabla P| \leq \frac{c}{r} \max_{F \cap B} |P|,$$

where ∇ denotes the gradient.

There are geometric characterizations of sets preserving Markov's inequality. One of them is that a set preserves Markov's inequality if, for some $\epsilon > 0$, no set $F \cap B(x_0, r)$, $x_0 \in F$, $0 < r < 1$, is contained in an $n-1$-dimensional band of width ϵr symmetric around x_0, see [9], Theorem 2, p. 38. Intuitively, this means that a set preserves Markov's inequality if the set is "nowhere flat".

Examples. The sets given above as examples of d-sets all preserve Markov's inequality. The union of the Sierpinski gasket, as described in Section 2, and the unit cube $\{(x,y) : 0 \leq x \leq 1, -1 \leq y \leq 0\}$ preserves Markov's inequality but is not a d-set. The unit circle in $\mathbb{R}^2$ is a d-set with $d = 1$, but does not preserve Markov's inequality.

We now return to the definition of Besov spaces on closed sets. If F is just a closed set, defining derivatives is a problem in the first place. Following a technique introduced by H. Whitney in [20], we consider collections of functions to be elements in the Besov spaces in the following way. Let α be non-integer and $k < \alpha < k+1$, and let $\{f^{(j)}\}_{|j|\leq k}$ be a family of functions defined μ-a.e. on F. Here j denotes an n-dimensional multi-index as above, and the functions $f^{(j)}$ should be thought of as a derivatives of $f^{(0)}$, although $f^{(j)}$ is in general not uniquely determined by $f^{(0)}$. Let R_j be formal Taylor remainders defined by

$$R_j(x,y) = f^{(j)}(x) - \sum_{|j+l|\leq k} \frac{f^{(j+l)}(y)}{l!}(x-y)^l,$$

and let $\|\{a_\nu\}\|_{l_q}$ be the sequence space norm $(\sum |a_\nu|^q)^{1/q}$. Different values of the positive constant c_0 in the definition below give rise to equivalent spaces.

Definition 4.4. *Let F be a d-set, $0 < d \leq n$. The family $\{f^{(j)}\}_{|j|\leq k}$ of functions defined on F belongs to* the Besov space $B_\alpha^{p,q}(F)$ *iff $f^{(j)} \in L^p(\mu), |j| \leq k$, and $\|\{a_{j\nu}\}_{\nu=0}^{\infty}\|_{l_q} < \infty$, $|j| \leq k$, where $a_{j\nu}$ is given by*

$$\left(2^{\nu d} \iint_{|x-y|<c_0 2^{-\nu}} |R_j(x,y)|^p d\mu(x) d\mu(y)\right)^{1/p} = 2^{-\nu(\alpha-|j|)} a_{j\nu}, |j| \leq k.$$

The norm of f in $B_\alpha^{p,q}(F)$ is $\sum_{|j|\leq k}(\|f^{(j)}\|_{p,\mu} + \|\{a_{j\nu}\}\|_{l_q})$.

These Besov spaces can be defined also for integer α but then the definition is more involved [9]. The definition looks less complicated if $0 < \alpha < 1$. Then there is just one sequence $\{a_{j\nu}\}$, and if we denote it by $\{a_\nu\}$, the expression for a_ν is given by formula (4) in Section 5.

Remark 4.1. The requirement that F is a d-set is less important and can be weakened considerably. For example, if $\alpha < 1$ and $\alpha p > 1$, then Besov spaces $B_\alpha^{p,q}(F)$ can be defined intrinsically on an arbitrary closed set F in such a way that the relevant trace theorem for traces to F of Besov spaces holds. See [7].

If F preserves Markov's inequality, the situation simplifies. Then the functions $f^{(j)}$ are uniquely determined by $f^{(0)}$ if $\{f^{(j)}\}_{|j|\leq k} \in B_\alpha^{p,q}(F)$, and so the family $\{f^{(j)}\}_{|j|\leq k}$ can be identified with the function $f^{(0)}$ which we then denote by f. If $F = \mathbb{R}^n$, then the Besov spaces defined above are equivalent to the classical Besov spaces on $\mathbb{R}^n$.

Recall that a function f belongs to the Sobolev space $W_k^p(\mathbb{R}^n)$ iff the norm $\sum_{|j|\leq k} \|D^j f\|_p < \infty$. The introduction of Besov spaces was partially motivated by the need when studying traces of Sobolev spaces. Trace problems were also the motivation when Besov spaces on subsets of $\mathbb{R}^n$ were introduced. We list here some basic well-known trace theorems. Since functions in the function spaces are in general defined only almost everywhere, and one takes the restriction to subsets of Lebesgue measure zero, one must explain what is meant with the restriction of a function. For f in $L^p(\mathbb{R}^n)$, define the strictly defined function $\tilde{f}$ by

$$\tilde{f}(x) = \lim_{r\to 0} \int_{B(x,r)} f(t)dt$$

at every point where the limit exists. If $F \subset \mathbb{R}^n$, we denote by $f|F$ the pointwise restriction to F of $\tilde{f}$. The interpretation of e.g. the statement $W_k^p(\mathbb{R}^n)|\mathbb{R}^m = B_\beta^{p,p}(\mathbb{R}^m)$ in Theorem 4.2 below is as follows: $\mathbb{R}^m$ denotes an m-dimensional subspace of $\mathbb{R}^n$. If $f \in W_k^p(\mathbb{R}^n)$, then $f|\mathbb{R}^m$ exists a.e. on $\mathbb{R}^m$ and belongs to $B_\beta^{p,p}(\mathbb{R}^m)$ and the restriction operator is continuous, and if $f \in B_\beta^{p,p}(\mathbb{R}^m)$, then there exists a function $Ef \in W_k^p(\mathbb{R}^n)$ such that $Ef|F = f$, and the extension is given by a linear and continuous operator.

Theorem 4.2. *Let k be a positive integer, $1 < p < \infty$, $0 < m < n$, and $\beta = k - (n-m)/p > 0$. Then*

$$W_k^p(\mathbb{R}^n)|\mathbb{R}^m = B_\beta^{p,p}(\mathbb{R}^m).$$

Theorem 4.3. *Let $1 \le p, q \le \infty$, and $\beta = \alpha - (n-m)/p > 0$. Then*

$$B_\alpha^{p,q}(\mathbb{R}^n)|\mathbb{R}^m = B_\beta^{p,q}(\mathbb{R}^m).$$

These theorems are generalized in [9] so that one can take traces to more general sets. In the following theorems, the meaning of the statements is immediate only if F preserves Markov's inequality, so that the functions in the Besov spaces on F can be identified with one function, see below for the general case. As indicated in Remark 4.1, the assumption that F is a d-set can be weakened.

Theorem 4.4. *Let F be d-set, $0 < d < n$, $1 < p < \infty$, let k be a positive integer, and $\beta = k - (n-d)/p > 0$. Then*

$$W_k^p(\mathbb{R}^n)|F = B_\beta^{p,p}(F).$$

Theorem 4.5. *Let F be a d-set, $0 < d \le n$, $1 \le p, q \le \infty$, and $\beta = \alpha - (n-d)/p > 0$. Then*

$$B_\alpha^{p,q}(\mathbb{R}^n)|F = B_\beta^{p,q}(F). \tag{3}$$

If F does not preserve Markov's inequality, the interpretation of e.g. the statement (3) is as follows. If $k < \beta \le k+1$ and $\{f^{(j)}\}_{|j|\le k} \in B_\beta^{p,q}(F)$, then there is a function $Ef \in B_\alpha^{p,q}(\mathbb{R}^n)$ such that $D^j f|F = f^{(j)}, |j| \le k$. Conversely, if $f \in B_\alpha^{p,q}(\mathbb{R}^n)$, then $\{D^j f|F\}_{|j|\le k} \in B_\beta^{p,q}(F)$. Also, if F does not preserve Markov's inequality, the linearity of the extension operator is lost if β is an integer.

5. Lipschitz spaces and the domain of the Dirichlet form

We next define the Lipschitz spaces on F. We do this for spaces not involving derivatives, since it is less clear what is the natural approach in the general case.

Definition 5.1. *Let F be a d-set, $0 < d \le n$, $\alpha > 0, 1 \le p, q \le \infty$. A function f belongs to* the Lipschitz space $\mathrm{Lip}(\alpha, p, q; F)$ *if $f \in L^p(\mu)$ and $\|\{a_\nu\}_{\nu=0}^\infty\|_{l_q} < \infty$, where*

$$2^{-\nu\alpha} a_\nu = 2^{\nu d} \int\int_{|x-y|<c_0 2^{-\nu}} |f(x) - f(y)|^p d\mu(x) d\mu(y))^{1/p}. \tag{4}$$

The norm of f in $\mathrm{Lip}(\alpha, p, q; F)$ *is $\|f\|_{p,\mu} + \|\{a_\nu\}\|_{l_q}$.*

For $p = \infty$ this means $2^{-\nu\alpha} a_\nu = \sup_{x,y \in F, |x-y| < c_0 2^{-\nu}} |f(x) - f(y)|$. The space $\mathrm{Lip}(\alpha, \infty, \infty; F)$ can be identified with the Lipschitz space of continuous bounded functions satisfying $|f(x) - f(y)| \le M|x-y|^\alpha$, $x, y \in F$, c.f. [9], p 137.

It is clear that if $0 < \alpha < 1$, then the spaces $\mathrm{Lip}(\alpha, p, q; F)$ and $B_\alpha^{p,q}(F)$ coincide, but if $\alpha \ge 1$, they are in general different. If $\alpha > 1$ and $F = \mathbb{R}^n$, then the spaces $\mathrm{Lip}(\alpha, p, q; F)$ are trivial, in the sense that they consist of constant functions, only. In general this is not true. For example, if $F = F_1 \cup F_2$, where F_1 and F_2 are disjoint compact sets, then $\mathrm{Lip}(\alpha, \infty, \infty; F)$ is clearly nontrivial for all $\alpha > 0$, since the function which equals 1 on F_1 and 0 on F_2, will be in $\mathrm{Lip}(\alpha, \infty, \infty; F)$. It was shown in [5] that if F is the Sierpinski gasket, then $\mathrm{Lip}(\alpha, 2, q; F)$ is non-trivial iff $\alpha < \ln 5/\ln 4$ or $\alpha = \ln 5/\ln 4$ and $q = \infty$, and in [8] that if F is the von Koch curve, $\mathrm{Lip}(\alpha, \infty, \infty; F)$ is nontrivial iff $\alpha \le \ln 4/\ln 3$.

Although the Lipschitz spaces and the Besov spaces are different if $\alpha \ge 1$, they are related. The following proposition is immediate from the definitions.

Proposition 5.1. *Let α be non-integer, and $k < \alpha < k+1$. Then $f \in \mathrm{Lip}(\alpha, p, q; F)$ iff the family $\{f^{(j)}\}_{|j| \le k}$, given by $f^{(0)} = f$ and $f^{(j)} = 0$ for $0 < |j| \le k$, is in $B_\alpha^{p,q}(F)$.*

If F preserves Markov's inequality, so that the functions in the Besov space can be identified with the function $f = f^{(0)}$, this can be formulated as follows. The space $\mathrm{Lip}(\alpha, p, q; F)$ is the subspace of $B_\alpha^{p,q}(F)$ consisting of functions whose "derivatives" $f^{(j)}$ of orders j, $0 < |j| \le k$ are zero μ-a.e. If $\alpha = k+1$, it can be seen that $\mathrm{Lip}(\alpha, p, q; F)$ is a subset of this subspace.

Besov spaces and Lipschitz spaces on $\mathbb{R}^n$ are also related to Sobolev spaces. It is known that $W_k^2(\mathbb{R}^n) = B_k^{2,2}(\mathbb{R}^n)$ and, of special interest in the present context, that the following theorem holds, see [15], Ch. 5, Section 3.5.

Theorem 5.2. *Let k be a positive integer and $1 < p < \infty$. Then*

$$W_1^p(\mathbb{R}^n) = \mathrm{Lip}(1, p, \infty; \mathbb{R}^n)$$

with equivalent norms.

In [5], the following result was proved, connecting the theory of function spaces on subsets of $\mathbb{R}^n$ and Dirichlet forms on the Sierpinski gasket. Since $\mathcal{D}$ plays, in the theory of Brownian motion on fractals, the same role as the Sobolev space $W_1^2(\mathbb{R}^n)$ in the classical theory, the theorem can be seen as a generalization of Theorem 5.2 for $p = 2$.

Theorem 5.3. *Let F denote the Sierpinski gasket, and let $\beta = \ln 5/\ln 4$. Then*

$$\mathcal{D} = \mathrm{Lip}(\beta, 2, \infty; F)$$

with equivalent norms.

This result has been generalized in [13] and [14] to treat not only the Sierpinski gasket, but the more general class of nested fractals. The result is then that the domain of the Dirichlet form on a nested fractal F is $\mathrm{Lip}(d_\omega/2, 2, \infty; F)$, where d_ω

is the walk dimension of F. Theorem 5.3 can be combined with the theorems in the previous section to give information about the space $\mathcal{D}$. For example we have the following corollary.

Corollary 5.4. *Let F and β be as in Theorem 5.3, and let $\alpha = \beta + (2-d)/2$. Then every $f \in \mathcal{D}$ can be extended to a function $Ef \in B_\alpha^{2,\infty}(\mathbb{R}^2)$, in such a way that $Ef|F = f$ and $D^j(Ef)|F = 0, |j| = 1$. Conversely, if $f \in B_\alpha^{2,\infty}(\mathbb{R}^2)$ with $D^j(Ef)|F = 0, |j| = 1$, then $f|F \in \mathcal{D}$.*

Proof. By Proposition 5.1, and the comments given after it, a function $f \in \mathcal{D} = \mathrm{Lip}(\beta, 2, \infty; \mathrm{F})$ can be identified with a function in $B_\beta^{2,\infty}(F)$ with "derivatives" $f^{(j)} = 0, |j| = 1$. The corollary then follows from Theorem 4.5. □

Remark 5.5. There is an analogy between the space $\mathcal{D}$ and a Sobolev space also concerning bases. For $i \geq 0$, let U_i be the set of dyadic points on the interval $[0,1]$ given by $U_i = \{k2^{-i}, k = 0, 1, \ldots 2^i\}$, let $U_{-1} = \emptyset$, and put $V_i = U_i \setminus U_{i-1}$. For $\xi \in V_i$, let ψ_ξ be the piecewise linear function defined on [0,1] which is equal to 1 at the point ξ, and equal to 0 at the other points of U_i. Then it is not difficult to show, using integration of the Haar series of the derivative of f, that $f \in W_1^2[0,1]$ iff f has a representation $f = \sum_{i=0}^\infty \sum_{\xi \in V_i} a_\xi \psi_\xi$, where $\sum_{i=0}^\infty \sum_{\xi \in V_i} a_\xi^2 2^i < \infty$. Thus we have, in a similar way as in Theorem 3.1, a representation in terms of a piecewise harmonic basis. We also remark that a representation of Besov spaces $B_\alpha^{(p,q)}(F)$ with low smoothness ($\alpha < 1$) on certain closed sets $F \subset \mathbb{R}^n$, including e.g. the Sierpinski gasket, in terms of a piecewise linear basis was given in [10].

6. Trace theorems

Let α and β be as in Corollary 5.4. If $f \in \mathcal{D}$, then the corollary gives that there is an extension $Ef \in B_\alpha^{2,\infty}(\mathbb{R}^2)$, and by Theorem 4.5, Ef restricted to the interval [0,1] on the x-axis is in $B_\gamma^{2,\infty}([0,1])$, where $\gamma = \alpha - (2-1)/2 = \beta - (d-1)/2$, which means the restriction of a function $f \in \mathcal{D}$ to the interval [0,1] (a "side" of the gasket) is in $B_\gamma^{2,\infty}([0,1])$. We shall see below that this result is not quite sharp. Also, no similar argument gives a converse result. If we extend a function $f \in B_\gamma^{2,\infty}([0,1])$ to a function $Ef \in B_\alpha^{2,\infty}(\mathbb{R}^2)$ using Theorem 4.5, and then restrict to F, we do not even come close to a correct result, the reason being that one can not expect that the derivatives of Ef are zero μ-a.e. on F, as required if one wants to use Corollary 5.4 in order to restrict Ef to F.

The following two theorems are proved in [4], using the characterization of the space $\mathcal{D}$ given in Theorem 3.1, the properties of harmonic functions on the gasket, and a discrete characterization of the Besov space $B_\alpha^{p,q}([0,1]$, valid for $1/p < \alpha < 1$, given in [11]. There is no place to give complete proofs here, but we give the extension operator explicitly which shows how Theorem 3.1 enters the picture. The proofs are different from the usual proofs of e.g. the Theorems 4.2-4.5, since new techniques seem to be required. The first theorem is a restriction theorem, the second an extension theorem.

Theorem 6.1. *Let F be the Sierpinski gasket, let $d = \ln 3/\ln 2$, $\beta = \ln 5/\ln 4$, and $\gamma = \beta - (d-1)/2$. Let Rf denote the pointwise restriction to $[0,1]$ of a continuous function f defined on F. Then R is a bounded linear operator from $\mathcal{D} = \mathrm{Lip}(\beta, 2, \infty; F)$ to $B_\gamma^{2,2}([0,1])$.*

To state the extension theorem, we start by describing an extension operator from $C([0,1])$ to functions defined on $\mathbb{R}^2$. Let U_i and V_i be the dyadic points in [0,1] defined in Remark 5.5, and let $\mathcal{U}_i$ and $\mathcal{V}_i$ denote the vertices of the Sierpinski gasket as in Section 2. We then have $U_i \subset \mathcal{U}_i$ and $V_i \subset \mathcal{V}_i$.

Let $f \in C([0,1])$. Denoting by s_0 the vertex of $\Delta \in \mathcal{T}_0$ not on the real axis, let the functions $E_n f$ be given by

$$E_n f = \sum_{i=0}^{n} \sum_{\xi \in V_i} a_\xi \psi_\xi + a_{s_0} \psi_{s_0},$$

where $a_{s_0} = (f(0) + f(1))/2$, and for $0 \leq i \leq n$, the coefficients a_ξ are given inductively by

$$a_\xi = f(\xi) - E_{i-1} f(\xi),\ \xi \in V_i,$$

where we put $E_{-1} f = 0$. Then $E_n f(\xi) = f(\xi)$, $\xi \in U_n$, and $E_n f$ belongs to H_n, in fact, $E_n f$ is harmonic in each $\Delta \in \mathcal{T}_i$, $i < n$ such that $\Delta \cap [0,1] = \emptyset$, and in each $\Delta \in \mathcal{T}_n$. We define an operator E on $C[0,1]$ by

$$Ef(x) = \lim_{n \to \infty} E_n f(x) = \sum_{i=0}^{\infty} \sum_{\xi \in V_i} a_\xi \psi_\xi(x) + a_{s_0} \psi_{s_0}(x), x \in F \setminus [0,1],$$

and $Ef(x) = f(x), x \in [0,1]$.

Theorem 6.2. *Let $d = \ln 3/\ln 2$, $\beta = \ln 5/\ln 4$, and $\gamma = \beta - (d-1)/2$. Then the operator E is a bounded linear operator from $B_\gamma^{2,2}([0,1])$ to $\mathcal{D} = \mathrm{Lip}(\beta, 2, \infty; F)$. In particular, E is an extension operator in the sense that the pointwise restriction to $[0,1]$ of the continuous function $Ef \in \mathcal{D}$ is f.*

These two theorems could be summarized as

$$\mathcal{D}|[0,1] = B_\gamma^{2,2}([0,1]).$$

As the space $\mathcal{D}$ plays, in the theory of Dirichlet forms on fractals, the role of a Sobolev space in the classical case, the present theorem could perhaps be considered as an analogy to the theorems 4.2 and 4.4 which characterize the trace of the Sobolev space $W_1^2(\mathbb{R}^n)$ as a Besov space. Note that in all cases, the loss of smoothness when restricting is half of the difference between the dimensions of the underlying sets.

Theorem 5.3 was (via Corollary 5.4) a useful tool in [6], and in a somewhat different way in [13], where it was shown that Brownian motion penetrating the Sierpinski gasket exists. Possibly more general trace theorems of the above type could be useful when studying Brownian motion moving between fractals as in [3].

References

[1] Barlow, M. T., Diffusions on fractals. *Lectures on probability theory and statistics (Saint-Flour, 1995), Lecture Notes in Math.*, vol 1690, 1–121, Springer, Berlin, 1998

[2] Fukushima M. and Shima T., On a spectral analysis for the Sierpinski gasket, *Potential Analysis* **1** (1992), 1–35

[3] Hambly, B.M. and Kumagai, T., Diffusion processes on fractal fields: heat kernel estimates and large deviations, prepint

[4] Jonsson A., A trace theorem for the Dirichlet form on the Sierpinski gasket, preprint, Dept of Math., Umeå Univ, **8** (2003)

[5] Jonsson A., Brownian motion on fractals and function spaces, *Math. Z.* **222** (1996), 495–504

[6] Jonsson A., Dirichlet forms and Brownian motion penetrating fractals, *Potential Analysis* **13** (2000), 69–80

[7] Jonsson, A., Besov spaces on closed subsets of $\mathbb{R}^n$, *Trans. Am. Math. Soc.* **341** (1994), 355–370

[8] Jonsson A., Haarwavelets of higher order on fractals and regularity of functions, to appear in *J. Math. Anal. and Appl.*

[9] Jonsson A. and Wallin H., Function spaces on subsets of $\mathbb{R}^n$, *Math. Reports* **2**, Part 1, Haarwood Acad. Publ., Chur, 1984

[10] Jonsson A. and Kamont A., Piecewise linear bases and Besov spaces on fractal sets, *Analysis Mathematica* **27** (2001), 77–117

[11] Kamont A., A discrete characterixation of Besov Spaces, *Approx. Th. Appl.* **13** (1997), 63–77

[12] Kigami J., *Analysis on fractals*, Cambridge Univ. Press, Cambridge, 2001

[13] Kumagai, T., Brownian motion penetrating fractals: an application of the trace theorem of Besov spaces. *J. Funct. Anal.* **170** (2000), 69–92.

[14] Pietruska-Paluba K., Some function spaces related to the Brownian motion on simple nested fractals, *Stochastics and Stochastics Reports* **67** (1999), 267–285

[15] Stein E.M., *Singular integrals and differentiability properties of functions*, Princeton Univ. Press, Princeton, 1970

[16] Strichartz R,S., Analysis on fractals, *Notices Amer. Math. Soc.* **46** (1999), 1199–1208

[17] Strichartz R.S., Function spaces on fractals, *J. Funct. Anal.* **198** (2003), 43–83.

[18] Triebel H., *Theory of function spaces*, Birkhäuser, Basel, 1983

[19] Triebel H., *Fractals and spectra*, Birkhäuser, Basel, 1997

[20] Whitney H., Analytic extension of differentiable functions defined in closed sets, *Trans. Amer. Math. Soc.* **36** (1934), 63–89

Alf Jonsson
Department of Mathematics
University of Umeå
90187 Umeå
Sweden
e-mail: `alf.jonsson@math.umu.se`

Progress in Probability, Vol. 57, 245–262

Spectral Zeta Function of Symmetric Fractals

Alexander Teplyaev

Abstract. This is an expository paper which includes several topics related to the spectral analysis on fractals. Such analysis on certain symmetric fractals can be completely described in terms of complex dynamics of a polynomial. The examples of such fractals include the Sierpiński gasket and an interval with a fractal measure. We discuss the spectral type of the Laplacian, complex spectral dimensions and spectral zeta function. The spectral zeta function has a product structure that involves a certain new zeta function of a polynomial and a "geometric" part. A similar product structure in the case of fractal strings was discovered by M. L. Lapidus. We give examples were the spectrum is singularly continuous on one dimensional fractals but is pure point on the infinite Sierpiński gasket, with a complete set of compactly supported eigenfunctions. We describe the spectrum of the Laplacian (in terms of the complex dynamics of a polynomial) and all the eigenfunctions with compact support.

Mathematics Subject Classification (2000). Primary 28A80; Secondary 05C25, 31C25, 35P05, 37F10, 47A10, 47A75, 47B39, 58J50, 60G18, 81Q50, 82D30, 90B10, 94C15.

Keywords. Fractal, Sierpiński gasket, self-similar graph, Dirichlet form, Laplacian, spectrum, Julia set, zeta function.

1. Introduction

There has been considerable attention given to the study of fractally structured media. These media have special scaling properties and, in this respect, may bridge the gap between crystalline structures and purely disordered materials. For instance, fractals may represent percolation clusters, quasicrystals, fractal electrical networks and porous materials. Recently fractal antennas were successfully tested in communication devices such as a GPS installed in a cellular phone ([21]). The Laplacian on fractals and fractal graphs play an important role in physics (see, for instance, [2, 18, 47, 48, 61] and references therein).

Physical processes are often described through differential equations involving the Laplace operator, such as wave and heat equations in classical physics and the Schrödinger equation in quantum physics. There is extensive mathematical literature devoted to fractals. However, the study of Laplacians on fractals started only a few years ago. First they were considered from the probabilistic point of view, and then an analytic approach was also developed (see, for example, an expository paper *Analysis on fractals* by R. Strichartz in the Notices of the AMS [63] and two recent books *Fractal geometry and number theory* by M. Lapidus and M. van Frankenhuysen [36] and *Analysis on fractals* by J. Kigami [30]). Fractal graphs and manifolds have found new geometric applications, such as in *Manifolds and graphs with slow heat kernel decay* by M. Barlow, T. Coulhon and A. Grigor'yan [3, 13], and groups with fractal spectrum were discovered by L. Bartholdi, R. Grigorchuk and A. Żuk [6, 7, 19].

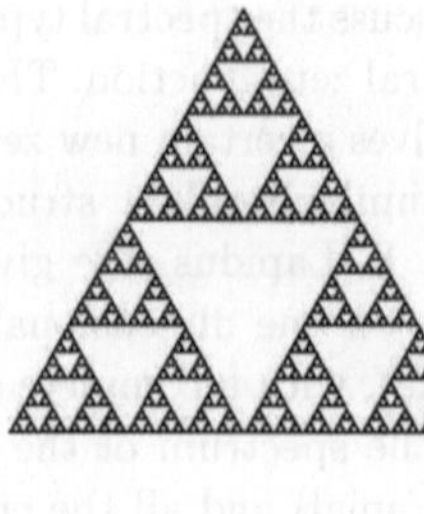

FIGURE 1. Sierpiński gasket.

The definition of a Dirichlet form on the Sierpiński gasket S uses a sequence of discrete approximations V_n to S that are defined inductively by

$$V_n = \bigcup_{j=1}^{3} \Psi_j(V_{n-1}),$$

where $V_0 = \{v_1, v_2, v_3\}$ is the set of vertices of S and $\Psi_j(x) = \frac{1}{2}(x + v_j)$ for $x \in \mathbb{R}^2$. The Sierpiński gasket is a unique compact set S such that $S = \bigcup_{j=1}^{3} \Psi_j(S)$. The union of V_n is dense in the Sierpiński gasket S.

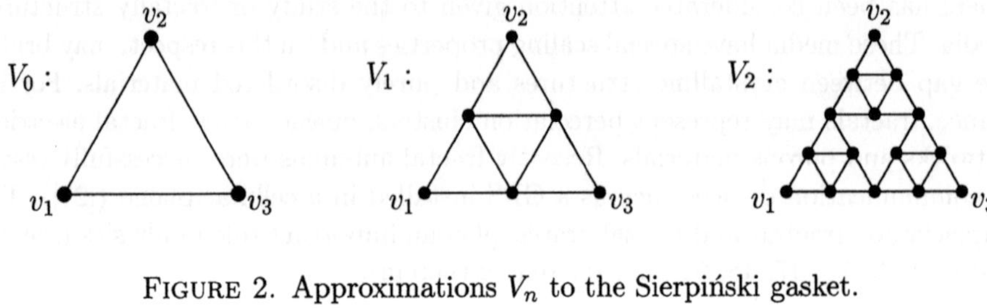

FIGURE 2. Approximations V_n to the Sierpiński gasket.

The **Dirichlet form** $\mathcal{E}$, which is local and regular by a result of J. Kigami [26, 27, 30], is defined on S by

$$\mathcal{E}(f,f) = \lim_{n\to\infty} \left(\tfrac{5}{3}\right)^n \sum_{\substack{y\sim x \\ x,y\in V_n}} (f(y)-f(x))^2$$

where $\sim$ means that y and x are neighbors in the finite lattice V_n. Note that this limit is nondecreasing for any function f. A continuous function h is **harmonic** if it minimizes $\mathcal{E}(h,h)$ given the three boundary values. If f is harmonic, then the sequence in the limit above is constant (does not depend on n).

Laplacian Δ_μ on the Sierpiński gasket is a densely defined linear operator satisfying the Gauss–Green formula

$$\mathcal{E}(f,f) = -\tfrac{3}{2}\int_S f\Delta_\mu f d\mu + \sum_{p\in\partial S} f(p)\partial_n f(p)$$

where μ is a normalized Hausdorff (Bernoulli) measure on the Sierpiński gasket, which is self-similar with weights $\frac{1}{3}, \frac{1}{3}, \frac{1}{3}$:

$$\mu = \tfrac{1}{3} \sum_{j=1,2,3} \mu\circ F_j.$$

This Laplacian also can be defined as a pointwise limit of normalized graph Laplacians on the sequence discrete approximations V_n. If the **discrete Laplacians** on V_n are

$$\Delta_n f(x) = \tfrac{1}{4}\sum_{\substack{y\in V_n \\ y\sim x}} f(y)-f(x), \quad x\in V_n\backslash V_0,$$

then Laplacian on S is

$$\Delta_\mu f(x) = \lim_{n\to\infty} 5^n \Delta_n f(x)$$

if the limit exists.

It is important to note that these definitions, and the analysis based on them, do not depend on the particular embedding of the Sierpiński gasket into $\mathbb{R}^2$, but can be formulated entirely in terms of its intrinsic topological structure (if the Sierpiński gasket is defined as a quotient of a shift space). In this project a number of different embeddings will play an important role, even though the energy form is the same.

The **spectral asymptotics** of Δ_μ is given by

$$0 < \liminf_{\lambda\to\infty} \frac{\rho(\lambda)}{\lambda^{d_s/2}} < \limsup_{\lambda\to\infty} \frac{\rho(\lambda)}{\lambda^{d_s/2}} < \infty$$

where $\rho(\lambda)$ is the **eigenvalue counting function** of the Dirichlet or Neumann Laplacian and the spectral dimension is

$$1 < d_s = \tfrac{\log 9}{\log 5} < 2.$$

We have, by an elementary computation, that $R(z) \in \sigma(\Delta_n)$ if and only if $z \in \sigma(\Delta_{n+1})$ where $z \neq -\frac{1}{2}, -\frac{3}{4}, -\frac{5}{4}$ and $R(z) = z(5+4z)$. Moreover, every eigenvalue of Δ_μ has a form $\lambda = \lim_{n\to\infty} 5^{n+m} R^{-n}(z_0)$ where $R^{-n}(z_0)$ is a preimage of $z_0 = -\frac{3}{4}, -\frac{5}{4}$ under the n-th iteration power of the polynomial $R(z)$. The multiplicity of such an eigenvalue is $C_1 3^m + C_2$ (see [14, 17, 47, 48, 57, 64]).

These results inspired a number of papers on the spectral analysis on fractals. So called "spectral decimation", first discovered by physicists in the case of Sierpiński gasket([2, 47, 48, 61]), is studied in [17, 57, 33, 38, 55, 70]. In [43] an analog of power series expansions on the Sierpiński gasketis considered. The spectrum of the Laplacian on the fractal pentagasket (a less symmetric analog of the Sierpiński gasket whose construction is based on a regular pentagon instead of a triangle) is studied in [1], and on an interval with positive measure in [54, 12, 67]. Note that the domain of the Laplacian can be very peculiar ([8, 45, 62, 65]). Some recent results on the Dirichlet forms and Laplacians are included in [23, 34, 72].

2. Spectral zeta functions of a polynomial

In the study of spectral zeta functions of the Sierpiński gasket and other fractals the first step is to define a new type of zeta function associated with a polynomial. Suppose that $R(z)$ is a polynomial of degree N such that $R(0) = 0$, $c = R'(0) > 1$, and the Julia set $\mathfrak{J}_R$ of $R(z)$ lies in $\mathbb{R}_+$. Although this condition can be relaxed, in our case it will be automatically satisfied because the spectrum of a nonnegative self-adjoint operator is contained in $\mathbb{R}_+$. Then the **zeta function of the polynomial** $R(z)$ is

$$\zeta_R^{z_0}(s) = \lim_{n\to\infty} \sum_{z\in R^{-n}\{z_0\}} (c^n z)^{-s/2}$$

for $\mathrm{Re}(s) > d_R = \frac{2\log N}{\log c}$. One of the reasons this new class of zeta functions is interesting is that the Riemann zeta function has a representation $\zeta(s) = \zeta_R^0(s)$ where $R(z)$ is a certain quadratic polynomial, which is associated with the Laplacian on an interval (considered as a p.c.f. self-similar set in the context of the analysis on fractals, see Section 4).

Theorem 2.1. *The zeta function $\zeta_R^{z_0}(s)$ of the polynomial $R(z)$ has a meromorphic continuation of the form*

$$\zeta_R^{z_0}(s) = \frac{f_1(s)}{1 - Nc^{-s/2}} + f_2^{z_0}(s)$$

where $f_1(s)$ is analytic in $\mathbb{C}$ and $f_2^{z_0}(s)$ is analytic for $Re(s) > 0$. The set of poles of $\zeta_R^{z_0}(s)$ is contained in $\left\{\frac{2\log N + 4in\pi}{\log c} : n \in \mathbb{Z}\right\}$, and there always is a pole at d_R. If $\mathfrak{J}_R$ is totally disconnected, then there exists $\varepsilon > 0$ such that $f_2^{z_0}(s)$ is analytic for $Re(s) > -\varepsilon$.

This theorem is proved in [67]. There are several open questions related to these zeta functions of polynomials. The first question is the existence of a meromorphic continuation to the complex plane. There are three, possibly related, approaches that one may try to use in order to obtain such continuation. The first

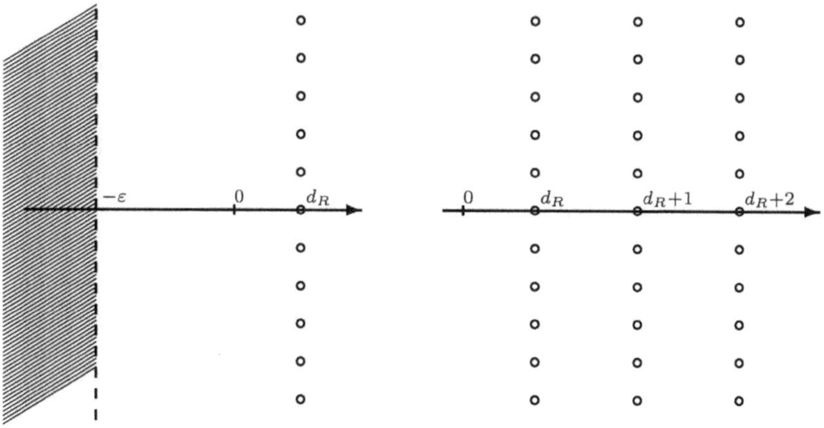

FIGURE 3. Poles of the zeta function of a polynomial (left) and of the Mellin transform (right).

approach is to find a functional equation which could be similar to the functional equation for the Riemann and some other zeta functions. Another approach is to find a relation of $\zeta_R(s)$ with a dynamical zeta function with weights (without weights this clearly doesn't work). The third approach is to try to relate the zeta function $\zeta_R(s)$ of a polynomial $R(z)$ to the **Mellin transform**

$$\mathcal{M}_R(s) = \lim_{n\to\infty} N^{-n} \sum_{z\in R^{-n}\{z_0\}} z^{-s/2} = \int_{\mathfrak{J}_R} z^{-s/2} d\rho(z)$$

of a balanced invariant measure ρ supported on the Julia set $\mathfrak{J}_R$ of $R(z)$. This Mellin transform is known to have a meromorphic continuation to $\mathbb{C}$ (see [9, 10] and references therein). The results of [9, 10] are not immediately applicable in our case. If it exists, the relation between $\zeta_R(s)$ and $\mathcal{M}_R(s)$ must be nontrivial since $\zeta_R(s)$ has one positive real pole by Theorem 2.1, while the Mellin transform of a balanced invariant measure of $R(z)$ has infinitely many poles on the positive real half line. For instance, in the case of the quadratic polynomial $R(z)$ for which $\zeta_R(s) = \zeta(s)$ is the Riemann zeta function, the Mellin transform is a ratio of two gamma functions. It seems plausible that there is a relationship between $\zeta_R(s)$ and the Mellin transform of a different measure on $\mathfrak{J}_R$. The measure most probably will be more complicated than the measures considered in [10] in that the density function involved is bounded neither above nor below.

3. Spectral zeta functions of the Sierpiński gasket

The **spectral zeta function** of a (Dirichlet) Laplacian L is defined as

$$\zeta_L(s) = \sum \lambda_j^{-s/2}$$

where $-\lambda_j$ are the eigenvalues of L. The poles of ζ_L are called the **complex spectral dimensions**. Based on Theorem 2.1 the following formula is proved in [67] for the spectral zeta function of the Laplacian Δ_μ on the Sierpiński gasket.

Theorem 3.1. *If* $R(z) = z(5-4z)$ *then*

$$\zeta_{\Delta_\mu}(s) = \tfrac{1}{2}\zeta_R^{\frac{3}{4}}(s)\left(\tfrac{1}{5^{s/2}-3} + \tfrac{3}{5^{s/2}-1}\right) + \tfrac{1}{2}\zeta_R^{\frac{5}{4}}(s)\left(\tfrac{3\cdot 5^{-s/2}}{5^{s/2}-3} - \tfrac{5^{-s/2}}{5^{s/2}-1}\right).$$

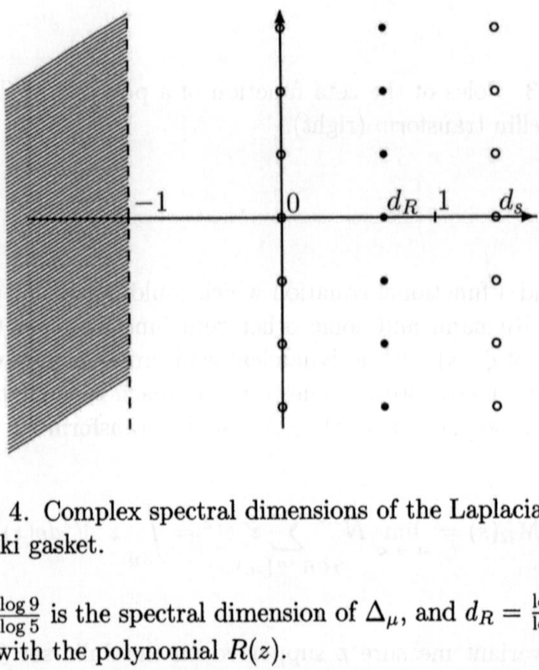

FIGURE 4. Complex spectral dimensions of the Laplacian on the Sierpiński gasket.

Here $d_s = \frac{\log 9}{\log 5}$ is the spectral dimension of Δ_μ, and $d_R = \frac{\log 4}{\log 5}$ is the dimension associated with the polynomial $R(z)$.

The representation of $\zeta_{\Delta_\mu}(s)$ in Theorem 3.1 is reminiscent of the spectral zeta function of a self-similar fractal string in the book by M. Lapidus and M. van Frankenhuysen [36]. A product formula

$$\zeta_L(s) = \zeta(s)\zeta_{\mathcal{L}}(s)$$

for fractal strings was first obtained by M. Lapidus in [35]. Here $\zeta(s)$ is the Riemann zeta function and $\zeta_{\mathcal{L}}(s) = \sum l_j^s$ is the geometric zeta function of a fractal string $\mathcal{L}$, a disjoint collection of intervals of lengths l_j. For example, if the fractal string $\mathcal{L}$ is the complement of the middle third Cantor set in $[0,1]$, then $\zeta_L(s) = \zeta(s)/(3^s - 2)$. Hence it is natural to say that in Theorem 3.1 we have products of the zeta functions of the polynomial and "geometric" zeta functions.

An interesting feature of the product formula in Theorem 3.1 is that the poles of the zeta functions of the polynomial are canceled by the zeros of the "geometric" part. This cancelation to be a general fact which can be explained by the geometry of symmetric p.c.f. fractals. Namely, one can use the self-similarity of the eigenprojectors in order to establish a direct relation between the "geometric" part of (3.1), the rational function $R(z)$, and the geometry of the self-similar structure that defines the fractal (see [64, 68]).

It would be very interesting to find the spectral zeta function in so called non lattice cases and in cases that do not allow the use of the spectral decimation technique (see [1, 4, 20, 32, 54, 55]), where the poles are expected to be not in an arithmetic progression, but scattered "at random" over a certain strip. Such spectral zeta functions were studied in [36] for non lattice fractal strings, but the situation for the Laplacian on a fractal seems to be more difficult. The renewal theory developed in [31, 30] is expected to play an important role. Another interesting and difficult question is to study spectral zeta function of the energy Laplacian Δ_ν. Heuristically speaking, this Laplacian corresponds to a fractal which is self-affine but not self-similar (see [28, 41, 65, 66, 71] and related results in [24, 25, 44, 46, 49, 50, 51, 52, 58]). Therefore the spectral analysis of the Laplacian Δ_ν is likely to involve analysis in several complex variables, unlike the simpler and more symmetric case of the Laplacian Δ_μ.

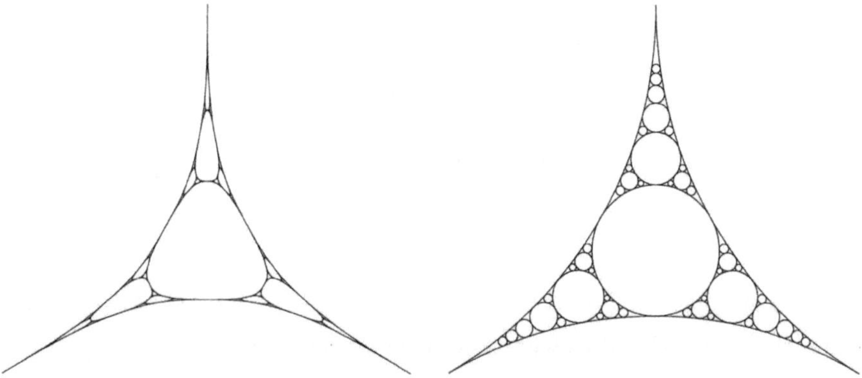

FIGURE 5. Sierpiński gasket in harmonic coordinates (left) and the residue set of the Apollonian packing (right).

In [42] there is a description of all, not necessarily self-similar, Dirichlet forms on the Sierpiński gasket. A general theory of Laplacians for such situations is developed in [30], but the spectrum is to be studied. In particular, it is shown in [66] that there exists a Dirichlet form on the Sierpiński gasket such that the residue set of the Apollonian packing (see [39, 16]) is the Sierpiński gasket in harmonic coordinates with respect to this form.

4. Laplacian on an interval and Riemann zeta function

We begin with defining the usual energy form of the unit interval by introducing a self-similar structure. In this way the interval can be considered as a p.c.f. fractal (see [27, 30]). Note that this construction is very similar to the one for the Sierpiński gasket, which is described in the Introduction. We define two contractions $F_1, F_2 : \mathbb{R}^1 \to \mathbb{R}^1$ with fixed points $p_j = 0, 1$ by $F_j(x) = \frac{1}{2}x + \frac{1}{2}p_j$. Then the interval $I=[0,1]$ is a unique compact set such that $I = \bigcup_{j=1,2} F_j(I)$. The **discrete approximations** to I are defined inductively by

$$V_n = \bigcup_{j=1,2} F_j(V_{n-1}) = \left\{\tfrac{k}{2^n}\right\}_{k=0}^{2^n}$$

where $V_0 = \partial I = \{0,1\}$ is the **boundary** of I. The **discrete Dirichlet (energy) form** on V_n is

$$\mathcal{E}_n(f) = 2^n \sum_{\substack{x,y\in V_n\\ y\sim x}} (f(y) - f(x))^2$$

and the **Dirichlet (energy) form** on I is

$$\mathcal{E}(f) = \lim_{n\to\infty} \mathcal{E}_n(f) = \int_0^1 |f'(x)|^2 dx.$$

A function h is **harmonic** if it minimizes the energy given the boundary values. We have that $\mathcal{E}_{n+1}(f) \geqslant \mathcal{E}_n(f)$ and $\mathcal{E}_{n+1}(h) = \mathcal{E}_n(h) = \mathcal{E}(h)$ for a harmonic h. This Dirichlet form on I is **self-similar** in the sense that

$$\mathcal{E}(f) = 2 \sum_{j=1,2} \mathcal{E}(f \circ F_j).$$

If we define the **discrete Laplacians** on V_n by

$$\Delta_n f(x) = \frac{1}{2} \sum_{\substack{y\in V_n\\ y\sim x}} f(y) - f(x), \quad x \in V_n \backslash V_0$$

then the Laplacian on I is $\Delta f(x) = \lim_{n\to\infty} 4^n \Delta_n f(x) = f''(x)$. This Laplacian satisfies the **Gauss–Green (integration by parts) formula**

$$\mathcal{E}(f) = -\int_0^1 f \Delta f dx + f f'\Big|_0^1.$$

It has **spectral asymptotics**

$$\lim_{\lambda\to\infty} \frac{\rho(\lambda)}{\lambda^{d_s/2}} = \frac{1}{\pi}$$

where $\rho(\lambda)$ is the **eigenvalue counting function** $\rho(\lambda) = \#\{j : \lambda_j < \lambda\}$ of the Dirichlet or Neumann Laplacian Δ and $d_s = 1$ is the spectral dimension. The self-similarity of this consruction allows to describe the spectrum in terms of the iterations of a quadratic polynomial $R(z) = 2z(2-z)$. Then one can easily obtain the following result (see [67]).

Theorem 4.1. *The Riemann zeta function $\zeta(s)$ has a representation*

$$\zeta(s) = \pi^s \lim_{n\to\infty} \sum_{\substack{z\in R^{-n}\{0\}\\ z\neq 0}} \left(4^n z\right)^{-s/2} = \pi^s \zeta_R(s)$$

where $Re(s) > d_R = 1$ and $R(z) = 2z(2-z)$.

5. Fractal Laplacian on an interval and its zeta function

We can modify the construction of the previous section to obtain a one parameter family of fractal Laplacians on the unit interval. We introduce a parameter $0 < p < 1$ and define $q = 1 - p$, which later will be shown to have a meaning of transitional probabilities of a random walk. Then we define contraction factors (or "resistance weights")

$$r_1 = r_3 = \frac{p + pq}{2 + pq} \quad \text{and} \quad r_2 = \frac{2q - pq}{2 + pq},$$

and "measure weights"

$$m_1 = m_3 = \frac{q}{1+q} \quad \text{and} \quad m_2 = \frac{p}{1+q}.$$

Note that up to a constant the resistance weights are reciprocals of the measure weights, and $m_1 + m_2 + m_3 = r_1 + r_2 + r_3 = 1$. We define three contractions $F_1, F_2, F_3 : \mathbb{R}^1 \to \mathbb{R}^1$ with fixed points $p_j = 0, \frac{1}{2}, 1$ by $F_j(x) = r_j x + (1 - r_j)p_j$. Then the interval I=[0, 1] is a unique compact set such that $I = \bigcup_{j=1,2,3} F_j(I)$. The **discrete approximations** to I are defined inductively by

$$V_n = \bigcup_{j=1,2,3} F_j(V_{n-1})$$

where $V_0 = \partial I = \{0, 1\}$ is the **boundary** of I. The **discrete Dirichlet (energy) form** on V_n is defined inductively

$$\mathcal{E}_n(f) = \sum_{j=1,2,3} \tfrac{1}{r_j} \mathcal{E}_{n-1}(f \circ F_j).$$

with $\mathcal{E}_0(f) = (f(1) - f(0))^2$, and the **Dirichlet (energy) form** on I is

$$\mathcal{E}(f) = \lim_{n\to\infty} \mathcal{E}_n(f).$$

A function h is **harmonic** if it minimizes the energy given the boundary values. We have that $\mathcal{E}_{n+1}(f) \geqslant \mathcal{E}_n(f)$ and $\mathcal{E}_{n+1}(h) = \mathcal{E}_n(h) = \mathcal{E}(h)$ for a harmonic h. The Dirichlet (energy) form on I is **self-similar** in the sense that

$$\mathcal{E}(f) = \sum_{j=1,2,3} \tfrac{1}{r_j} \mathcal{E}(f \circ F_j).$$

The **μ–Laplacian** Δ_μ is defined by the following Gauss–Green (integration by parts) formula

$$\mathcal{E}(f) = -\int_0^1 f \Delta_\mu f d\mu + f f' \big|_0^1.$$

where μ is a unique probability self-similar measure with weights m_1, m_2, m_3, that is

$$\mu = \sum_{j=1,2,3} m_j \mu \circ F_j.$$

We also can define Δ_μ as a limit

$$\Delta_\mu f(x) = \lim_{n\to\infty} \left(1+\tfrac{2}{pq}\right)^n \Delta_n f(x)$$

where the discrete Laplacians

$$\Delta_n f(x_k) = \begin{cases} pf(x_{k-1}) + qf(x_{k+1}) - f(x_k) \\ \qquad\qquad\qquad\qquad\qquad\qquad \text{or} \\ qf(x_{k-1}) + pf(x_{k+1}) - f(x_k) \end{cases}$$

are defined as the generators of the nearest neighbor random walks on V_n with transitional probabilities p and q assigned according to the weights of the corresponding intervals. Note that $p = \frac{m_2}{m_1+m_2}$, $q = \frac{m_1}{m_1+m_2}$. Also transitional probabilities p and q can be assigned inductively as as shown on Figure 6.

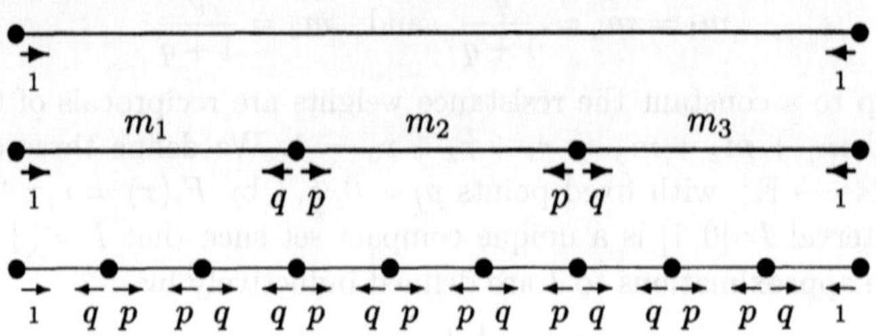

FIGURE 6. Random walks corresponding to the discrete Laplacians Δ_n.

The construction described above follows the lines of the general theory of p.c.f. fractals in [30]. Note that the standard Laplacian and Dirichlet form on $I = [0, 1]$, defined in the previous section, correspond to $p = \frac{1}{2}$. If $p \neq \frac{1}{2}$ then we can make a change of variable on the unit interval I so that either the Dirichlet form becomes the standard one, or the measure becomes the Lebesgue measure, but not both. Thus for different values of p the μ–Laplacians are different even up to a change of variable. By the result of Kigami and Lapidus [31], the Dirichlet or Neumann Laplacian Δ_μ has the **spectral asymptotics**

$$0 < \liminf_{\lambda\to\infty} \frac{\rho(\lambda)}{\lambda^{d_s/2}} \leqslant \limsup_{\lambda\to\infty} \frac{\rho(\lambda)}{\lambda^{d_s/2}} < \infty$$

where $\rho(\lambda)$ is the eigenvalue counting function. The spectral dimension is

$$d_s = \frac{\log 9}{\log\left(\frac{2+pq}{pq}\right)} \leqslant 1.$$

All the inequalities are strict if and only if $p \neq q$.

Similarly to the case of the Sierpiński gasket, one can prove the following simple "spectral decimation" result (see [68]). If $z \neq -1 \pm p$ then $R(z) \in \sigma(\Delta_n)$ if and only if $z \in \sigma(\Delta_{n+1})$, where $R(z) = z(z^2 + 3z + 2 + pq)/pq$. This implies, together with Theorem 2.1, the following theorem.

Theorem 5.1. *There are constants c_1, c_2, z_1, z_2 that depend on the boundary conditions such that*

$$\zeta_{\Delta_\mu}(s) = c_1^s \zeta_R^{z_1}(s) + c_2^s \zeta_R^{z_2}(s)$$

and $d_R = d_s$. If $p = q = \frac{1}{2}$ then $\zeta_{\Delta_\mu}(s) = \pi^s \zeta(s)$ and $d_R = 1$. If $p \neq \frac{1}{2}$ then $d_R < 1$ and $\zeta_{\Delta_\mu}(s)$ has a meromorphic continuation to $Re(s) > -\varepsilon$. If $p \to 0$ or $p \to 1$ then $\varepsilon \to 2$.

6. Self-similarity and nature of the spectrum

A good example of a fractal graph is a Sierpiński lattice which is an increasing union of copies of graphs V_n defined above. There are uncountably many non isomorphic Sierpiński lattices with empty boundary, but essentially one with nonempty boundary (which consists of one "corner" point). The following theorem on the spectrum of the discrete Laplacian Δ on a Sierpiński latticeis proved in [64].

Theorem 6.1. *The spectrum of Δ is $\sigma(\Delta) = \mathfrak{J} \cup \mathfrak{D}$, where $\mathfrak{D} =$ is a set of isolated eigenvalues of infinite multiplicity accumulating to the Julia set $\mathfrak{J}$ of the polynomial $R(z) = z(5 + z)$. This Julia set is a Cantor set of Lebesgue measure zero. The spectrum of Δ is pure point, eigenfunctions with compact support are complete. The set of eigenvalues is*

$$\mathcal{E} = \{-\tfrac{3}{2}\} \bigcup \left(\bigcup_{m=0}^{\infty} R^{-m}\{-\tfrac{3}{4}\} \right) \bigcup \left(\bigcup_{m=0}^{\infty} R^{-m}\{-\tfrac{5}{4}\} \right).$$

It would be interesting to extend this result for non symmetric fractals and in the so called non lattice case. Existence and completeness of localized eigenfunctions is established in [5, 29, 37, 53, 64] in many situations when the boundary is empty. The completeness for the Neumann Laplacian on the Sierpiński lattice and infinite Sierpiński gasket is proved in [64]. Note, however, that if the Sierpiński lattice has nonempty boundary, then the compactly supported eigenfunctions of the Dirichlet Laplacian are **not** complete.

The description of the spectrum as $\sigma(\Delta) = \mathfrak{J} \cup \mathfrak{D}$ where $\mathfrak{J}$ is a Julia set of a rational function, and $\mathfrak{D}$ is a set of isolated eigenvalues, is obtained in [38] for a large class of fractal graphs. However, one needs more information to establish the nature of the spectrum.

Based on the fractal Laplacian on an interval one can obtain a simple one–dimensional discrete model with purely singular continuous spectrum ([69]). The Laplacian is a discrete probabilistic Laplacian corresponding to a nearest neighbor random walk on $\mathbb{Z}_+$. For this walk, zero is a repulsive boundary, and the other transition probabilities assume two values p and $q = 1 - p$ that are assigned in a

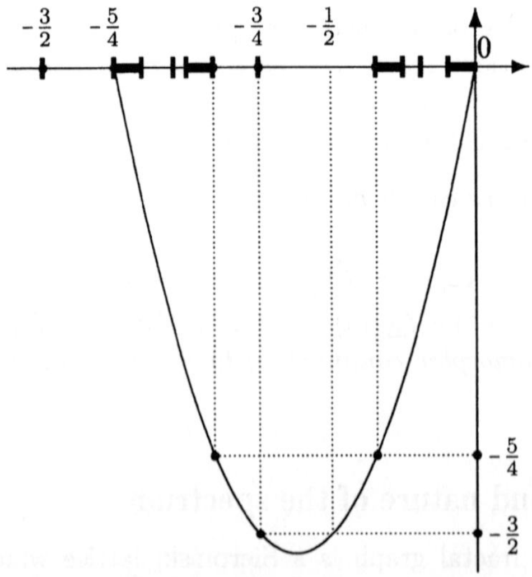

FIGURE 7. The complex dynamics of the polynomial $R(z)$ and the spectrum of the Sierpiński lattice.

self-similar fashion as shown on Figure 9. The spectrum can be analyzed in terms of the complex dynamics of the cubic polynomial $R(z) = z(z^2 + 3z + 2 + pq)/pq$.

Theorem 6.2. *The spectrum of the Laplacian is the Julia set of $R(z)$. If $p = q = \frac{1}{2}$, then the spectrum is an interval and is absolutely continuous. If $p \neq q$, then the spectrum is singularly continuous and is a Cantor set of Lebesgue measure zero.*

On Figure 10 we give a sketch that describes the complex dynamics of the family of cubic polynomials associated with the fractal Laplacians on the interval. The curved dotted line corresponds to the case when $p = \frac{1}{2}$ and the Julia set is the interval $[-2, 0]$. For any other value of p (we still assume $0 < p < 1$), the graph of the polynomial $R(z)$ behaves like the shown solid curved line. It is easy to see that then the Julia set of $R(z)$ is a Cantor set of Lebesgue measure zero. Note that the change $p \mapsto 1 - p$ does not change the polynomial $R(z)$, although the Laplacians Δ_μ are different.

The result in Theorem 6.2 is very sensitive since the nature of the spectrum of the Dirichlet Laplacian is not known, even though the spectrum as a set is still the Julia set of $R(z)$. The analysis of this model is related to results of J. Béllissard, D. Bessis, J. Geronimo, P. Moussa *et al.* on orthogonal polynomials (see, for instance, [10]) and of D. Damanik, S. Jitomirskaya, H. Schulz-Baldes, B. Simon, B. Solomyak *et al.* on spectral analysis of ergodic random and almost periodic Schrödinger operators (see [15, 22, 56, 59, 60] and other works). In fact, the difficulty in analyzing this one dimensional self-similar Dirichlet Laplacian is the same as for the Dirichlet Laplacian on the Sierpiński lattice. In particular, one can establish a certain nonlinear self–similarity property of the spectral measure.

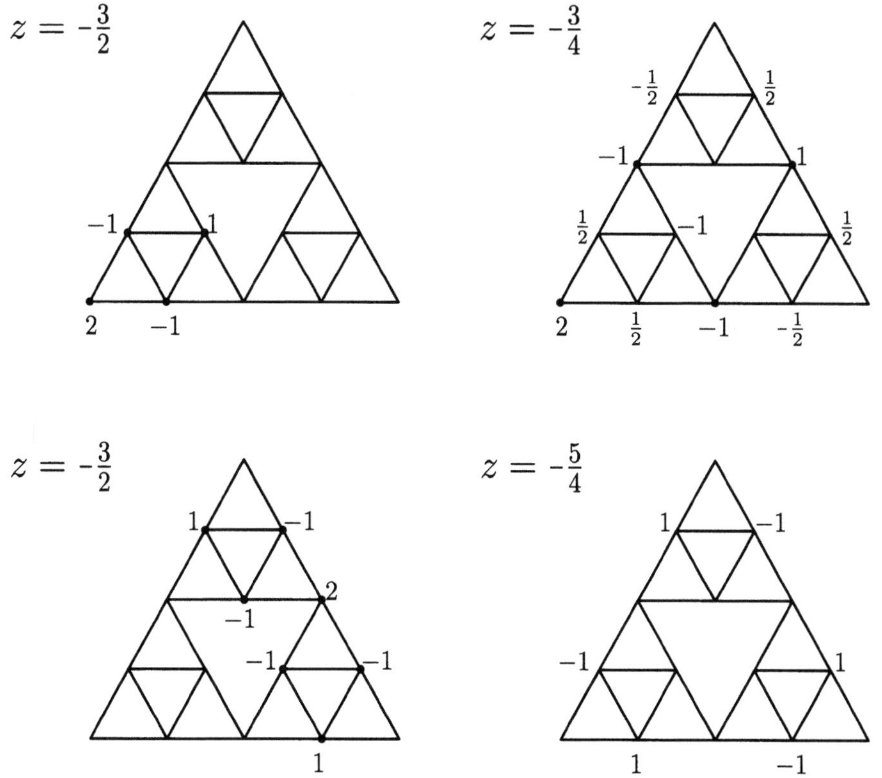

FIGURE 8. Localized eigenfuctions of the Sierpiński lattice.

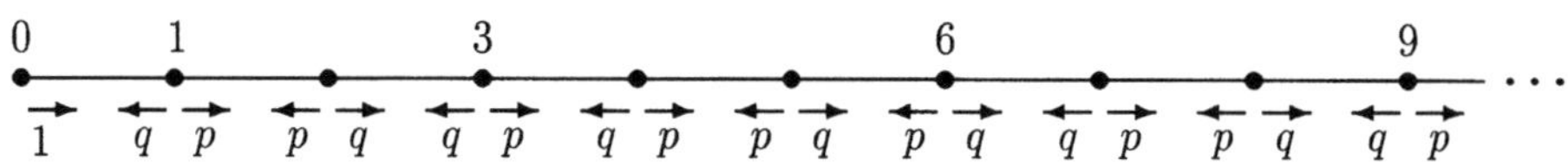

FIGURE 9. Transition probabilities of a self-similar random walk

In the case of the fractal Laplacian on an interval this identity is

$$\frac{d(\nu \circ R_{1,2,3})}{d\nu}(z) = \frac{3z^2 + 6z + 2 + pq}{(z+1)^2 - p^2}$$

where $R_{1,2,3}$ are the branches of the polynomial R. Note that this density is identically equal to 3 if $p = q = \frac{1}{2}$, but otherwise is not bounded and is not separated from zero on $\mathfrak{J}_R$.

There exists a number of works of B. Simon et al. on the relation between pure point and singularly continuous spectra of self-adjoint operators (see, for instance, [59, 60] and references therein). It is often possible to prove that the spectrum is pure point or singular continuous "generically" in one sense or another, but it might be hard to give a definitive answer for a fixed perturbation.

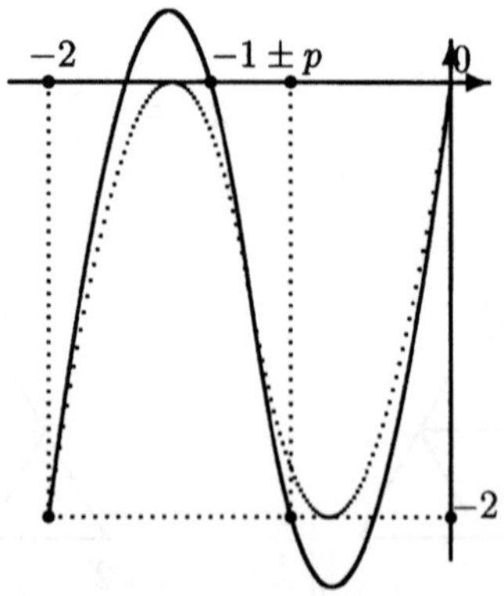

FIGURE 10. Sketch of the cubic polynomial $R(z)$ associated with the fractal Laplacians on the interval.

An interesting question is the relation of the fractal graphs, like the ones we described here, with the graphs of fractal groups recently discovered by L. Bartholdi, R. Grigorchuk and A. Żuk (see [6, 7, 19]).

References

[1] B. Adams, S.A. Smith, R. Strichartz and A. Teplyaev, *The spectrum of the Laplacian on the pentagasket.* Fractals in Graz 2001, Trends Math., Birkhäuser Basel (2003) 1–24.

[2] S. Alexander, *Some properties of the spectrum of the Sierpiński gasket in a magnetic field.* Phys. Rev. B **29** (1984), 5504–5508.

[3] M.T. Barlow, T. Coulhon and A. Grigor'yan, *Manifolds and graphs with slow heat kernel decay.* Invent. Math. **144** (2001), 609–649.

[4] M. T. Barlow and B. M. Hambly, *Transition density estimates for Brownian motion on scale irregular Sierpinski gaskets.* Ann. Inst. H. Poincaré Probab. Statist., **33** (1997), 531–557.

[5] M. T. Barlow and J. Kigami, *Localized eigenfunctions of the Laplacian on p.c.f. self-similar sets.* J. London Math. Soc., **56** (1997), 320–332.

[6] L. Bartholdi and R. Grigorchuk, *On parabolic subgroups and Hecke algebras of some fractal groups.* Serdica Math. J. **28** (2002), 47–90.

[7] L. Bartholdi, R. Grigorchuk and V. Nekrashevich, *From fractal groups to fractal sets.* Fractals in Graz 2001, Trends Math., Birkhäuser Basel (2003) 25–118.

[8] O. Ben-Bassat, R. S. Strichartz and A. Teplyaev, *What is not in the domain of the Laplacian on Sierpinski gasket type fractals.* J. Funct. Anal. **166** (1999), 197–217.

[9] D. Bessis, J. Geronimo and P. Moussa, *Mellin transforms associated with Julia sets and physical applications.* J. Statist. Phys. **34** (1984), 75–110.

[10] D. Bessis, J. S. Geronimo and P. Moussa, *Function weighted measures and orthogonal polynomials on Julia sets.* Constr. Approx. **4** (1988), 157–173.

[11] D. Bessis, J. Geronimo and P. Moussa, *Erratum: "Function weighted measures and orthogonal polynomials on Julia sets" [Constr. Approx. 4 (1988), no. 2, 157–173].* Constr. Approx. **6** (1990), 335–336.

[12] E.J. Bird, S.-M. Ngai and A. Teplyaev, *Fractal Laplacians on the Unit Interval*, to appear in Annales des sciences mathématiques du Québec (2003).

[13] T. Coulhon and A. Grigor'yan, *Pointwise estimates for transitional probabilities of random walks in infinite graphs.* Fractals in Graz 2001, Trends Math., Birkhäuser Basel (2003) 119–134.

[14] K. Dalrymple, R. S. Strichartz and J. P. Vinson, *Fractal differential equations on the Sierpinski gasket.* J. Fourier Anal. Appl., **5** (1999), 203–284.

[15] D. Damanik and B. Solomyak, *Some high-complexity Hamiltonians with purely singular continuous spectrum.* Ann. Henri Poincaré **3** (2002), 99–105.

[16] K. J. Falconer, *The Geometry of Fractal Sets.* Cambridge Tracts in Mathematics, 85. Cambridge University Press, 1986.

[17] M. Fukushima and T. Shima, *On a spectral analysis for the Sierpiński gasket.* Potential Analysis **1** (1992), 1-35.

[18] Y. Gefen, A. Aharony and B. B. Mandelbrot, *Phase transitions on fractals. I. Quasi-linear lattices. II. Sierpiński gaskets. III. Infinitely ramified lattices.* J. Phys. A **16** (1983), 1267–1278; **17** (1984), 435–444 and 1277–1289.

[19] R. Grigorchuk and A. Żuk, *On the asymptotic spectrum of random walks on infinite families of graphs.* Random walks and discrete potential theory (Cortona, 1997), 188–204, Sympos. Math., XXXIX, Cambridge Univ. Press, Cambridge, 1999.

[20] B. M. Hambly, *On the asymptotics of the eigenvalue counting function for random recursive Sierpinski gaskets.* Probab. Theory Related Fields, **117** (2000), 221–247.

[21] R. G. Hohlfeld and N. Cohen, *Self-Similarity and the Geometric Requirements for Frequency Independence in Antennae.* Fractals **7** (1999), 79–84.

[22] S. Jitomirskaya, *Nonperturbative localization.* Proceedings of the International Congress of Mathematicians, Vol. III (Beijing, 2002), 445–455, Higher Ed. Press, Beijing, 2002.

[23] A. Jonsson, *The trace of functions in the domain of the Dirichlet form.* Fractal Geometry and Stochastics III, Birkhäuser (2004).

[24] V. A. Kaimanovich, *Random walks on Sierpiński graphs: hyperbolicity and stochastic homogenization.* Fractals in Graz 2001, Trends Math., Birkhäuser Basel (2003) 145–184.

[25] V. A. Kaimanovich and A. M. Vershik, *Random walks on discrete groups: boundary and entropy.* Ann. Probab., **11** (1983), 457–490.

[26] J. Kigami, *A harmonic calculus on the Sierpiński spaces.* Japan J. Appl. Math. **6** (1989), 259–290.

[27] J. Kigami, *Harmonic calculus on p.c.f. self-similar sets.* Trans. Amer. Math. Soc. **335** (1993), 721–755.

[28] J. Kigami, *Harmonic metric and Dirichlet form on the Sierpiński gasket.* Asymptotic problems in probability theory: stochastic models and diffusions on fractals (Sanda/Kyoto, 1990), 201–218, Pitman Res. Notes Math. Ser., **283**, Longman Sci. Tech., Harlow, 1993.

[29] J. Kigami, *Distributions of localized eigenvalues of Laplacians on p.c.f. self–similar sets.* J. Funct. Anal. **156** (1998), 170–198.

[30] J. Kigami, *Analysis on fractals.* Cambridge Tracts in Mathematics **143**, Cambridge University Press, 2001.

[31] J. Kigami and M. L. Lapidus, *Weyl's problem for the spectral distribution of Laplacians on p.c.f. self-similar fractals.* Comm. Math. Phys. **158** (1993), 93–125..

[32] J. Kigami and M. L. Lapidus, *Self–similarity of volume measures for Laplacians on p.c.f. self–similar fractals.* Comm. Math. Phys. **217** (2001), 165–180.

[33] B. Krön, *Green functions on self-similar graphs and bounds for the spectrum of the Laplacian.* Ann. Inst. Fourier (Grenoble) **52** (2002), 1875–1900.

[34] T. Kumagai, *Function spaces and stochastic processes on fractals: heat kernel estimates for jump-type processes.* Fractal Geometry and Stochastics III, Birkhäuser (2004).

[35] M. L. Lapidus, *Spectral and fractal geometry: from the Weyl-Berry conjecture for the vibrations of fractal drums to the Riemann zeta-function.* Differential equations and mathematical physics (Birmingham, AL, 1990), 151–181, Math. Sci. Engrg., **186**, Academic Press, Boston, MA, 1992.

[36] M. L. Lapidus and M. van Frankenhuysen, *Fractal geometry and number theory. Complex dimensions of fractal strings and zeros of zeta functions.* Birkhäuser, Boston, 2000.

[37] L. Malozemov and A. Teplyaev, *Pure point spectrum of the Laplacians on fractal graphs.* J. Funct. Anal. **129** (1995), 390–405.

[38] L. Malozemov and A. Teplyaev, *Self-similarity, operators and dynamics.* Math. Phys. Anal. Geom. **6** (2003), 201–218.

[39] R. D. Mauldin and M. Urbański, *Dimension and measures for a curvilinear Sierpinski gasket or Apollonian packing.* Adv. Math., **136** (1998), 26–38.

[40] V. Metz, *Self-similar fractals and self-similar energy.* Fractals in Graz 2001, Trends Math., Birkhäuser Basel (2003) 225–240.

[41] V. Metz and K.-T. Sturm, *Gaussian and non-Gaussian estimates for heat kernels on the Sierpiński gasket.* Dirichlet forms and stochastic processes (Beijing, 1993), 283–289, de Gruyter, Berlin, 1995.

[42] R. Meyers, R. Strichartz and A. Teplyaev, *Dirichlet forms on the Sierpinski gasket,* to appear in Pacific J. Math. (2003).

[43] J. Needleman, R. Strichartz, A. Teplyaev and P.-L. Yung, *Calculus on the Sierpiński gasket: polynomials exponentials and power series,* submitted (2003).

[44] P. Nikitin, *The Hausdorff dimension of the harmonic measure on a de Rham curve.* (Russian) Zap. Nauchn. Sem. S.-Peterburg. Otdel. Mat. Inst. Steklov. (POMI) **283** (2001), Teor. Predst. Din. Sist. Komb. i Algoritm. Metody. **6**, 206–223.

[45] A. Öberg, R.S. Strichartz and A.Q. Yingst, *Level sets of harmonic functions on the Sierpinski gasket.* Ark. Mat. **40** (2002), 335–362.

[46] Y. Peres and B. Solomyak, *Problems on self-similar sets and self-affine sets: an update.* Fractal geometry and stochastics, II (Greifswald/Koserow, 1998), 95–106, Progr. Probab., **46**, Birkha"user, Basel, 2000.

[47] R. Rammal, *Spectrum of harmonic excitations on fractals.* J. Physique **45** (1984), 191–206.

[48] R. Rammal and G. Toulouse, *Random walks on fractal structures and percolation clusters.* J. Physique Letters **44** (1983), L13–L22.

[49] G. de Rham, *Un peu de mathématiques à propos d'une courbe plane.* Elemente der Math., **2** (1947), 73–76, 89–97.

[50] G. de Rham, *Sur une courbe plane.* J. Math. Pures Appl., **35** (1956), 25–42.

[51] G. de Rham, *Sur quelques courbes definies par des equations fonctionnelles.* Univ. e Politec. Torino. Rend. Sem. Mat., **16** (1956/1957), 101–113.

[52] G. de Rham, *Sur les courbes limites de polygones obtenus par trisection.* Enseignement Math., **5** (1959), 29–43.

[53] C. Sabot, *Pure point spectrum for the Laplacian on unbounded nested fractals.* J. Funct. Anal. **173** (2000), 497–524.

[54] C. Sabot, *Integrated density of states of self-similar Sturm-Liouville operators and holomorphic dynamics in higher dimension.* Ann. Inst. H. Poincaré Probab. Statist. **37** (2001), 275–311.

[55] C. Sabot, *Spectral properties of self-similar lattices and iteration of rational maps.* Mém. Soc. Math. Fr. (N.S.) **92** (2003).

[56] H. Schulz-Baldes, *Anomalous transport in aperiodic media.* XIIIth International Congress on Mathematical Physics (London, 2000), 199–205, Int. Press, Boston, MA, 2001.

[57] T. Shima, *On eigenvalue problems for Laplacians on p.c.f. self-similar sets.* Japan J. Indust. Appl. Math. **13** (1996), 1–23.

[58] N. Sidorov and A. M. Vershik, *Ergodic properties of the Erdös measure, the entropy of the golden shift, and related problems.* Monatsh. Math., **126** (1998), 215–261.

[59] B. Simon, *Singular spectrum: recent results and open questions.* XIth International Congress of Mathematical Physics (Paris, 1994), 507–512, Internat. Press, Cambridge, MA, 1995.

[60] B. Simon and T. Wolff, *Singular continuous spectrum under rank one perturbations and localization for random Hamiltonians.* Comm. Pure Appl. Math. **39** (1986), 75–90.

[61] R. B. Stinchcombe, *Fractals, phase transitions and criticality.* Fractals in the natural sciences. Proc. Roy. Soc. London Ser. A **423** (1989), 17–33.

[62] R. S. Strichartz, *Taylor approximations on Sierpiński type fractals.* J. Funct. Anal. **174** (2000), 76–127.

[63] R. S. Strichartz, *Analysis on fractals.* Notices AMS, **46** (1999), 1199-1208.

[64] A. Teplyaev, *Spectral Analysis on Infinite Sierpiński Gaskets.* J. Funct. Anal. **159** (1998) 537-567.

[65] A. Teplyaev, *Gradients on fractals.* J. Funct. Anal. **174** (2000) 128–154.

[66] A. Teplyaev, *Energy and Laplacian on the Sierpiński gasket,* to appear in Mandelbrot Jubilee, Proc. Sympos. Pure Math., AMS (2003).

[67] A. Teplyaev, *Spectral zeta functions of fractals,* preprint (2003).

[68] A. Teplyaev, *Self-similarity of the spectrum of fractal graphs,* preprint.

[69] A. Teplyaev, *Fractal Laplacians with singular continuous spectrum,* preprint.

[70] E. Teufl, *The average displacement of the simple random walk on the Sierpiński graph.* Combin. Probab. Comput. **12** (2003) 203–222.

[71] E. Teufl, *On the Hausdorff dimension of the Sierpiński gasket with respect to the harmonic metric.* Fractals in Graz 2001, Trends Math., Birkhäuser Basel (2003) 263–270.

[72] H. Triebel, *The fractal Laplacian and multifractal quantities.* Fractal Geometry and Stochastics III, Birkhäuser (2004).

Acknowledgment

The author is very grateful to R. Strichartz for his invaluable guidance during the author's work in analysis on fractals, and to M. Lapidus for mentoring, encouragement and support.

Alexander Teplyaev
Department of Mathematics
University of Connecticut
Storrs CT 06269
USA
e-mail: `teplyaev@math.uconn.edu`

MIX
Papier aus verantwortungsvollen Quellen
Paper from responsible sources
FSC® C105338

If you have any concerns about our products,
you can contact us on
ProductSafety@springernature.com

In case Publisher is established outside the EU,
the EU authorized representative is:
Springer Nature Customer Service Center GmbH
Europaplatz 3, 69115 Heidelberg, Germany

Printed by Libri Plureos GmbH
in Hamburg, Germany